Urban Forestry and Arboriculture in Malaysia

Sreetheran Maruthaveeran · Wendy Y. Chen ·
Justin Morgenroth
Editors

Urban Forestry and Arboriculture in Malaysia

An Interdisciplinary Research Perspective

Springer

Editors
Sreetheran Maruthaveeran
Department of Landscape Architecture
Faculty of Design & Architecture
Universiti Putra Malaysia
Serdang, Malaysia

Wendy Y. Chen
Department of Geography
Faculty of Social Sciences
University of Hong Kong
Hong Kong, Hong Kong

Justin Morgenroth
School of Forestry
University of Canterbury
Christchurch, New Zealand

ISBN 978-981-19-5417-7 ISBN 978-981-19-5418-4 (eBook)
https://doi.org/10.1007/978-981-19-5418-4

This Springer imprint is published by the registered company Springer Nature Singapore Pte Ltd.
The registered company address is: 152 Beach Road, #21-01/04 Gateway East, Singapore 189721, Singapore

Preface

The term *urban forestry* was first used in the 1960s in North America. Though initially the term was not used widely, soon it became accepted by a broader group of experts. Later, urban forestry research, policy and practice developed with high-level research being undertaken at the universities and the federal and state agencies in America. Soon this concept spread to Europe, where Britain was among the first European countries to embrace the concept of urban forestry by involving in urban tree planting. Nevertheless, in most Asian countries, this concept arrived quite late, particularly in Malaysia. In Malaysia, the term *landscaping* was used more commonly than *urban forestry*, particularly among the government and private institutions, politicians, stakeholders, academicians and the public. The term 'landscaping' in Malaysia refers to beautifying the cityscapes, e.g., planting flowering shrubs, planting trees for shades, creating more green spaces like parks, etc.

In Malaysia, the term *urban forestry* was used more broadly after the involvement of the Malaysia-Denmark Twinning Programme: Multipurpose Forestry in a Changing Society funded by DANIDA (2003–2006). The Danish counterparts who took the lead in the project were Dr. Kjell Nilsson, Prof. Dr. Cecil Konijnendijk van den Bosh and Prof. Dr. Thomas B. Randrup who basically introduced the concept of *urban forestry and urban greening* and *arboriculture*. Through this project, many initiatives were taken such as research collaboration with several institutions in Malaysia such as Universiti Putra Malaysia (UPM), Forest Research Institute Malaysia (FRIM) and Kuala Lumpur City Hall (DBKL) with universities in Denmark such as the Royal Veterinary and Agricultural University (KVL) which later merged with the University of Copenhagen, Denmark. It was also during this project that the 1st Arborist Certification under the International Society of Arboriculture (ISA), USA, was organised in Malaysia by the Forest Research Institute Malaysia (FRIM) in June 2005 with the support of the Danish counterparts. I, personally involved as one of the Malaysian counterparts in this twinning programme, personally feel privileged to be part of this project. Through this networking with my Danish counterparts, I also had the opportunity to pursue my Ph.D. at the University of Copenhagen under the supervision of Prof. Dr. Cecil Konijnendijk van den Bosh.

Later, many local universities and research institutions in Malaysia were involved in research on urban forestry and urban greening from the environmental, social, health and well-being and economic perspective. Some of these studies also focus on arboricultural studies. In addition, the number of ISA Certified Arborist has also increased from 23 (from the 1st batch in 2005) up to 92 certified arborists at present. A society under the name of the Malaysian Society of Arborist (PArM) was also established to serve as an umbrella body for arboriculture practitioners who are involved in the maintenance and management of urban forest trees with a membership of about 300 as of 2021. Nevertheless, there are no publications that document the research work on urban forestry and arboriculture in Malaysia as a reference for the local researchers, academicians, policy-makers and students. As a researcher and academician, I feel it's time to have a reference on urban forestry and arboricultural research in Malaysia though not many aspects of urban forestry research were covered. With this, as the coordinator of the Urban Forestry Section (6.07.00) for the International Union of Forest Research and Organizations (IUFRO) and also an ISA Certified Arborist, I initiated the attempt to coordinate the compilation work on urban forestry and arboricultural research in Malaysia together with the help of my deputies, Professor Dr. Wendy Yan Chen from University of Hong Kong and Associate Professor Dr. Justin Morgenroth from the University of Canterbury, New Zealand, who also agreed to be the editors of this book. I gratefully acknowledge the help of Wendy and Justin in making this effort a success.

This book is divided into two parts. The first part aims at reviewing the main topics in the field of urban forestry and arboriculture in Malaysia, while the second part of the book provides a set of case studies around Malaysia that illustrates some specific studies on urban forestry and arboriculture.

I hope this book can be a good reference and inspire more research on urban forestry and arboriculture in Malaysia. It is also hoped that more research collaboration can be initiated between the local and international institutions so that Malaysia will sustain a strong position in scientific and technical research on urban forestry and arboriculture in this region particularly in Southeast Asia.

I am grateful to our contributors from several local universities, research institutions and tree care organisations for their articles which consist of diverse collections of issues from different aspects of urban forestry and arboriculture. Finally, I must thank Springer and the entire team behind it for their continuous support in publishing this book.

Serdang, Malaysia

Sreetheran Maruthaveeran, Ph.D.
Associate Professor, Coordinator
Urban Forestry Section (6.07.00)
International Union of Forest Research
Organizations (IUFRO)

Contents

Editors and Contributors

About the Editors

Sreetheran Maruthaveeran is an Associate Professor in the Department of Landscape Architecture, Faculty of Design and Architecture, Universiti Putra Malaysia (UPM). He also has served as a Senior Researcher at the Forest Research Institute Malaysia (FRIM) since 2003 before joining the academic line in 2016. His main research interest lies in the social aspect of urban forestry, arboriculture, recreation and leisure sciences. Currently, he serves as the Editorial Board Member for *Journal of Outdoor Recreation and Tourism, World Leisure Journal and Leisure Studies*. He is also the Editor-in-Chief for *ALAM CIPTA International Journal of Sustainable Tropical Design Research*. Sreetheran also serves as the Coordinator for the Urban Forestry Unit, 6.07.00 of the International Union of Forest Research Organizations (IUFRO). He is also a Professional Technologist (Ts) registered under the Malaysian Board of Technologists (MBOT) and a Certified Arborist (MY0300A) under the International Society of Arboriculture (ISA), USA, since 2007.

Wendy Y. Chen is a Professor in the Department of Geography, serving as the director of the International Centre for China Development Studies, the University of Hong Kong. She serves as the Editor-in-Chief for Urban Forestry and Urban Greening since January 2019, a top international journal in the field of urban forestry. Her research interests include urban forestry, natural resource evaluation and management, urban nature and urban ecology, ecosystem services and ecosystem service economics, environmental management and sustainable development. She is currently the Deputy Coordinator for the Urban Forestry Unit, 6.07.00 of the International Union of Forest Research Organizations (IUFRO).

Justin Morgenroth is an Associate Professor at the School of Forestry, University of Canterbury. His areas of research interest include urban forest dynamics, as well as explorations of tree response to a variety of urban stresses. He is a former Chair of the International Society of Arboriculture's Science and Research Committee and Associate Editor for top scientific journals such as *Urban Forestry and Urban Greening* and *Arboriculture and Urban Forestry*. He is currently the Deputy Coordinator for the Urban Forestry Unit, 6.07.00 of the International Union of Forest Research Organizations (IUFRO).

Contributors

Abdu Arifin Department of Forestry Science and Biodiversity, Universiti Putra Malaysia (UPM), Serdang, Selangor Darul Ehsan, Malaysia

Abdul Aziz Nor Akmar Department of Recreation and Ecotourism, Faculty of Forestry and Environment, Universiti Putra Malaysia (UPM), Serdang, Malaysia

Abdullah Rohayu Universiti Teknologi Malaysia, Skudai, Malaysia

Ahmad Mohd Farid Forest Research Institute Malaysia (FRIM), Kepong, Malaysia

Ahmad Syazwan Samsuddin Mycology and Pathology Branch Forest Biodiversity Division, Forest Research Institute Malaysia (FRIM), Kuala Lumpur, 52109 Selangor Darul Ehsan, Malaysia;
Department of Forest Science and Biodiversity, Universiti Putra Malaysia (UPM), Serdang, 43400 Selangor Darul Ehsan, Malaysia

Ariffin Wan Tarmeze Wan Forest Research Institute Malaysia (FRIM), Kepong, Malaysia

Cheong Yew Loong Genetic & Agriculture (G&A) Research Centre, Kabupaten Pelalawan, Riau, Indonesia

Darus Nadirah Department of Environmental, School of Civil Engineering, Universiti Sains Malaysia, George Town, Penang, Malaysia

Dorairaj Deivaseeno Institute for Environment and Development (LESTARI), Universiti Kebangsaan Malaysia, Bangi, Selangor, Malaysia

Foo Chee Hung Property Business Development, MKH Berhad, Kuala Lumpur, Malaysia

Govender Nisha Institute of Systems Biology (INBIOSIS), Universiti Kebangsaan Malaysia (UKM), Bangi, Selangor Darul Ehsan, Malaysia

Halim Herni Department of Environmental, School of Civil Engineering, Universiti Sains Malaysia, George Town, Penang, Malaysia

Hamzah Helmi Centre of Studies for Landscape Architecture, Department of Built Environment Studies and Technology, Universiti Teknologi MARA (UiTM), Perak, Seri Iskandar, Malaysia

Haron Zaiton Department of Environmental, School of Civil Engineering, Universiti Sains Malaysia, George Town, Penang, Malaysia

Hasan Ramly Department of Landscape Architecture, Faculty of Architecture and Ekistics, Universiti Malaysia Kelantan (UMK), Bachok, Kelantan Darul Naim, Malaysia

Hishamuddin Muhammad Syahmi Forest Research Institute Malaysia (FRIM), Kepong, Malaysia

Jahya Zanariah Department of Environmental, School of Civil Engineering, Universiti Sains Malaysia, George Town, Penang, Malaysia

Jasmani Zanariah Faculty of Built Environment and Surveying, Universiti Teknologi Malaysia, UTM Skudai, Johor Bahru, Johor, Malaysia

Kanniah Kasturi Devi Universiti Teknologi Malaysia, Skudai, Malaysia

Karam Daljit Singh Department of Land Management, Universiti Putra Malaysia (UPM), Serdang, Selangor Darul Ehsan, Malaysia

Lamit Hasanuddin Faculty of Built Environment and Surveying, Universiti Teknologi Malaysia, UTM Skudai, Johor Bahru, Johor, Malaysia

Lee Seng Hua Institute of Tropical Forestry and Forest Products (INTROP), Universiti Putra Malaysia (UPM), Serdang, Selangor Darul Ehsan, Malaysia

Lim Wing-Shen Faculty of Tropical Forestry, Universiti Malaysia Sabah (UMS), Kota Kinabalu, Sabah, Malaysia

Lum Wei Chen Institute for Infrastructure Engineering and Sustainable Management (IIESM), Universiti Teknologi MARA (UiTM), Shah Alam, Selangor Darul Ehsan, Malaysia

Maruthaveeran Sreetheran Department of Landscape Architecture, Faculty of Design and Architecture, Universiti Putra Malaysia (UPM), Serdang, Malaysia

Mohd Roseli Ahmad Nazarudin Forestry and Environment Division, Forest Research Institute Malaysia (FRIM), Kepong, Selangor, Malaysia

Mojiol Andy Russel Faculty of Tropical Forestry, Universiti Malaysia Sabah (UMS), Kota Kinabalu, Sabah, Malaysia

Nath Tapan Kumar University of Nottingham Malaysia, Semenyih, Selangor, Malaysia

Ong Su Ping Forest Research Institute Malaysia, Kepong, Selangor, Malaysia

Osman Normaniza Institute of Biological Sciences, Faculty of Science, University of Malaya (UM), Kuala Lumpur, Malaysia

Othman Noriah Department of Landscape Architecture, Faculty of Architecture Planning and Surveying, Universiti Teknologi MARA (UiTM), Bandar Puncak Alam, Selangor Darul Ehsan, Malaysia

Pathil Saifful Tree Care Safety™, Cheras, Malaysia

Rajoo Keeren Sundara Department of Forestry Science, Universiti Putra Malaysia Bintulu Sarawak Campus (UPMKB), Bintulu, Sarawak, Malaysia;
Institute of Ecosystem Science Borneo, Universiti Putra Malaysia Bintulu Sarawak Campus (UPMKB), Bintulu, Malaysia

Sajap Ahmad Said Subang Jaya, Selangor Darul Ehsan, Malaysia

Salim Shahzarimin Legal and Regulatory Planning Division, PLANMalaysia (Department of Town and Country Planning), Putrajaya, Malaysia

Shahidan Mohd Fairuz Department of Landscape Architecture, Faculty of Design and Architecture, Universiti Putra Malaysia (UPM), Serdang, Selangor Darul Ehsan, Malaysia

Shyong Yap Zhen Department of Environmental, School of Civil Engineering, Universiti Sains Malaysia, George Town, Penang, Malaysia

Siong Ho Chin Centre for Environmental Sustainability and Water Security (IPASA), Research Institute for Sustainable Environment (RISE), Universiti Teknologi, Johor Bahru, Malaysia

Sukri Nik Adlin Nik Mohamed Forest Research Institute Malaysia (FRIM), Kepong, Malaysia

Tan Li Peng Department of Paraclinical, Universiti Malaysia Kelantan (UMK), Kota Bharu, Kelantan, Malaysia

van den Bosch Cecil C. Konijnendijk Department of Forest Resources and Management, Faculty of Forestry, University of British Columbia, Vancouver, BC, Canada

Vijayanathan Jeyanny Forest Research Institute Malaysia (FRIM), Kepong, Malaysia

Yahya Khairulzan Department of Environmental, School of Civil Engineering, Universiti Sains Malaysia, George Town, Penang, Malaysia

Zaki Sheikh Ahmad Malaysia-Japan International Institute of Technology, Universiti Teknologi Malaysia (UTM), Kuala Lumpur, Malaysia

List of Figures

List of Tables

Chapter 1
Defining Urban Forestry and Arboriculture in Malaysia

Sreetheran Maruthaveeran

Abstract The concept of urban forestry emerged in the 1960s in North America, however in most Asian countries including Malaysia, this concept arrived in the 1980s or perhaps even later. Taking the concept of urban forestry and arboriculture into real-world practice in Malaysia was brought by the Twinning Programme: Multipurpose Forestry in a Changing Society, a Denmark-Malaysia collaboration funded by DANIDA during 2003–2006. Many initiatives were brought into realisation in terms of the scholarly research of urban forestry and arboriculture and in terms of practices among the practitioners, academicians and researchers. Many important milestones from this Twinning Programme were achieved, notably the 1st Arborist Certification in Malaysia in 2005 and training of personnel from Malaysia into the higher education organisations (PhD studies) in Denmark. This Twinning Programme further generated studies on urban forestry and arboriculture which later were branched out into varieties of scope e.g., environment, social, health and well-being, etc.—There is still a lack of research from other perspectives such as the valuation (economics), governance, urban soil, carbon sequestration and tree risk management in Malaysia. New knowledge in the context of tropical urban forestry and arboriculture needs to be developed and practised. It is therefore necessary and timely to develop bachelor programmes in urban forestry and arboriculture at the higher education level in Malaysia.

Keywords Urban forest · Arborist · Tree care · Urban tree management · Malaysia

The term *urban forestry* was first coined in 1965 at the University of Toronto, Canada as a title to a graduate student's study of the success and failures of municipal tree planting projects in an area of Metropolitan Toronto (Johnston, 1996). Later on, the term was defined in detail in 1970 by Professor Erik Jorgensen as 'a specialised branch of forestry and has in its objectives the cultivation and management of trees for their present and potential contribution to the physiological, sociological and economic

---

S. Maruthaveeran (✉)
Department of Landscape Architecture, Faculty of Design and Architecture, Universiti Putra Malaysia (UPM), Serdang, Malaysia
e-mail: sreetheran@upm.edu.my

S. Maruthaveeran et al. (eds.), *Urban Forestry and Arboriculture in Malaysia*,
https://doi.org/10.1007/978-981-19-5418-4_1

well-being of urban society' (Jorgensen, 1986, p. 173). Deneke (1993) expanded on the term as a sustained planning, planting, protection, maintenance, and care of trees, forests, greenspace and related resources in and around cities and communities for economic, environmental, social and public health benefits for people. Nonetheless one of the most commonly cited definition of 'urban forestry' was developed by the Society of American Foresters, which defines urban forestry as 'the art, science and technology of managing trees and forest resources in and around urban community ecosystems for the physiological, sociological, economic and aesthetic benefits tree provide society' (Helms, 1998, p. 193). Whereas Harris et al. (1999, p. 1) defined 'urban forestry is the management of planted and naturally occurring trees in urban and urban-interface areas.' This has led to the need to take care of our urban trees which has promoted the concept and profession of arboriculture (Randrup et al., 2005). Arboriculture is regarded as one of the branches of horticulture that concerns the cultivation of woody plants, particularly trees, shrubs and vines (Harris et al., 1999). This branch of study in essence focuses on single trees or small groups of trees in urban areas. The goals of arboriculture are to establish and maintain healthy, aesthetic and safe trees (Dujesiefken et al., 2005). Urban forestry and arboriculture are interrelated disciplines that are frequently treated synonymously by professionals and the public in the same way (Miller et al., 2015). Arboriculture is an inherent aspect of urban forestry but should be viewed distinctly in terms of professional preparation and core competencies (O'Herrin et al., 2020).

After the emergence of the concept of urban forestry in North America, it later spread to Europe gradually, where Britain was among the first European countries to embrace the concept by involving it in urban tree planting (Konijnendijk, 2003). In the 1990s the promotion and advancement of urban forestry in Europe started with a network of green space researchers from 22 European countries financed by the European Union (COST Action E12 Urban Forests and Trees) (Konijnendijk et al., 2006). A series of seminars and conferences were held through this network, based on which the status of research and higher education on urban forestry in Europe were reviewed and compiled into the first European reference book on urban forestry titled 'Urban Forests and Trees' (Konijnendijk et al., 2005). Another direct spin-off of COST Action E12 was also the launching of a new scientific journal, *Urban Forestry and Urban Greening* in 2002 by Urban and Fischer Publishers to cater to the urban forestry research community (Konijnendijk, 2003).

Nevertheless, in most Asian countries including Malaysia, this concept arrived in the 1980s or even later. Despite this belated conceptualization, urban forestry activities such as urban tree planting have been practised for more than a century in Malaysia. Koening (1894) has mentioned that *Pterocarpus indicus* or the Angsana have been planted in 1778 in Malacca and in 1802 in Penang (Burkill, 1966). With its spreading crown and ease of propagation, *P. indicus* became a popular tree for urban greening/landscaping in Malaysia in the 1990s (Philip, 1999). With this being said, the term *landscaping* has been used more commonly rather than *urban forestry* and/or *arboriculture*, particularly among the government and private institutions, politicians, stakeholders, academicians and the public (Sreetheran et al., 2006). The greening of urban Malaysia in the 1970s has focused primarily on landscape beautification and has mainly been the province of horticulturists, landscapers, nursery workers, town planners and architects, with negligible inputs from foresters. The term 'landscaping' in Malaysia refers to activities mainly on beautifying the cities e.g., planting flowering shrubs, planting trees for shades, creating more green spaces like parks etc. In short, trees were planted for the sake of beautifying the cityscape. Perhaps, for this reason, the term 'landscaping' has been used more widely than *urban forestry* and/or *arboriculture*.

In Malaysia, particularly in Kuala Lumpur, the term *urban forestry* was proposed initially by Clive L. Justice, a prominent landscape architect coming from Canada to serve at the Department of Parks and Recreation under the Kuala Lumpur City Hall (DBKL) in the 1980s. Justice was invited by DBKL to provide an improved urban environment that will make working and living in the city pleasant and enjoyable (Justice, 1986). One of the key ideas proposed by Justice to DBKL was to form a special section or division under the Parks and Recreation Department to administer standards of tree care and management under urban conditions, undertake programs of tree selection, tree planting and urban forest management (Justice, 1986). He also emphasised on the need to establish a team of qualified Landscape Architects, Arborists, Foresters and Urban Forest Ecologists under this special section, who should develop and administer urban forest and tree programmes throughout the city of Kuala Lumpur.

Apart from DBKL, other government organisation such as the Forest Research Institute Malaysia (FRIM) has also played a vital role in the beautification of the cities in the earlier days. In fact, an Urban Forestry Unit was established in FRIM during the service of the first director-general of FRIM, Tan Sri Dato' Dr. Salleh Haji Mohd Nor. This unit has involved actively in many massive landscape projects at the national level including in the development of a mini forest inside the main terminal of the Kuala Lumpur International Airport (KLIA). The main terminal building area was designed using the concept of 'Airport in the Forest, Forest in the Airport', in which the airport is surrounded by lush green elements. This concept was given by a renowned Japanese architect, Kisho Kurokawa when KLIA was opened for business in 1998.

Until the early 2000s, the term *urban forestry* or *arboriculture* became used broadly after the Malaysia-Denmark Twinning Programme: Multipurpose Forestry in a Changing Society funded by DANIDA (2003–2006). The Danish counterparts who took the lead in the project were Dr. Kjell Nilsson, Prof. Dr. Cecil Konijnendijk van den Bosh and Prof. Dr. Thomas B. Randrup who introduced the concept of 'urban forestry and urban greening' and 'arboriculture' to Malaysia. Through this Twinning Programme, a series of initiatives were taken through the collaboration between the Denmark and Malaysian counterparts. On the Malaysian side, the Forest Research Institute Malaysia (FRIM) took the lead and joined forces with other organisations such as Universiti Putra Malaysia (UPM) and DBKL. The Malaysian participants collaborated with a university in Denmark, the Royal Veterinary and Agricultural University (KVL) which later merged with the University of Copenhagen, Denmark in 2007. The author was also invited for a PhD workshop on urban forest governance

Fig. 1.1 Under the Twining Programme, the author was invited to attend a PhD course on urban forest governance in the summer of 2005 at the Royal Veterinary and Agricultural University (KVL) (later merged with the University of Copenhagen in 2007), Frederiksberg C, Copenhagen, Denmark

in the summer of 2005 at the Royal Veterinary and Agricultural University (KVL), Copenhagen, Denmark (Fig. 1.1). A number of researchers and academicians from Malaysia were also trained at the University of Copenhagen for their PhD studies under the supervision of several renowned professors in urban forestry and urban greening and landscape architecture such as Prof. Dr. Cecil Konijnendijk van den Bosh (Fig. 1.2) Prof. Dr. Kjell Nilsson, Prof. Dr. Thomas B. Randrup and Prof. Dr. Ulrika K. Stigsdotter.

Fig. 1.2 The author with his PhD supervisor, Prof. Dr. Cecil Konijnendijk van den Bosch upon completion of his PhD *viva-voce* on 29 May 2015 at the Section for Landscape Architecture and Planning, Department of Geosciences and Nature Management, University of Copenhagen, Frederiksberg Campus, Frederiksberg C, Denmark

Through this Twinning Programme, the Danish side with the Malaysian counterparts, particularly the Faculty of Forestry, UPM initiated urban forestry as a bachelor's degree programme in 2005. Though it was not an accomplishment, the faculty offers a bachelor's degree programme in Forestry majoring in Urban Forestry. Those students who are interested to major in urban forestry will be offered an elective course package (21 credits) which consists of 13 courses related to urban forestry and arboriculture such as Urban Forest Management, Integrated Pest Management of Urban Forests, Urban Tree Health, Risk Management in Urban Forestry, Arboriculture, etc. These courses are also offered at the postgraduate level, particularly for the programme of Master in Sustainable Landscape Management at the Faculty of Design and Architecture, UPM as an elective course (Fig. 1.3). Urban forestry and arboriculture courses were also provided in other universities in Malaysia such as Universiti Malaysia Sabah (UMS) and University College of Agroscience Malaysia (UCAM). In UMS, similar courses are designed and adopted for bachelor's degree programmes such as International Tropical Forestry and Nature Parks and Recreation. The same goes in UCAM for those students under the Bachelor Programme of Landscape Management, who can find elective courses such as Urban Forestry and Tree Risk Management. Besides, some Polytechnics or Community Colleges similarly offer arboriculture courses for their diploma programmes such as Diploma in Landscape and Horticulture. Albeit still fragmented, urban forestry education has been increasingly emphasised particularly in those programmes related to forestry, landscape and parks and recreation. With an increasing demand from the industry, perhaps more students should be trained under well-established urban forestry or arboricultural bachelor programmes at the higher education level in Malaysia, so that well-rounded personnel could be nurtured to integrate urban forestry and arboriculture with its roots in forestry and parks and recreation.

It was also during the implementation of this project, that the 1st Arborist Certification course under the International Society of Arboriculture (ISA), US was organised in Malaysia by the Forest Research Institute Malaysia (FRIM) in June 2005 with the support of the Danish counterparts. This 1st Arborist Certification course was conducted by William M. Fountain, a professor of Arboriculture from the University of Kentucky who was involved in training the arborists for the ISA certification exam. Professor Fountain has played a significant role in inspiring the local tree care practitioners, researchers and academicians in their work and has contributed tremendous effort in educating and training the Malaysian arborist/urban foresters. Since then, the number of ISA Certified Arborists has increased from 23 (from the 1st batch in 2005) up to 92 certified arborists at present. Later, this first batch of certified arborists took the initiative to form a society named as the Malaysian Society of Arborist (PArM) which was established in 2005 to serve arboriculture practitioners in Malaysia. It has attracted about 300 members as of 2021. Resultantly, the term 'arborist' has become much more common among property managers, developers, landscape contractors, landscape architects, town planners, architects, engineers, etc. Many arborists in Malaysia were also hired as consultants in dealing with trees specifically on tree risk assessment, tree planting, transplanting, tree pest and diseases and many more (Fig. 1.4).

Fig. 1.3 The author with his first batch of students from the Tropical Arboriculture class under the programme of Master in Sustainable Landscape Management at the Faculty of Design and Architecture, Universiti Putra Malaysia (UPM)

Further, due to a high enthusiasm among arborists and tree climbers across the country, Tree Climbers Malaysia (TCM) was initiated to promote tree climbing as a recreational activity for the public and tree lovers. It is also one of the organisations listed by Tree Climbers International (TCI) as resources outside the U.S for recreational tree climbing. TCM also organises a series of courses for recreational purposes and professional training. Currently, TCM works closely with the National Institute of Occupational Safety & Health (NIOSH) for tree worker safety training. PArM also took an effort to organise the Malaysia Tree Climbing Championship (MTCC) recognized by the ISA in 2018 and 2019. Tree climbers who receive credits from TCM were also qualified to compete in the Asia Pacific Tree Climbing (APTCC) in Christchurch, New Zealand, in 2019. One of the climbers from Malaysia was appointed as the scoring judge at the APTCC. Taken together, the concept of urban forestry and arboriculture has been gradually accepted by researchers and practitioners in Malaysia. It has also captured the interest of the local town councils, landscape contractors, landscape architects, nurseries, etc., particularly those who are directly involved with the urban tree care industry. A similar evolutionary path was also noticed in other parts of the world because arboriculture predates urban forestry as a profession and has a much larger professional footprint because of the large scale of the commercial tree care industry (O'Herrin et al., 2020). There is a growing number of ISA-certified arborists in Malaysia, the majority of these certified arborists are those who are directly involved with the urban tree care industry such as

Fig. 1.4 As a certified arborist (MY-0300A), the author has been involved actively in tree risk assessment in golf courses, schools, embassies, government and private organisations and private houses since 2007

landscape contractors, landscape architects, developers, and nursery owners besides academicians and researchers.

Apart from the practices of urban forestry and arboriculture, substantial research in this field has been carried out in Malaysia. According to a recent systematic review on urban forestry research in Malaysia by Sundara Rajoo et al. (2021), there is a continuous increase in the number of scholarly studies from 2007–2021. This is closely related to the country's rapid urbanisation, with Malaysia being considered one of the most urbanised countries in Southeast Asia (Alias et al., 2010). It is also estimated by 2025, the Malaysian urban population will increase from 34.3% in 1971 to almost 80% by 2025 (National Urbanisation Policy, 2016). Consequently, there is a growing awareness of the need for sustainable urbanisation in Malaysia, similar to other countries of the global south. However, the majority of the research themes of urban forestry so far focus on the environmental perspective, covering issues of biodiversity, urban ecology, microclimate and pollution. For example, many studies were conducted on the urban biodiversity, particularly on birds in Malaysia (e.g. Idilfitri & Mohamad, 2012; Jasmani et al., 2017; Karuppannan et al., 2013; Lim & Mojiol, 2019). Another group of studies focuses on urban heat island and urban trees' functions (e.g. Hamid et al., 2019; Harun et al., 2020; Morries et al., 2015; Salleh et al., 2013; Shahidan et al., 2012; Wang et al., 2019), attributed to the fact that urban heat island in this region has brought many serious issues e.g. heat waves, flash floods since 2004 (Ramakreshnan et al., 2018), similar to other mega-urban areas in Asia that are known to suffer from significant urban heating (Hunt et al., 2018) with important implications on human health and wellbeing (Hajat et al., 2014). Furthermore, this is also due to the international environmental agendas, namely United Nation's Sustainable Development Goals (SDGs) and the Malaysian government's short-term development objectives under the 11th Malaysian Development Plan (2016–2020) which emphasises sustainable development and increasing environmental conservation (Ngu et al., 2020). Unfortunately, issues related to carbon sequestration of urban forestry, which could also contribute to urban climate change mitigation, are still lacking. Most of the relevant research concentrates on natural forests and plantations. Not much has been done on carbon sequestration in the context of urban forests. A couple of empirical studies simply compute the volume of carbon sequestration of trees planted in urban green reserves and the open space in Malaysian cities (e.g. Kanniah & Ho, 2018; Misni et al., 2016).

Special attention has also been given to the social-cultural values of urban forests, covering studies investigating urban residents' behaviours in and around urban forests, such as Sreetheran (2010, 2015, 2017), Malek et al. (2014), Kerishnan and Sreetheran (2021), and Anak Siba et al. (2020). These studies focused on various issues such as urban green space usage patterns, motives, fear of crime in urban green spaces, park safety etc.

Currently, a new research trend has been witnessed: health and wellbeing and the role of urban forests in maintaining public wellbeing. Many studies in Malaysia looked into how urban green spaces (e.g. parks, forest reserves, and green corridors) could cater to the well-being of the public in general (e.g. Foo, 2016; Mansor & Harun, 2014; Nath et al., 2018; Rajoo et al., 2019). The increasing rate of urbanisation is also associated with increased obesity, heart diseases and poorer mental health, owing to the increased exposure to pollution, as well as changes in behaviours and lifestyles that directly influence disease risk like smoking, sedentary time, social behaviour etc. (Ewing et al., 2006; McLafferty & Wang, 2009). In Malaysia, the National Health and Morbidity Survey, conducted by the Ministry of Health (MOH), indicates that the occurrence of mental health problems among people aged 16 years and above accounts for 29.2%, approximately 4.2 million. This indicates that one in three Malaysians has experienced mental health problems. And the current situation is very worrying as there is an increase in the reported cases related to mental health problems over the past 10 years, from 10.6% in 1996 to 11.2% in 2006. With these issues at the national level, more knowledge is needed in experimental research, especially on stress reduction and attention restoration. In addition, more studies on health and wellbeing require theoretical frameworks in the tropical context with more robust assessment and monitoring methods.

Besides empirical studies about urban forests' environmental and cultural benefits, a number of studies and publications focus mainly on urban tree inventory, tree vandalism, tree risk assessment, tree health assessment, and management policies. For example, a pictorial guide to identify hazardous trees was written by Sreetheran (2009) (Fig. 1.5); a historical perspective of urban forestry in Malaysia by Sreetheran et al. (2006); developing criteria and indicators to evaluate hazardous trees in Kuala Lumpur by Sreetheran and Yaman (2010); street tree inventory in Kuala Lumpur by Sreetheran et al. (2011); developing criteria for urban street tree vandalism assessment (Hamzah et al., 2020); assessing the level of pruning knowledge among tree maintenance workers (Badrulhisham & Othman, 2016); tree species selection model for street trees (Hasan et al., 2018) and on urban tree phenology and flower induction (Ahmad Nazarudin et al., 2012). Several studies have also focused on urban tree pests and diseases particularly on ornamental trees (Farid et al., 2018; Hua et al., 2015; Lee, 2014; Philip, 1999). To further enhance this group of research, more rigorous methods should be applied particularly in the field of tree risk assessment, apart from just using instruments such as Picus® Sonic Tomograph or the Resistograph®, in the field to determine internal defects such as cavities or decay, so that robust assessment and knowledge about species failure and defect profile among Malaysian trees can be generated.

The last strand of studies looked specifically into urban heritage trees in Malaysia (Lau et al., 2017; Mohd et al., 2019; Wan Ali et al., 2016). These studies were initiated because the rapid development of urban areas in Malaysia has transformed the land use from the natural environment into a built environment. This situation causes a lot of urban heritage trees to be felled to make way for urban development (Mohd et al., 2019), which had led many researchers to look at the challenges of implementing tree preservation order in Malaysia (Ibrahim et al., 2019; Nik Mohamed Sukri et al., 2020). Realising that more trees need to be cut down to accommodate the rapid urban expansion, the government of Malaysia has introduced the Town and Country Planning Act 1976 (Act 172) and Federal Territory (Planning) Act 1982 (Act 267), both of which include the preservation of trees, however the implementation remains challenging.

Fig. 1.5 The pictorial guidebook 'Hazardous Trees' was published by the Forest Research Institute Malaysia (FRIM) in 2009. This book was written by the author as an aid to identify the criteria and indicators to evaluate hazardous trees and the elements in hazard tree management. It was one of the earliest references on arboriculture published in Malaysia particularly on tree risk assessment

Overall, there is still a limited number of research focusing on urban tree governance, though recently an emerging group of researchers, academicians and practitioners is looking into relevant issues such as the implementation of the tree preservation order in Malaysia. Besides that, research on urban soil science and the valuation (economics) related to urban forests is still missing in Malaysia. More educational materials could be compiled and published related to urban forestry and arboriculture in the context of the tropics or Malaysia particularly.

This book aims to provide a review and also empirical discussion on the issues related to urban forestry and arboriculture in Malaysia. It also highlights the potential and limitations within different research disciplines when dealing with urban forestry and arboriculture. For the practitioners, this book could help set out a more realistic agenda for what can be planned and achieved within this field of urban forestry and arboriculture. Hopefully, this first compilation of research work on urban forestry and arboriculture in Malaysia could not only serve as a reference for the academicians, researchers, practitioners and students but also inspire more research on urban forestry and arboriculture, particularly in those fields which are still lacking in the context of tropical urban forestry and arboriculture.

References

Ahmad Nazarudin, M. R., Tsan, F. Y., Normaniza, O., & Adzmi, Y. (2012). Phenological growth stages of the golden penda tree (Xanthostemon chrysanthus). *Annals of Applied Biology, 161*(1), 12–15. https://doi.org/10.1111/j.1744-7348.2012.00546.x

Alias, A., Ali, A. S., & Wai, C. K. (2010). New urbanism and township developments in Malaysia. *Urban Design International, 16*, 76–93.

Anak Siba, M. A. J., Sreetheran, M., & Ujang, N. (2020). Investigating the use and constraints associated with green outdoor environment at workplaces: What do the office employees in Kuala Lumpur say? *Urban Forestry & Urban Greening, 51*, 126692. https://doi.org/10.1016/j.ufug.2020.126692

Badrulhisham, N., & Othman, N. (2016). Knowledge in tree pruning for sustainable practices in Urban setting: Improving our quality of life. *Procedia-Social and Behavioral Sciences, 234*, 210–217. S1877042816314902. https://doi.org/10.1016/j.sbspro.2016.10.236

Burkill, I. H. (1966). *A dictionary of the economic products of the Malay Peninsula* (2 vols.). Ministry of Agriculture and Cooperation.

Farid, A. M., Wan Muhd Azrul, W. A., Syazwan, S. A., Lee, S. Y., & Mohamed, R. (2018). First record of basal stem rot of foxtail palm *Wodyetia bifurcata* caused by *Ganoderma boninense* in Malaysia. *Plant Disease, 102*(7), 461–1461. https://doi.org/10.1094/pdis-12-17-2027

Deneke, F. (1993, May 30–June 2). Urban forestry in North America: Towards a global ecosystem perspective. In G. Blouin & R. Comeau (Eds.), *Proceedings of the First Canadian Urban Forests Conference* (pp. 4–8, 151 p). Winnipeg MB.

Dujesiefken, B., Drenou, C., Oven, P., & Stobbe, H. (2005). Arboricultural practices. In C. Kjonijnendijk, K. Nilsson, & T. B. Randrup (Eds.), *Urban forests and trees* (pp. 419–441). Springer.

Ewing, R., Brownson, R. C., & Berrigan, D. (2006). Relationship between urban sprawl and weight of United States youth. *American Journal of Preventive Medicine, 31*(6), 464–474. S074937970600331X. https://doi.org/10.1016/j.amepre.2006.08.020

Foo, C. H. (2016). Linking forest naturalness and human wellbeing—A study on public's experiential connection to remnant forests within a highly urbanized region in Malaysia. *Urban Forestry & Urban Greening. 16*, 13-24. S1618866716000054. https://doi.org/10.1016/j.ufug.2016.01.005

Hajat, S., Vardoulakis, S., Heaviside, C., & Eggen, B. (2014). Climate change effects on human health: Projections of temperature-related mortality for the UK during the 2020s, 2050s and 2080s. *Journal of Epidemiology and Community Health, 68*, 641–648.

Hamid, H. H. A., Latif, M. T., Nadzir, M. S. M., Uning, R., Khan, M. F., & Kannan, N. (2019). Ambient BTEX levels over urban suburban and rural areas in Malaysia. *Air Quality Atmosphere & Health, 12*(3), 341–351. https://doi.org/10.1007/s11869-019-00664-1

Hamzah, H., Othman, N., & Mohd Hussain, N. H. (2020). Setting the criteria for urban tree vandalism assessment. *Planning Malaysia, 18*(14). https://doi.org/10.21837/pm.v18i14.8

Harris, R. W., Clark, J. R., & Matheny, N. P. (1999). *Arboriculture: Integrated management of trees, shrubs and vines*, 4th ed. Englewood Cliffs, Prentice Hall.

Harun, Z., Reda, E., Abdulrazzaq, A., Abbas, A. A., Yusup, Y., & Zaki, S. A. (2020). Urban heat island in the modern tropical Kuala Lumpur: Comparative weight of the different parameters. *Alexandria Engineering Journal, 59*(6), 4475–4489. S111001682030377X. https://doi.org/10.1016/j.aej.2020.07.053

Hasan, R., Othman, N., & Ismail, F. (2018). Developing Malaysian roadside tree species selection model in urban areas. *Planning Malaysia Journal, 16*(7). https://doi.org/10.21837/pmjournal.v16.i7.515

Helms, J. A. (1998). *The dictionary of forestry*. Society of American Foresters.

Hua, L. S., Chen, L., Sajap, A. S., Peng, T., & Ashaari, Z. (2015). Development of Pteroma Pendula Joannis (Lepidoptera: Psychidae) feeding on selected landscape trees in Peninsular Malaysia. *The Malaysian Forester, 78*(1), 87–96.

Hunt, J. C., Aktas, Y. D., Mahalov, A., Moustaoui, M., Salamanca, F., & Georgescu, M. (2018). Climate change and growing megacities: Hazards and vulnerability. In *Proceedings of Institution of Civil Engineers—Engineering sustainability* (pp. 314–332).

Idilfitri, S., & Mohamad, N. H. N. (2012). Role of ornamental vegetation for birds' habitats in Urban parks: Case study FRIM Malaysia. *Procedia-Social and Behavioral Sciences, 68*, 894–909. S1877042812057588. https://doi.org/10.1016/j.sbspro.2012.12.275

Ibrahim, P. I., Zahrull Pauzi, H. F., & Mohd Masri, N. N. (2019). The implementation of tree preservation order in urban environment: Public and local authority perception. *Journal of Architecture, Planning & Construction Management, 9*(1), 1–18.

Jasmani, Z., Ravn, H. P., & van den Bosch, C. C. K. (2017). The influence of small urban parks characteristics on bird diversity: A case study of Petaling Jaya Malaysia. *Urban Ecosystems, 20*(1), 227–243. https://doi.org/10.1007/s11252-016-0584-7

Johnston, M. (1996). A brief history of urban forestry in the United States. *Arboricultural Journal, 20*, 257–278.

Jorgensen, E. (1986). Urban forestry in the rearview mirror. *Arboricultural Journal, 10*, 177–190.

Justice, C. L. (1986). The concept of urban forest as applied to Kuala Lumpur, Malaysia. *Journal of Arboriculture, 12*, 178–181.

Kanniah, K. D., & Ho, C. (2018). Urban forest cover change and sustainability of Malaysian cities. *Chemical Engineering Transactions, 56*, 673–678.

Karuppannan, S., Baharuddin, Z. M., Sivam, A., & Daniels, C. B. (2013). Urban green space and Urban biodiversity: Kuala Lumpur Malaysia. *Journal of Sustainable Development, 7*(1), 1. https://doi.org/10.5539/jsd.v7n1p1

Kerishnan, P. B., & Sreetheran, M. (2021). Factors contributing to the usage of pocket parks-A review of evidence. *Malaysia. Urban Forestry & Urban Greening, 58*, 126985. https://doi.org/10.1016/j.ufug.2020.126647

Koening, J. G. 1894. Journal of a voyage from India to Siam and Malacca in 1779. *Journal of the Straits Branch of the Royal Asiatic Society*, (26), 58–201.

Konijnendijk, C. C. (2003). A decade of urban forestry in Europe. *Forest Policy and Economics, 5*(3), 173–186.

Konijnendijk, C. C., Nilsson, K., Randrup, T. B., & Schipperijn, J. (Eds.). (2005). *Urban forests and trees*. Springer.

Konijnendijk, C. C., Randrup, T. B., & Nilsson, K. (2000). Urban forestry research in Europe: An overview. *Journal of Arboriculture, 26*(3), 152–161.

Konijnendijk, C. C., Ricard, R. M., Kenney, A., & Randrup, T. B. (2006). Defining urban forestry—A comparative perspective in North America and Europe. *Urban Forestry & Urban Greening, 4*, 93–103.

Lau, B., Jonathan, Y., & Alias, M. (2017). Heritage tree expert assessment and classification: Malaysian perspective. *World Academy of Science, Engineering and Technology, Open Science Index 128, International Journal of Biological and Ecological Engineering, 11*(8), 629–635.

Lee, C. Y. (2014). Urban forest insect pests and their management in Malaysia. *Formosan Entomologist, 33*, 207–214.

Lim, W. S., & Mojiol. A. R. (2019). A preliminary assessment on avian community in the urban forest of universiti Malaysia sabah. *Transaction on Science and Technology, 6*(3), 292–297.

Malek, N. A., Mariapan, M., & Rahman, N. I. A. A. (2014). Community participation in quality assessment for green open spaces in Malaysia. *Procedia Social and Behavioral Sciences, 168*, 219–228.

Mansor, M., & Harun, N. Z. (2014). Health issues and awareness and the significant of green space for health promotion in Malaysia. *Procedia-Social and Behavioral Sciences, 153*, 209–220. S1877042814054974. https://doi.org/10.1016/j.sbspro.2014.10.055

McLafferty, S., & Wang, F. (2009). Rural reversal?. *Cancer, 115*(12), 2755–2764. https://doi.org/10.1002/cncr.24306

Miller, R. W., Hauer, R. J., & Werner, L. P. (2015). *Urban forestry: Planning and managing urban greenspaces* (3rd ed.). Waveland Press.

Misni, A., Jamaluddin, S., & Kamaruddin, S. M. (2015). Carbon sequestration through urban green reserve and open space. *Planning Malaysia Journal, 13*(5), https://doi.org/10.21837/pmjournal.v13.i5.142

Mohd Ariffin, N. F., Abdul Aziz, N. A., & Mohd Yunos, M. Y. (2019). The significance of heritage trees conservation for urban development in Taiping Lake Garden, Malaysia. *International Journal of Engineering & Technology. Alam Cipta: International Journal of Sustainable Tropical Design Research & Practice, 12*, 39–46.

Morris, K. I., Salleh, S. A., Chan, A., Ooi, M. C. G., Abakr, Y. A., Oozeer, M. Y., & Duda, M. (2015). Computational study of urban heat island of Putrajaya. *Malaysia, Sustainable Cities and Society, 19*, 359–372. https://doi.org/10.1016/j.scs.2015.04.010

Nath, T. K., Han, S. S. Z., & Lechner, A. M. (2018). Urban green space and well-being in Kuala Lumpur Malaysia. *Urban Forestry & Urban Greening, 36*, 34–41. S1618866718300815. https://doi.org/10.1016/j.ufug.2018.09.013

National Urbanisation Policy II. (2016). Federal department of town and country planning, Peninsular Malaysia.

Ngu, H. K., Lee, M. D., & Osman, M. S. B. (2020). Review on current challenges and future opportunities in Malaysia sustainable manufacturing: Remanufacturing industries. *Journal of Cleaner Production, 273*, 123071.

O'Herrin, K., Wiseman, P. E., Day, S. D., & Hauer, R. J. (2020). Professional identity of urban foresters in the United States. *Urban Forestry & Urban Greening, 54*, 126741. https://doi.org/10.1016/j.ufug.2020.126741

Philip, E. (1999). Wilt disease of angsana (Pterocarpus inducus) in Peninsular Malaysia and its possible control. *Journal of Tropical Forest Science, 11*(3), 519–527.

Rajoo, K. S., Karam, D. S., & Aziz, N. A. A. (2019). Developing an effective forest therapy program to manage academic stress in conservative societies: A multi-disciplinary approach. *Urban Forestry & Urban Greening, 43*126353. S1618866719300123. https://doi.org/10.1016/j.ufug.2019.05.015

Ramakreshnan, L., Aghamohammadi, N., Fong, C. S., Ghaffaraianhoseini, A., Wong, L. P., Hassan, N., & Sulaiman, N. M. A. (2018). A critical review of urban heat Island phenomenon in the context of greater Kuala Lumpur Malaysia. *Sustainable Cities, 39*, 99–113.

Randrup, T. B., Konijnendijk, C. C., Kaennel Dobbertin, M., & Prüller, R. (2005). The concept of urban forestry in Europe. In C. C. Konijnendijk, K. Nilsson, T. B. Randrup, & J. Schipperijn (Eds.), *Urban forests and trees* (pp. 9–21). Springer.

Salleh, S. A., Latif, Z. A., Mohd, W. M. N. W., & Chan, A. (2013). Factors Contributing to the Formation of an Urban Heat Island in Putrajaya Malaysia. *Procedia-Social and Behavioral Sciences, 105*, 840–850. S1877042813044613. https://doi.org/10.1016/j.sbspro.2013.11.086

Shahidan, M. F., Jones, P. J., Gwilliam, J., & Salleh, E. (2012). An evaluation of outdoor and building environment cooling achieved through combination modification of trees with ground materials. *Building and Environment, 58*, 245–257. https://doi.org/10.1016/j.buildenv.2012.07.012

Sreetheran, M. (2009). *Hazardous Trees* (52 p). Forest Research Institute Malaysia (FRIM).

Sreetheran, M. (2010). Establishing performance indicators from the user perspective as tools to evaluate the safety aspects of urban parks in Kuala Lumpur. *Pertanika Journal of Social Sciences & Humanities, 18*, 199–207.

Sreetheran, M. (2015). *A socio-ecological approach of fear of crime in urban green spaces—A case in Kuala Lumpur, Malaysia* [Unpublished Doctoral Dissertation]. University of Copenhagen.

Sreetheran, M. (2017). Exploring the urban park use, preference and behaviours among the residents of Kuala Lumpur, Malaysia. *Urban Forestry & Urban Greening, 25*, 85–93. https://doi.org/10.1016/j.ufug.2017.05.003

Sreetheran, M., Adnan, M., & Azuar, A. K. K. (2011). Street tree inventory and tree risk assessment of selected major roads in Kuala Lumpur, Malaysia. *Arboriculture & Urban Forestry, 37*, 226–235.

Sreetheran, M., Philip, E., Adnan, M., & Siti Zakiah, M. (2006). A historical perspective of urban tree planting in Malaysia. *An International Journal of Forestry and Forest Industries (Unaslyva), 57*, 6.

Sreetheran, M., & Yaman, A. R. (2010). The identification of criteria and indicators to evaluate hazardous street trees of Kuala Lumpur, Malaysia: A Delphi study. *Journal of Forestry, 108*(7), 360–364.

Sukri, N. A. N. M., Ismail, Z., & Ariffin, W. T. W. (2020). Developing a model of tree preservation model (Act 172) implementation in construction projects. *International Journal of Sustainable Construction Engineering and Technology, 11*(1), 18–30.

Sundara Rajoo, K., Karam, D. S., Abdu, A., Rosli, Z., & Gerusu, G. J. (2021). Urban forest research in Malaysia: A systematic review. *Forets, 12*, 903. https://doi.org/10.3990/f12070903

Wan Ali, A., Hassan, N., & Hassan, K. (2016). The morphology of heritage trees in colonial town: Taiping Lake. *Procedia—Social and Behavioral Sciences, 222*, 621–630.

Wang, K., Aktas, Y. D., Stocker, J., Carruthers, D., Hunt, J., & Malki-Epshtein, L. (2019). Urban heat island modelling of a tropical city: Case of Kuala Lumpur. *Geoscience Letters, 6*, 4. https://doi.org/10.1186/s40562-019-0134-2

Sreetheran Maruthaveeran is an Associate Professor at the Department of Landscape Architecture in the Faculty of Design and Architecture, Universiti Putra Malaysia (UPM). He was a Senior Researcher at the Forest Research Institute Malaysia (FRIM) for 15 years before joining the academic line in 2016. He received his Ph.D. in Landscape Architecture and Planning from the University of Copenhagen, Denmark in 2015 under the supervision of Prof. Dr. Cecil Konijnendijk van den Bosch. Sreetheran's research interest lies in the social aspects of urban forestry and urban greening, recreation and leisure sciences. He teaches courses such as landscape ecology, tropical arboriculture, landscape maintenance and research methods at the bachelor's and master's levels in UPM. He is also a Professional Technologist (Ts) registered under the Malaysian Board of Technologists (MBOT) and a Certified Arborist (MY0300A) under the International Society of Arboriculture (ISA), USA since 2007. Currently, he is one of the Editorial Board Members for the *Journal of Outdoor Recreation and Tourism*, Leisure Studies and World Leisure Journal. Sreetheran is also the Editor-in-Chief for *ALAM CIPTA International Journal on Sustainable Tropical Design Research and Practice*. Currently he is the Coordinator of the Urban Forestry Unit (6.07.00) for the International Union of Forest Research Organizations (IUFRO).

Part I
Review

Chapter 2
The Status and Future of Urban Forestry in Sabah, Malaysia

Andy Russel Mojiol and Wing-Shen Lim

Abstract A wide range of urban green spaces can be entered by the public in Sabah. These urban green spaces are dominated by indigenous tree species, although certain exotic trees are incorporated into the urban landscapes of Sabah. A combination of indigenous and exotic tree species is applied in local urban landscaping, to maximise the ecosystem services provided to a certain area by the urban trees and shrubs. To date, various perspectives of urban forestry in Sabah have been explored by researchers, which include the species composition and ecology of the urban wildlife and vegetation, conditions of the urban trail, health condition of urban tree stands, prospects and functions of various urban habitats, and the perception and satisfaction of public to the urban green spaces. The impacts of the urbanisation and introduction of exotic tree species to the indigenous tree community, and the implementation of GIS in monitoring the change in urban green cover, are yet to be conducted in the urban landscapes of Sabah. With this, further research is required to provide a more comprehensive understanding of the current conditions of every existing urban area and green space presented in Sabah.

Keywords Ecosystem service · Urban biodiversity · Urban green space · Urban landscape

Introduction

Sabah is one of the Thirteen Federal States of Malaysia, and it is situated on Borneo Island, as shown in Fig. 2.1. Presently, it is home to approximately 3.88 million population in 2020, which has increased by 1.4% of the population every year since 2017 (Department of Statistics Malaysia, 2019). Moreover, three cities have been established in Sabah, and then the human populations in these three urban areas are higher than those in the suburban and rural areas of the state. This phenomenon actually highlights the fact that the urban population has steadily increased with

A. R. Mojiol (✉) · W.-S. Lim
Faculty of Tropical Forestry, Universiti Malaysia Sabah (UMS), Jalan UMS, 88400 Kota Kinabalu, Sabah, Malaysia
e-mail: andy@ums.edu.my

S. Maruthaveeran et al. (eds.), *Urban Forestry and Arboriculture in Malaysia*,
https://doi.org/10.1007/978-981-19-5418-4_2

each passing year, thus triggering the occurrence of rapid urbanisation in Sabah, in response to the given matter. Nevertheless, the acceleration of urban development and expansion can reduce the accessibility of the public to nature (Chaudhry & Tewari, 2011). Therefore, "Urban Forestry" was introduced and incorporated into the process of developing and managing the urban landscapes for Malaysia in 1995, to ensure that the green spaces were available in every urban area in the country. Presently, Sabah has successfully blended nature into the local urban areas, thus resulting in the presence of many green cities and towns in the state.

Generally, urban forestry focuses on maintaining the tangible and intangible benefits provided by the urban trees to the environment and dwellers of a certain urban area. Currently, the urban green spaces in Sabah can be entered by the public, may they be situated within or outside an urban area, because they are established to provide ecosystem services to the surrounding urban environment. Costanza et al. (2017) and Nair et al. (2018) emphasised that the ecosystem services provided by an urban green space were grouped under four different categories, namely the provisioning, regulating, supporting, and cultural services, and then the quality of services were close to those provided by the natural forested areas (see Table 2.1). The ecosystem services provided by the urban forests included providing tranquillity to the urban dwellers (e.g. Lee et al., 2004; Mojiol, 2018a), increasing the aesthetic value (e.g. Mojiol, 2001; Mojiol & Bürger-Arndt, 2012), reducing the disturbance and pollution impacts (e.g. Chaudhry & Tewari, 2011; Mojiol, 2018b), supporting the house urban biodiversity (e.g. Lim & Mojiol, 2019; Wells et al., 2014), regulating the environmental condition at an urban area (e.g. Konijnendijk et al., 2013; Nair et al., 2018), providing tourism, education, research and recreation opportunities (e.g. Lim et al., 2019; Mojiol, 2018a), and maintaining the urban nutrient and hydrological cycles (e.g. Konijnendijk et al., 2013; Nair et al., 2018). Table 2.1 shows the contributions of ecosystem services that are provided by an urban green space to the surrounding urban environment.

Urban Forests in Sabah

Different types of urban green spaces can be found in Sabah, particularly in major cities such as Kota Kinabalu, Sandakan and Tawau. Based on its size, setting and location, the facilities, infrastructures, and outdoor recreational activities provided at a certain urban green space can be varied from those of other urban green spaces. Currently, the number of established urban green spaces in Kota Kinabalu is more than those in Sandakan and Tawau. Since 'Nature' is the main element of the urban forestry concept applied in local urban landscaping, thus Kota Kinabalu has earned the nickname 'Rainforest City', while Sandakan and Tawau are referred as the 'Nature City' and 'Planters' Land' respectively (Lee et al., 2004).

Konijnendjik et al. (2013) defined urban green structures as any areas or structures that incorporated naturally-grown or cultivated trees into urban landscapes, hence including both the natural and cultivated forested areas. Presently, major urban forests

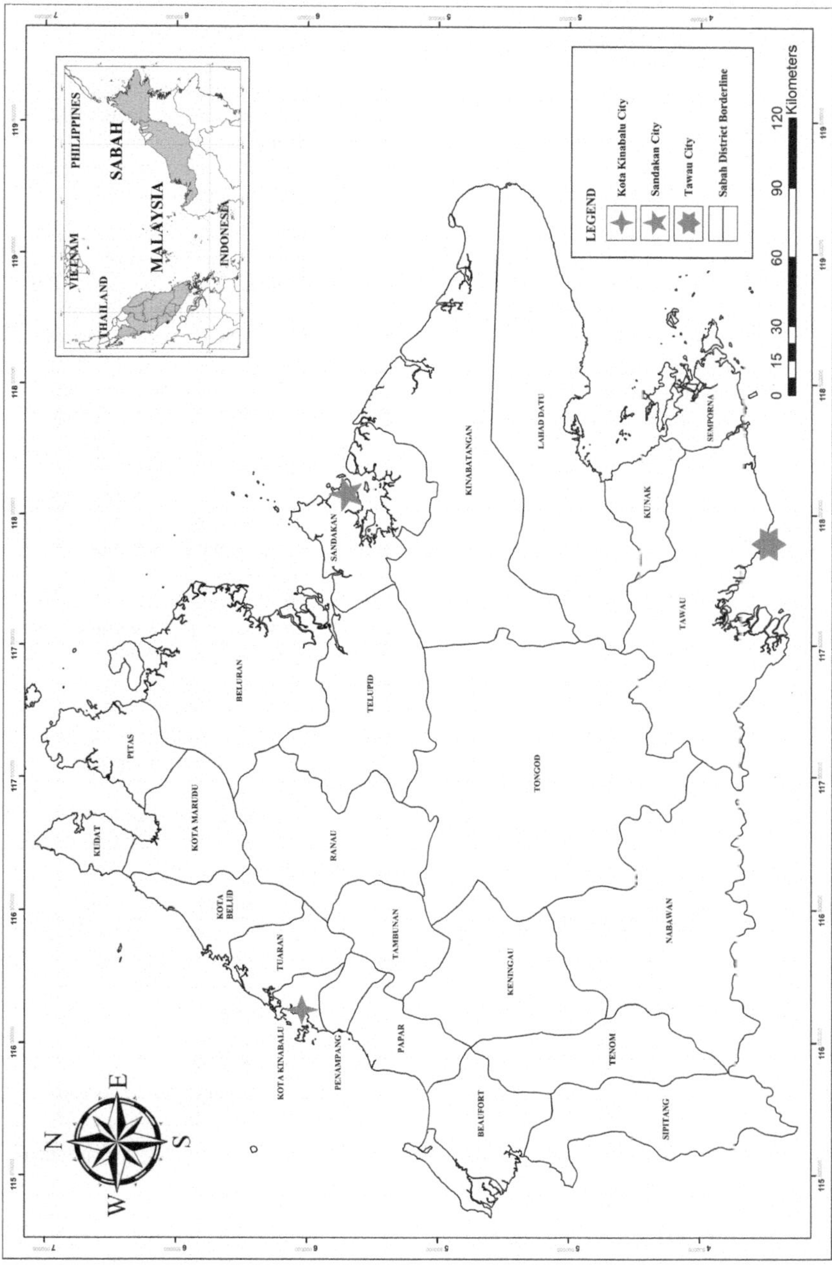

Fig. 2.1 Location of Sabah and its three major cities

Table 2.1 Contributions of the ecosystem services that are provided by an urban green space to the surrounding urban environment

Category	Ecosystem service	Example
Provisioning	Consumable Food Tangible Resource Genetic Resource	Provisioning of edible plant-based foods to the urban dwellers. Providing consumable water, fuel, lumber, and other non-timber forest products to the urban dwellers Functioning as the source of biological materials and products in the urban environment
Regulating	Pollution Control Climate Control Air Purification Carbon Sequestration Shading Effect Disturbance Regulation Natural Pollination Biological Control Erosion Control and Sediment Retention	Regulating the impacts of various pollution in urban areas.. Regulating the surrounding solar radiation, ambient temperature and humidity of the urban environment Removing air pollutants and providing clean oxygen supply to the urban area Functioning as carbon sinks for the surrounding urban areas. Providing additional shade to the urban dwellers at daytime. Regulating the impacts of certain environmental disturbance (e.g. wind damage, storm, flood, and landslide) in the urban areas Providing pollinators that are crucial for the reproduction of urban plant populations Maintaining the distribution and abundance of urban wildlife in a state of equilibrium through the prey-predator relationship Reducing the impacts of sediment loss, coastal and soil erosion in the urban environments
Supporting	Habitat Provisioning Nutrient Cycling Soil Formation	Functioning as habitat for both the indigenous and exotic species of plants, and also for both the resident and migratory species of wildlife in the urban environments Maintaining equilibrium in nutrient cycling of the urban areas Accumulating organic matters as soil nutrients for urban plants

(continued)

Table 2.1 (continued)

Category	Ecosystem service	Example
Cultural	Recreation Opportunity Intangible Resource	Supplying green landscapes with natural resources suitable for the conducting of outdoor recreation and tourism activities Providing intangible products that can benefit the well-being of urban dwellers (e.g. aesthetic, spiritual, scientific, and educational values)

Source Adapted from Mojiol and Lim (2020)

that can be entered by the public in Sabah include pocket parks, urban parks and forest reserves. A pocket park can be defined as a smaller-sized urban park that is often being established in a highly-urbanised area. Because of its limited space, only trees and seats are given at a pocket park, majorly for providing the urban dwellers with a safe place to rest under shade during the daytime, travel from one place to another and socialise with others (Peschardt et al., 2012), such as in the Lintasan Deasoka Pocket Park and Kampung Air Pocket Park in Kota Kinabalu (see Fig. 2.2).

Both urban parks and pocket parks have in fact provided similar ecosystem services to the surrounding environments and urban inhabitants. However, the aesthetic value remains to be the major contribution of pocket parks, while the urban parks mainly provide outdoor activity opportunities for the urban dwellers in Sabah (Mojiol, 2018a). Therefore, different types of infrastructures and facilities have been constructed at the urban park, to allow a wide range of outdoor recreation activities to be implemented by various individuals at the given park. The provided infrastructures and facilities range from a playground to a nature trail, kiosk, public toilets, table and chair, bench, and jogging and cycling tracks (Lee et al., 2004; Lim et al., 2019).

There will be at least one urban park in every city in Sabah. Currently, Kota Kinabalu has the highest number and variations of parks compared to Sandakan and Tawau. For example, in Kota Kinabalu, there is a protected wetland park (Kota Kinabalu Wetland Centre), public parks (e.g. Tun Fuad Stephen Park, Teluk Likas Public Park and Prince Philip Public Park), a memorial park (Tugu Petagas Public Park), community forest (Kionsom Recreation Centre), and secondary tropical inland forest (UMS Peak) (Lim et al., 2019; Mojiol, 2018a). Whereas in Sandakan, we have the Sandakan Rainforest Park and Sandakan Memorial Park as the urban parks (Damit, 2009; Sugau et al., 2017), whereas the Muhibbah Public Park and Tawau Sports Complex are examples of urban parks in Tawau.

Apart from the pocket parks and urban parks, forest protected areas can also be found as urban green spaces in Sabah. Generally, forest protected area is a totally protected forest that provides various types of ecosystem services to its surrounding environment, at regional and global scales in long-term (Costanza et al., 2017; Nair et al., 2018), hence different from the urban parks and pocket parks that mainly supply cultural services to the urban inhabitants in Sabah (Mojiol, 2018a). Some of

Fig. 2.2 The three types of urban forests that can be found within and outside of the three major cities of Sabah, such as the: (**a**) Lintasan Deasoka Pocket Park (Kota Kinabalu); (**b**) Kampung Air Pocket Park (Kota Kinabalu); (**c**) Teluk Likas Public Park (Kota Kinabalu); (**d**) Sandakan Rainforest Park (Sandakan); (**e**) Heritage Amenity Forest Reserve (Sandakan), and (**f**) Tawau Hill Park (Tawau)

these forest protected areas are considered colossal in size than the other types of urban green spaces (Gobilik & Limbawang, 2010; Nair et al., 2018). Additionally, these protected forests are located at the periphery of the cities, therefore highly accessible for public use (Lee et al., 2004).

At the moment, Kawang Recreation Centre (Papar) and Heritage Amenity Forest Reserve (Sandakan) are some of the forest protected areas that can be accessed by the public near Kota Kinabalu and Sandakan respectively (Damit, 2009; Nair et al., 2018). Then, Tawau Hill Park and Gemok Hill Forest Reserve are forest protected areas that serve as urban forests of Tawau (Gobilik & Limbawang, 2010; Lee et al., 2004). The Infrastructures and facilities provided at forest protected areas are more diverse than those at pocket parks and urban parks, and then unique tourist attractions are available in every forest protected area as well. As an example, nature trails are provided for visitors to experience the nature of Kawang Recreation Centre and Gemok Hill Forest Reserve (Lee et al., 2004), while the Heritage Amenity Forest Reserve has a deer farm, fish pond and suspension bridge for urban inhabitants of Sandakan to interact with the tropical forest ecosystem of Sabah (Sugau et al., 2017). Table 2.2 shows the summary on the three types of urban green spaces in Sabah. Then, Fig. 2.2 displays the three types of urban forests that can be found within and outside of the three major cities of Sabah.

Table 2.2 Summary of the descriptions, roles and examples for the three types of urban green spaces in Sabah

Urban Forest	Description	Role	Example
Pocket Park	Urban green space that is smaller than 0.5 ha and established at the city centre with limited free space. Highly accessible to urban inhabitants by walking	Providing a safe zone to rest, walk and socialise with others	Kampung Air Pocket Park (KK) & Lintasan Deasoka Pocket Park (KK)
Urban Park	Urban green space that is larger than 0.5 ha and established outside of the city centre, but still within the city border. Accessible to urban inhabitants by walking, more accessible with transportation	Allowing many people to do outdoor recreation activities together at the same time	Teluk Likas Public Park (KK); Sandakan Rainforest Park (SDK); & Tawau Sport Complex (TWU)
Forest Protected Area	Total protected forests managed mainly to provide ecosystem services, but open for public use. Accessible only to urban inhabitants with transportation	Providing provisioning, regulating, cultural, and supporting services to the surrounding area	Kawang Recreation Centre (KK); Heritage Amenity Forest Reserve (SDK); & Gemok Hill Forest Reserve (TWU)

Note KK = Kota Kinabalu; SDK = Sandakan; TWU = Tawau
Source Adapted from Mojiol and Lim (2020)

Urban Tree Diversity in Sabah

In Sabah, both indigenous and exotic tree species are employed in urban landscape design (National Landscape Department, 2008). According to Skinner (2019), the cultivation of trees in urban landscaping was first practised in Sabah during the nineteenth century by the British North Borneo Company. More than 100 trees that were "huge and shady" with ornamental values were planted in Kota Kinabalu, when the North Borneo War Monument was erected to remember the fallen British soldiers from World War I in 1923. Nonetheless, Kota Kinabalu was largely damaged during World War II (1940–1946), and then the local tree planting activity ceased until Kota Kinabalu was taken over by the British once again (1947–1956). Since then, the local tree planting activity continued even after the independence of Malaya (1957) and the establishment of the Federation of Malaysia (1963). In 2006, about one million indigenous trees of Sabah were cultivated within the city centre of Kota Kinabalu and its surrounding areas by the Kota Kinabalu City Hall (DBKK), to further increase the greenery in the local cityscape. Later, the Roads Landscape Beautification Project was initiated in 2012 to increase the shaded areas in the city centre (Lajiun, 2012). This project was financed by the Sabah State Government, as part of the initiatives in the 9th Malaysia Plan (2006–2010), and this was continued during the 10th Malaysia Plan (2011–2015).

Currently, many of the decade-old urban trees in Kota Kinabalu are declared by DBKK as heritage trees. These include the Angsana (*Pterocarpus indicus*) that are mainly found along Gaya Street, as well as the Rain Tree (*Samanea saman*) found mostly near Sabah Museum, Wisma Muis and Sabah State Mosque, as shown in Fig. 2.3. The Rain Tree is indigenous to northern South America, while Angsana is indigenous to south-eastern Asia (Gardner et al., 2011). However, both species can produce beautiful inflorescences and grow into tall trees with broad crowns. Hence, both species are highly valued as urban tree species that can provide shade and aesthetic values to urban areas of Sabah. Moreover, combinations of indigenous and exotic tree species are also applied in the local urban landscaping, in which more indigenous species are cultivated compared to the exotic species (Mojiol & Bürger-Arndt, 2012; Mojiol, 2018b). While at Sandakan, the indigenous dipterocarps and nepenthes species remain intact as part of the local urban landscape, along with other disturbed and pristine forest species (Damit, 2009; Sugau et al., 2017), as shown in Fig. 2.4. Nevertheless, certain indigenous tree species like the Pulai (*Alstonia angustiloba*) and Parasol Leaf Tree (*Macaranga tanarius*) are well-distributed throughout the urban secondary-regrowth forests of Sabah, thus becoming common to be sighted here (Damit, 2009; Sugawara et al., 2009). In other words, a wide range of indigenous tree species remain viewable in the urban green spaces in Sabah, and then particular exotic tree species are cultivated to further enhance the quality of ecosystem services provided to the surrounding urban regions, such as aesthetic value enhancement, education and research opportunities provisioning and wind-breaking (Mojiol, 2018a, 2018b).

Fig. 2.3 The oldest planted tree in Kota Kinabalu, Sabah, the Rain Tree (*Samanea saman*) (Location: Wisma Muis, Kota Kinabalu)

A tree species is used in urban landscaping when it can provide key functions and can be maintained easily in the urban landscape of Sabah for the long term (National Landscape Department, 2008). These key functions include providing shade, visual and acoustic screening, pollution and erosion control, wind-breaking, regulating urban climate, and enhancing aesthetic value (Mojiol, 2018a; Nair et al., 2018). Since the three major cities of Sabah only have limited spaces for trees and shrubs planting, tree species that can contribute multiple key functions to the surrounding urban areas are prioritised. First of all, the different combinations of trees and shrub species are cultivated together as hedging (visual screening) and to absorb the urban noise (acoustic screening), which can subsequently create a sense of serenity in the local urban areas and green spaces (Chaudhry & Tewari, 2011; Konijnendijk, et al., 2013; Mojiol, 2018a). The selected species are comprised of both indigenous and exotic species, which include the Pink Trumpet Tree (*Tabebuia pallida*), Royal Palm (*Roystonea regia*), Clove (*Syzygium aromaticum*), Sealing-wax Palm (*Cyrtostachys renda*), Red Coondoo (*Mimusops elengi*), False Asoka (*Monoon longifolia*), Tembusu (*Cyrtophyllum fragrans*), and the Weeping Fig (*Ficus benjamina*). This is followed by tree species with unique features (e.g. leaves, flower and stem/trunk) are selected and cultivated to enhance the aesthetic values of urban areas and green spaces in Sabah. Some of the tree species that are commonly cultivated for this purpose are the Pong-pong (*Cerbera odollam* and *C. manghas*), Bauhinia (*Bauhinia blakeana* and *B. purpurea*), Pink Poui (*Tabebuia rosea*), Flame of the Forest (*Delonix regia*), and Indian Rose Chestnut (*Mesua ferrea*) trees. A list

Fig. 2.4 Nepenthes species that can be found at the urban forest of Sandakan, Sabah (Location: Sandakan Rain Forest Park)

of urban tree species that are frequently observed in the urban landscapes of Sabah is displayed in Table 2.3.

Since the three major cities in Sabah are situated near the seashores, therefore the local infrastructures, vegetation and coastlines are highly susceptible to damage and erosion caused by the strong coastal wind throughout the year. Due to this, trees that can withstand the kinetic force of coastal wind and adapt to the coastal environment are suitable to be planted for wind- breaking and controlling the local soil and coastal erosions (Mojiol, 2018a; Mojiol & Bürger-Arndt, 2012). Tree species that are primarily selected for erosion control and wind- breaking are the River Tamarind (*Leucaena leucocephala*), Casuarina (*Casuarina equisetifolia*), Coconut (*Cocos nucifera*), Indian Almond (*Terminalia catappa*), and the Pong- pong (*Cerbera odollam*) trees. Furthermore, trees with broad crown like the Mahogany (*Khaya senegalensis*), Sandbox Tree (*Hura crepitans*), Yellow Flame (*Peltophorum pterocarpum*), Shrubby Simpoh (*Dillenia suffroticosa*), Weeping Fig (*Ficus benjamina*), Golden Shower (*Cassia fistula*), and Alexandrian Laurel (*Calophyllum inophyllum*), are good shade provider, particularly in urban areas (Mojiol, 2018a, 2018b). In addition, these species are beneficial in pollution control and climate regulation when

Table 2.3 A list of urban tree species that are frequently observed in the urban landscapes of Sabah

Family	Species	Common name	TA	MHA
Exotic species				
Annonaceae	*Monoon longifolia*	False Asoka	√	–
Bignoniaceae	*Tabebuia pallida*	Pink Trumpet Tree	√	–
	Tabebuia rosea	Pink Poui	√	√
Caricaceae	*Carica papaya*	Papaya	–	√
Euphorbiaceae	*Hevea brasiliensis*	Rubber Tree	–	√
	Hura crepitans	Sandbox Tree	√	–
Fabaceae	*Acacia auriculiformis*	Ear-leaf Acacia	√	√
	Acacia mangium	Mangium	√	√
	Albizia chinensis	Albizia	–	√
	Andira inermis	Brown-heart	√	–
	Andira surinamensis	Kedondong	√	–
	Cassia biflora	Twin Flowered Cassia	√	–
	Cassia siamea	Kassod Tree	√	–
	Delonix regia	Flame of the Forest	√	√
	Erythrina crista-galli	Coral Tree	√	–
	Leucaena leucocephala	River Tamarind	–	√
	Pithecellobium dulce	Manila Tamarind	√	–
	Samanea saman	Rain tree	√	√
Meliaceae	*Khaya senegalensis*	Mahogany	√	–
Moraceae	*Ficus elastica*	Indian Rubber	√	√

(continued)

Table 2.3 (continued)

Family	Species	Common name	TA	MHA
Myrtaceae	*Syzygium aromaticum*	Clove	√	–
Palmae	*Roystonea regia*	Royal Palm	√	√
Sapindaceae	*Filicium decipiens*	Fern Tree	–	√
Indigenous species				
Anacardiaceae	*Mangifera caesia*	Beluno	–	√
	Mangifera indica	Mangga	√	√
	Mangifera pajang	Bambangan	–	√
Apocynaceae	*Alstonia angustiloba*	Pulai	–	√
	Alstonia macrophylla	Hard Milkwood	–	√
	Alstonia spatulata	Hard Milkwood	–	√
	Cerbera manghas	Pong-pong	√	–
	Cerbera odollam	Pong-pong	√	√
Araucariacea	*Agathis borneensis*	Borneo Kauri	–	√
Bignoniaceae	*Spathodea campanulata*	African Tulip	–	√
Bombacaceae	*Ceiba pentandra*	Kapok Tree	–	√
	Durio acutifolius	Tupaloh	–	√
	Durio zibethinus	Durian	–	√
Calophyllaceae	*Mesua ferrea*	Indian Rose Chestnut	√	–
Casuarinaceae	*Casuarina equisetifolia*	Aru	√	√
Clusiaceae	*Calophyllum inophyllum*	Alexandrian Laurel	√	√
	Garcinia malaccensis	Forest Mangosteen	–	√

(continued)

Table 2.3 (continued)

Family	Species	Common name	TA	MHA
	Garcinia mangostana	Mangosteen	–	√
Combretaceae	*Terminalia catappa*	Indian Almond	√	√
	Terminalia subspathulata	Malayan Terminalia	–	√
Datiscaceae	*Octomeles sumatrana*	Binuang	–	√
	Dillenia excelsa	Purple Simpoh	–	√
	Dillenia suffroticosa	Shrubby Simpoh	√	√
Elaeocarpaceae	*Elaeocarpus stipularis*	Benzoin Oil-fruit	–	√
	Muntingia calabura	Malayan Cherry	–	√
Euphorbiaceae	*Baccaurea angulata*	Red Angle Tampoi	–	√
	Baccaurea motleyana	Rambai	–	√
	Macaranga gigantea	Elephant's Ear Tree	–	√
	Macaranga tanarius	Parasol Leaf Tree	–	√
	Macaranga triloba	Mahang	–	√
	Wetria insignis	Forest Rambai	–	√
Fabaceae	*Bauhinia x blakeana*	Bauhinia	√	√
	Bauhinia purpurea	Bauhinia	√	–
	Cassia alata	Candle Bush	–	√
	Cassia fistula	Golden Shower	√	√
	Cassia spectabilis	Weeping Cassia	√	–
	Dalbergia sissoo	Indian Rosewood	–	√
	Paraserianthes falcataria	White Albizia	–	√
	Peltophorum pterocarpum	Yellow Flame	√	√

(continued)

Table 2.3 (continued)

Family	Species	Common name	TA	MHA
	Pterocarpus indicus	Angsana	√	√
	Millettia pinnata	Indian Beech	√	–
	Tamarindus indica	Tamarind	–	√
Gentianaceae	*Cyrtophyllum fragrans*	Tembusu	√	√
	Cyrtophyllum gigantea	Giant Tembusu	–	√
	Fagraea volubilis	Todopon Puok	–	√
Lauraceae	*Alseodaphne bancana*	Medang	–	√
	Cinnamomum camphora	Camphor Tree	–	√
	Cinnamomum iners	Wild Cinnamon	√	–
	Cinnamomum zeylanicum	Cinnamon	–	√
Lythraceae	*Lagerstroemia indica*	Crepe Myrtle	√	–
	Lagerstroemia speciosa	Crepe Myrtle	√	√
Melastomataceae	*Melastoma malabatricum*	Indian Rhododendron	–	√
Meliaceae	*Azadirachta excelsa*	Giant Neem	–	√
	Lansium domesticum	Langsat	–	√
	Melia indica	Neem Tree	–	√
Moraceae	*Artocarpus anisophyllus*	Entawak	–	√
	Artocarpus altilis	Breadfruit	–	√
	Artocarpus elasticus	Terap	–	√
	Artocarpus heterophyllus	Jackfruit	–	√
	Artocarpus integer	Cempedak	–	√
	Ficus benjamina	Weeping Fig	–	√

(continued)

Table 2.3 (continued)

Family	Species	Common name	TA	MHA
	Ficus microcarpa	Malayan Banyan	–	√
	Ficus speciosa	Fig	√	–
Musaceae	*Musa speciosa*	Dwarf Banana	√	–
	Musa violascens	Wild Banana	–	√
Myrsinaceae	*Ardisia elliptica*	Seashore Aridisa	–	√
Myrtaceae	*Syzygium cerasiformis*	Eugenia	–	√
	Syzygium grandis	Sea Apple	√	√
	Syzygium jambos	Rose Apple	–	√
	Syzygium malaccensis	Malay Apple	–	√
	Syzygium oleina	Eugenia	–	√
	Syzygium papillosa	Eugenia	–	√
	Psidium guajava	Guava	–	√
	Syzygium cumini	Java Plum	–	√
Oxalidaceae	*Sarcotheca diversifolia*	Tabarus	–	√
	Sarcotheca glauca	Aremajuh	–	√
Palmae	*Areca catechu*	Pinang	–	√
	Calamus subinermis	Rattan	–	√
	Cocos nucifera	Coconut	√	√
	Cyrtostachys rendu	Sealing-wax Palm	√	–
	Nypa fructicans	Nipah Palm	–	√
	Oncosperma tigillarium	Nibong palm	–	√
	Ptychosperma macarthurii	MacArthur Palm	√	–

(continued)

Table 2.3 (continued)

Family	Species	Common name	TA	MHA
	Veitchia merrillii	Christmas Palm	√	–
Phyllanthaceae	*Phyllanthus urinaria*	Shatterstone	–	√
Rubiaceae	*Morinda citrifolia*	Indian Mulberry	–	√
	Pleiocarpidia sandakanica	Buloh-buloh	–	√
Rutaceae	*Citrus microcarpa*	Calamondin	–	√
Sapindaceae	*Guioa pleuropteris*	Senyamuk	–	√
	Nephelium lappaceum	Rambutan	–	√
Sapotaceae	*Mimusop elengi*	Red Coondoo	√	√
	Palaquium gutta	Gutta Percha	–	√
Sterculiaceae	*Commersonia bartramia*	Brown Kurrajong	–	√
Ulmaceae	*Trema cannabina*	Poison Peach	√	–
Verbenaceae	*Vitex pinnata*	Malayan Teak	–	√

Note TA = Town Area, and; MHA = Mixed Horticulture Area. The listed tree species only represent a portion of the overall urban species that are used in urban landscaping at town and mixed horticulture areas of Sabah
Sources Mojiol and Bürger-Arndt (2012) and Mojiol (2018a, 2018b)

compared to shrubs, which further emphasises the importance of these species in urban areas (Mojiol, 2001).

Presently, trees that are commonly cultivated around the cityscapes of Sabah include the False Asoka (*Monoon longifolia*), Flame of the Forest (*Delonix regia*), Pink Poui (*Tabebuia pallida*), Sandbox Tree (*Hura crepitans*), Royal Palm (*Roystonea regia*), Rain tree (*Samanea saman*), Pong-pong (*Cerbera odollam* and *C. manghas*), Coconut (*Cocos nucifera*), Malayan Banyan (*Ficus microcarpa*), Bauhinia (*Bauhinia x* blakeana and *B. purpurea*), Clove (*Syzygium aromaticum*), Aru (*Casuarina equisetifolia*), Golden Shower (*Cassia fistula*), Indian Almond (*Terminalia catappa*), Red Coondoo (*Mimusop elengi*), Angsana (*Pterocarpus indicus*), Crepe Myrtle (*Lagerstroemia indica* and *L. speciosa*), Yellow Flame (*Peltophorum pterocarpum*), and Wild Cinnamon (*Cinnamomum iners*). Likewise, these tree species possess various unique characteristics that allow them to provide various key functions to the local urban areas, and then they can be easily maintained for long term. Some of the key ecosystem services that are provided by these urban tree species include regulating the urban climate, providing habitat to the urban wildlife, wind-breaking, and providing recreation and education opportunities to the urban inhabitants (Mojiol & Lim, 2020). Contrarily, long-term maintenance of certain tree species is difficult, which makes them unsuitable for urban landscaping, although they can provide various key functions. In Sabah, the Acacias (*A. mangium* and *A. auriculiformis*) are identified as some of the exotic tree species that are unsuitable for urban landscaping because these trees produce a ton of dry foliage throughout the day, hence the long-term maintenances of these species are both costly and time-consuming (Mojiol & Bürger-Arndt, 2012). In sum, indigenous tree species are much more preferred than exotic species in urban landscaping in Sabah. Figure 2.5 shows the tree species that are frequently cultivated in the urban landscapes of Sabah.

Research Trends, Challenges and Future of Urban Forestry in Sabah

To date, various perspectives of urban forestry in Sabah have been explored, including the species composition and ecology of urban flora and fauna (e.g. Chin et al., 2014; Lim & Mojiol, 2019; Wells et al., 2014), urban trail conditions (Lim et al., 2019), health condition of urban tree stands (e.g. Mojiol, 2018b; Mojitol & Bürger-Arndt, 2012), potentials and functions of different urban habitats (e.g. Lee et al., 2004; Mojiol & Bürger-Arndt, 2012), and lastly the public perception and satisfaction towards urban green spaces (e.g. Hilmi & Mojiol, 2017; Mojiol, 2018a; Mojiol et al., 2017). The summary of several published studies that have explored the different perspectives of urban forestry in the three major cities of Sabah is tabulated and displayed as shown in Table 2.4.

Presently, information on the species composition of urban trees at Kota Kinabalu are more available (e.g. Majuakim et al., 2018; Mojiol & Bürger-Arndt, 2012; Mojiol et al., 2016; Sugawara et al., 2009) compared to Sandakan (e.g. Damit, 2009; Chin

Fig. 2.5 Tree species that are frequently cultivated tree species at the urban landscapes of Sabah, such as the: (**a**) Malayan Banyan (*Ficus microcarpa*); (**b**) Crepe Myrtle (*Lagerstroemia* sp.); (**c**) Royal Palm (*Roystonea regia*); (**d**) Yellow Flame (*Peltophorum pterocarpum*), and (**e**) Rain tree (*Samanea saman*)

Table 2.4 Summary of the published studies that have investigated the various perspectives of urban forestry in the three major urban areas of Sabah, Malaysia

Field of study	Citation	Study area	Type of urban forest
Potentials and Functions of Urban Forests	Lee et al. (2004)	Kota Kinabalu, Sandakan and Tawau	Urban Parks and Forest Reserves
	Mojiol and Bürger-Arndt (2012)	Kota Kinabalu	Town Area
Perception and Satisfaction of Public	Mojiol and Bürger-Arndt (2012)	Kota Kinabalu	Town Area
	Mojiol et al. (2013)	Kota Kinabalu	Forest Reserve
	Hilmi and Mojiol (2017)	Kota Kinabalu	Town Area
	Mojiol et al. (2017)	Kota Kinabalu	Urban Park
	Mojiol (2018a)	Kota Kinabalu	Urban Parks and Pocket Parks
	Nair et al. (2018)	Kota Kinabalu	Forest Reserve
Recreation Ecology	Lim et al. (2019)	Kota Kinabalu	Urban Park
Tree Species Composition and Ecology	Gobilik and Limbawang (2010)	Tawau	Forest Reserve
	Mojiol and Bürger-Arndt (2012)	Kota Kinabalu	Town Area
	Chin et al. (2014)	Sandakan	Urban Park
	Majuakim et al. (2018)	Kota Kinabalu	Urban Parks
Tree Health Assessment	Mojiol and Bürger-Arndt (2012)	Kota Kinabalu	Town Area
	Mojiol (2018b)	Kota Kinabalu	Town Area
Wildlife Monitoring and Management	Mojiol et al. (2008)	Kota Kinabalu	Urban Park
	Wells et al. (2014)	Kota Kinabalu	Residential areas and the adjacent forests at urban, suburban and rural areas
	Lim and Mojiol (2019)	Kota Kinabalu	Urban Park

Note This table only lists out some of the published research studies that are conducted at the three major cities of Sabah, and which referred to their respective field of study. Therefore, this table does not show every past research study that has focused on urban forestry in the three major cities, as well as in other urban areas of Sabah

et al., 2014; Sugau et al., 2017) and Tawau (e.g. Gobilik & Limbawang, 2010). Therefore, urban tree conditions at Kota Kinabalu are more well-defined than those at Sandakan and Tawau. Generally, different urban habitats have different densities, sizes, species composition and diversity of tree stands, where exotic trees like rubber (*Hevea brasiliensis*), Mangium and Ear-leaf Acacia are commonly found at different urban habitats in Kota Kinabalu (Mojiol & Bürger-Arndt, 2012). Overall, the health condition of local urban tree stands is considered as moderate to good, based on the fact that a majority of these urban trees exhibit moderate growth performance and controllable tree disease transmission issues, as reported by Mojiol (2018b).

A higher variation of wildlife species is observed at the least-disturbed urban habitats of Sabah (Damit, 2009; Lim & Mojiol, 2019; Mojiol et al., 2008; Wells et al., 2014). Whereas certain exotic wildlife species like the Asian House Shrew (*Suncus murinus*) and Asian Black Rat (*Rattus rattus*) were determined as invasive urban pests of Kota Kinabalu. However, only a few of such studies were conducted in Sandakan and Tawau, at the same time a majority of past studies were focusing on urban birds (e.g. Lim & Mojiol, 2019; Mojiol et al., 2008). Therefore, further research into the non-bird community (e.g. reptiles, insects, mammals, and amphibians) in the urban environment of Sabah is required in the future. Additionally, the main interests of researchers on urban wildlife are generally to the species composition and ecology of urban wildlife, such as the habitat preference of respective wildlife species across different urban green spaces (e.g. Lim & Mojiol, 2019; Mojiol et al., 2008).

Most of the urban dwellers in Sabah realise the differences in functions between various forms of urban green spaces, and also the function of urban forestry in planning and managing of local urban tree stands (Mojiol, 2018a). Presently, willingness to pay (WTP) is frequently being applied in quantifying values perceived by the public on the ecosystem services provided by urban green spaces in Sabah (e.g. Hilmi & Mojiol, 2017; Mojiol et al., 2017; Nair et al., 2018). Visitors from different gender and age groups, as well as distances, travel to an urban green space, value ecosystem services differently, although urban ecosystem services like soil fertility and erosion control, provisioning of clean water, recreational opportunities, and aesthetic value are highly regarded to be important (Mojiol et al., 2017; Nair et al., 2018). Therefore, most of them are willing to provide funding in conserving these biological assets, with a present value of about RM 1,086,373.87 (Hilmi & Mojiol, 2017). Furthermore, recreational opportunities can provide numerous benefits to both the visitors and the surrounding local community, thus urban green spaces are promotable as tourist destinations in Sabah (Lee et al., 2004).

Classification of urban green spaces at Kota Kinabalu using geographical information system (GIS) is also being explored by local experts in Sabah. Digital mapping using secondary data is a cost and time-effective way to evaluate the current conditions of different urban habitats at Kota Kinabalu (Mojiol & Bürger-Arndt, 2012). For example, Sabah Soil Map is used to map out the soil associations throughout Kota Kinabalu, which is useful in assessing the soil distribution and agricultural capability across the given urban region. Then, the application of GIS in mapping out the boundaries of different urban habitats can ease the related authority in field demarcation and monitoring of urban habitats that are contributing to urban ecosystem services and sensitive to human disturbance.

In summary, urban conditions at Sandakan and Tawau are rarely examined by both international and local experts, therefore limited information is available regarding the existing condition of urban green spaces there. Urban trail condition at Sabah is hardly being examined, same goes for the management of urban green spaces at

Sandakan and Tawau. Species diversity and composition of urban flora and fauna at Tawau require further research in the future, together with urban non-avian communities presented at Kota Kinabalu and Sandakan, due to very limited information available on given matters. Other than these, there are two other aspects of urban forestry that have yet to be explored in Sabah. Urbanisation and the introduction of exotic species into isolated urban forest patches can impact the species richness and abundance of indigenous tree communities negatively, as well as minimise the connectivity and seed dispersal through wildlife and wind between different urban forest patches (Olejniczak et al., 2018). On the other hand, the application of remote sensing techniques in monitoring long-term change in the urban green cover of a particular city and urban green space (e.g. Kanniah, 2017) has yet been done here as well. Henceforth, further research should be conducted on these two topics, so that a holistic understanding of the current conditions of urban green spaces at Sabah can be verified in the future.

Conclusion

Urban forestry in Sabah has been explored by past researchers from different perspectives, but then a majority of past studies have been conducted at Kota Kinabalu. Because of that, the information related to the species composition and ecology of urban flora and fauna, urban trail conditions, health condition of the urban tree stands, potentials and functions of different urban habitats, and the public perception and satisfaction with the urban green spaces at Kota Kinabalu, are generally much well-defined, when compared to Sandakan and Tawau. From the economic perspective, the present value of biological assets at urban green spaces of Kota Kinabalu has been examined only once, while a relevant study has yet to be conducted at Sandakan and Tawau at this moment. Then, the social aspect of urban forestry in Sabah is mainly examined, in terms of the perceptions of urban dwellers towards the difference in ecosystem services provided by different urban green spaces, as well as the importance of urban forestry in urban tree planning and management. Similarly, the given research topics are mainly examined for the urban green spaces and urban tree stands at Kota Kinabalu, whereas those at Sandakan and Tawau have yet to be investigated at the same time.

The environmental aspect of urban forestry in Sabah can be evaluated through different perspectives, and yet a majority of the past studies were conducted at Kota Kinabalu. Even so, the existing conditions of urban trails and urban tree stands at Kota Kinabalu are evaluated only once at this moment, despite that management is the key to the maintenance of the biological, physical and cultural aspects of an urban trail and urban tree stand from being degraded along with time. Then, species composition and habitat preference, are verified as the main interests of researchers when it comes to studying urban flora and fauna presented in different urban green spaces in Sabah. Furthermore, the lack of relevant studies on urban non-avifauna communities emphasises the need to conduct further research into this topic in the future. Currently, the impact of urbanisation and the introduction of exotic tree species toward indigenous tree diversity, as well as the impact of green cover change in isolated urban green spaces, are yet to be explored by researchers at the urban landscapes of Sabah. Therefore, these topics are determined to be the research gaps in urban forestry in Sabah, and then further research should be conducted on the given topics, not only in the three major cities but also in other urban areas of Sabah.

References

Chaudhry, P., & Tewari, V. P. (2011). Urban forestry in India: Development and research scenario. *Interdisciplinary Environmental Review, 12*(1), 80–93.

Chin, L., Chung, A. Y. C., & Clarke, C. (2014). Interspecific variation in prey capture behavior by co-occurring nepenthes pitcher plants: Evidence for resource partitioning or sampling- scheme artifacts? *Plant Signaling and Behavior, 9*, 1–16.

Costanza, R., de Groot, R., Braat, L., Kubiszewski, I., Fioramonti, L., Sutton, P., Farber, S., & Grasso, M. (2017). Twenty years of ecosystem services: How far have we come and how far do we still need to go? *Ecosystem Services, 28*, 1–16.

Damit, A. (2009). *Sandakan Rainforest Park: A scientifically & historically important forest*. Forest Research Centre.

Department of Statistics Malaysia. (2019). *Department of Statistics Malaysia, Official Portal*. Retrieved 25 October 2019, form https://www.dosm.gov.my/v1/

Endreny, T. A. (2018). Strategically growing the urban forest will improve our world. *Nature Communications, 9*(1), 10–12.

Gardner, S., Sidisunthorn, P., & Lai, E. M. (2011). *Heritage trees of Penang*. Areca Books.

Gobilik, J., & Limbawang, S. (2010). Notes on species composition and ornamental gingers in Tawau Hills Park, Sabah. *Journal of Tropical Biology and Conservation, 7*(1), 31–48.

Hilmi, M. A., & Mojiol, A. R. (2017). Contingent valuation on urban trees in city of Kota Kinabalu, Sabah. *Transactions on Science and Technology, 4*(2), 166–173.

Kanniah, K. D. (2017). Quantifying green cover change for sustainable urban planning: A case of Kuala Lumpur, Malaysia. *MIT-UTM Malaysia Sustainable Cities Program* (Working Paper Series 1). Universiti Teknologi Malaysia.

Konijnendijk, C. C., Annerstedt, M., Nielsen, A. B., & Maruthaveeran, S. (2013). Benefits of Urban Parks—A systematic review. *A Report for IFPRA*. World Urban Parks.

Lajiun, J. (2012, June 20). City Hall spending RM1 million for tree planting. *The Borneo Post*. Retrieved 5 May 2020, from https://www.theborneopost.com/2012/06/20/city-hall-spending-rm1-million-for-tree-planting/

Lee, Y. F., Ligunjang, J., & Yong, S. C. (2004). *Urban forestry and its relevance to tourism development in Sabah*. Paper presented in the Asia Europe Meeting (ASEM) Symposium on Urban Forestry.

Lim, W. S., & Mojiol, A. R. (2019). A preliminary assessment on avian community in the urban forest of Universiti Malaysia Sabah. *Transactions on Science and Technology, 6*(3), 292–297.

Lim, W. S., Nisa, S., & Mojiol, A. R. (2019). Rapid observational assessment on urban forest trails established at UMS Peak of Universiti Malaysia Sabah. *Tropical Forest Journal, 14*(1), 18–31.

Majuakim, L., Lee, A. M. L., & Gisil, J. (2018). An inventory of flora in urban forests of Universiti Malaysia Sabah Campus, Sabah, Malaysia. *Journal of Tropical Biology & Conservation, 15*, 173–188.

Mojiol, A. R. (2001). *Pokok-Pokok Hiasan Bandar*. Penerbit Universiti Malaysia Sabah.

Mojiol, A. R. (2018a). Public awareness on the importance of urban forest parks in Kota Kinabalu City, Sabah. *Borneo Science, 39*(1), 39–47.

Mojiol, A. R. (2018b). Tree health assessment for roadside tree in Kota Kinabalu City Centre, Sabah. *Borneo Science, 39*(2), 104–113.

Mojiol, A. R., & Bürger-Arndt, R. (2012). *Urban green areas management in Kota Kinabalu, Sabah-Malaysia: Concept and sustainability*. Lambert Academic Publishing.

Mojiol, A. R., & Lim, W. S. (2020). Urban forestry in Sabah, Malaysia: A perspective review. In G. N. Tanjina Hasnat & M. K. Hossain (Eds.), *Examining international land use policies, changes, and conflicts*. IGI Global Publishers.

Mojiol, A. R., Hassan, A., Maluda, J., & Immit, S. (2008). Rapid assessment on the abundance of bird species utilising the Kota Kinabalu Wetland Centre mangroves. *Journal of Tropical Biology and Conservation, 4*(1), 99–107.

Mojiol, A. R., Lo, M. W., Lintangah, W., & Amat, R. (2016). *A guide to the plants of Kota Kinabalu Wetlands*. Sabah Wetland Conservation Society.

Mojiol, A. R., Sompud, J., Igau, A. O., & Yong, Y. S. (2013). Contributions of Kawang Forest Reserve to the local community. *Sepilok Bulletin, 45*(17 and 18), 35–45.

Mojiol, A. R., Zamri, Z., Hilmi, M. A., & Gitom, M. (2017). Visitors' Willingness to Pay (WTP) at Kionsom Recreation Centre, Inanam, Kota Kinabalu Sabah. *Transactions on Science and Technology, 4*(42), 174–182.

Nair, G. V. G., Mojiol, A. R., Kamlun, K. U., & Lintangah, W. (2018). The contribution of forest ecosystem services toward the local community living vicinity to the forest protected area: The case of Kawang Forest Reserve, Sabah, Malaysia. *Transactions on Science and Technology, 5*(1), 25–30.

National Landscape Department. (2008). *National Landscape Guideline* (2nd ed.). National Landscape Department.

Olejniczak, M. J., Spiering, D. J., Potts, D. L., & Warren, R. J. (2018). Urban forests form isolated archipelagos. *Journal of Urban Ecology, 4*(1), 1–8.

Peschardt, K. K., Schipperijn, J., & Stigsdotter, U. K. (2012). Use of small public urban green spaces (SPUGS). *Urban Forestry & Urban Greening, 11*(3), 235–244.

Skinner, S. (2019, June 9). Protecting our heritage tree. *The Daily Express*. Retrieved 4 May 2020, from http://www.dailyexpress.com.my/read/2992/protecting-our-heritage-trees/

Sugau, J. B., Pereira, J. T., & Damit, A. (2017). *A guide to the trees in Heritage Amenity Forest Reserve (HQ), Sandakan: A Revised and Expanded Edition*. Sabah Forestry Department.

Sugawara, A., Sudin, M., Said, I. M., Suleiman, M., Gisil, J., & Sundaling, D. (2009). *Buku Panduan Hutan Bukit UMS: Tumbuh-tumbuhan*. Pernebit Universiti Malaysia Sabah.

Wells, K., Lakim, M. B., & O'Hara, R. B. (2014). Shifts from native to invasive small mammals across gradients from tropical forest to urban habitat in Borneo. *Biodiversity and Conservation, 23*(9), 2289–2303.

Andy Russel Mojiol is an Associate Professor at the Faculty of Tropical Forestry, Universiti Malaysia Sabah (UMS), and has been lecturing at UMS since 1998. He obtained his bachelor's and master's degree at the Universiti Putra Malaysia (UPM) and his doctorate degree in Park Planning and Management from the University of Goettingen, Germany in 2006. Andy is actively involved in research in the field of urban forestry and zoology and has contributed several articles in research journals and chapters in books. At present, he is an active member of the International DAAD Scholar and Alumni, Germany.

Wing-Shen Lim completed his M.Sc. in Forestry Science at the Faculty of Science and Natural Resources (FSSA), Universiti Malaysia Sabah (UMS) in May 2021. Currently, he serves as a demonstrator for the Zoology and Wildlife Management course at the Faculty of Tropical Forestry, UMS. Lim is actively involved in research related to Wildlife Monitoring and Management. Simultaneously, he partakes in research studies in the field of Urban Forestry and Recreation Ecology. He is currently pursuing his Ph.D. in Forestry Science at the Faculty of Tropical Forestry, UMS.

Chapter 3
Tree Preservation Order of Act 172: A Malaysian Legislation Towards Sustainable Urban Forests

Nik Adlin Nik Mohamed Sukri, Wan Tarmeze Wan Ariffin, and Shahzarimin Salim

Abstract An urban forest has to be sustained for it to continuously benefit the urban dwellers and the environment. The essence of urban forest sustainability is to have good urban tree management, and one means to achieve this state of affairs is through a legislation called Tree Preservation Order (TPO). The Malaysian TPO legislation comes under the Town and Country Planning Act 1976 or Act 172, hence known as TPO (Act 172) and it is meant to be implemented by the local planning authorities (LPAs) in the states of Peninsular Malaysia. TPO (Act 172) was introduced as a reaction to a disaster in which a high-rise building collapsed and believed to be the result of widespread lack of caring for the environment in development projects. This chapter thoroughly discusses on the TPO (Act 172) provisions, the tree felling prohibition provisions and the procedures for an LPA to make a TPO It also spells out the current status of TPO (Act 172) implementation which is considered very unsatisfactory and needs reviewing. Finally, the way forward for ensuring effective TPO (Act 172) implementation is mentioned.

Keywords Tree Preservation Order · Act 172 · Sustainable · Urban forest · Local planning authority

Introduction

Trees are an important part of any forest, including urban forests. Without trees, the forest will not sustain and animals who depend on it will disappear. In a city, town or a suburb, trees that have been planted or grown naturally and other floras and faunas around make up an urban forest. While we want to sustain the urban forest as much as possible, there will always be requests to remove some of the trees to give way to

N. A. N. M. Sukri (✉) · W. T. W. Ariffin
Forest Research Institute Malaysia (FRIM), Kepong, Malaysia
e-mail: nikadlin@frim.gov.my

S. Salim
Legal and Regulatory Planning Division, PLANMalaysia (Department of Town and Country Planning), Putrajaya, Malaysia

S. Maruthaveeran et al. (eds.), *Urban Forestry and Arboriculture in Malaysia*,
https://doi.org/10.1007/978-981-19-5418-4_3

development. The local planning authority (LPA) should evaluate the tree removal application thoroughly, and carefully consider every opportunity to preserve the trees before approving the removal.

Nevertheless, as far as urban forest sustainability is concerned, the tree removal may not be detrimental at all and may even be beneficial in the long run if it is wisely planned and properly executed. The statement may seem excessive, but that is not impossible. Every removal of trees provides a new chance to make-up for any mistake done before, for example, the planting of wrong tree species. In the situation where the trees must be replaced, taking lessons from the previous mistakes, better tree species can be proposed.

The bottom line is that the essence of urban forest sustainability is to have good urban tree management, and one means to achieve this state of affairs is by making the trees so important that they have to be preserved. Hence, many countries in the world practise a legislation called Tree Preservation Order (TPO). This chapter elaborates on the TPO legislation of Malaysia, how it began, its provisions, rules, TPO making procedures and the status of its implementation.

Tree Preservation Order (TPO)

There are many versions of the TPO definition that can be found in the literature. In this chapter, we would like to define TPO as "a written order made by a LPA, which in general, makes it a criminal offence to remove, fell (cut down), top, lop, uproot, wilfully damage or wilfully destroy a tree protected by that order, or to cause or permit such actions, without the authority's permission".

The power to make TPO and other similar legislation provisions is provided to LPAs through various acts, for instance, the *Town and Country Planning Act* in England, the *Planning Act* in Ireland, the *Parks and Trees Act* in Singapore, and the *Environmental Planning and Assessment Act* in Australia. In Malaysia, the TPO making power provisions are conferred under acts such as the *Town and Country Planning Act 1976* (Act 172) for the states in the Peninsular, and the *Federal Territory (Planning) Act 1982* (Act 267) for Kuala Lumpur. Since the principle of TPO making power is the same regardless of the Act and for the sake of representing Malaysia, the rest of this chapter will elaborate only on the provisions of Act 172 (referred to as TPO [Act 172]). TPO (Act 172) applies to most of the states in Malaysia, hence the progress of its implementation could reflect the efforts made by the Malaysian government toward urban forest sustainability.

Town and Country Act (Act 172)—Amendment 1995 (Act A933)

The collapse of Highland Towers Block A in Kuala Lumpur on the 11th of December 1993 has been perceived by many as the impetus for the inclusion of TPO provisions in Act 172. The Highland Towers tragedy was a wake-up call for the nation that if we did not take care of the environment, disasters would be inevitable.

Following the tragedy, a special cabinet committee headed by the Honourable Deputy Prime Minister came up with a proposal that Act 172 had to be amended in response to the growing concerns over the need to handle environmental issues more seriously and with better initiatives such as conserving and preserving the natural resources, greening the environment and protecting historical buildings which are directly affected by the town and country planning activities (JPBDSM, 2002). The proposed amendment to Act 172 was brought up and approved as a bill in the 42nd meeting of *Majlis Negara bagi Kerajaan Tempatan* (National Council for Local Governments) or MNKT held on the 30th of September 1994. About a year later, the bill was passed by the Parliament and on the 20th of October 1995, it was consented by the Yang di-Pertuan Agong. The law is referred to as the *Town and Country Planning Act (Amendment) 1995* or Act A933.

The amendment to Act 172 was intended, among others, to ensure that trees are to be preserved as much as possible for the benefit of people and the environment. Hence, a new part, namely Part VA (TREE PRESERVATION ORDER) was included after section 35 (Power to make rules) of Act 172 which gives power to LPAs in the states of Peninsular Malaysia to make TPO as they wish. In addition to the power to make an order to preserve trees, A933 also incorporates provisions that allow LPAs to prohibit the removal or felling of trees in development projects.

Tree Preservation Order of Act 172—TPO (Act 172)

The provisions of TPO (Act 172) are detailed in section 35A to 35H (Laws of Malaysia, 2014). Principally, TPO (Act 172) can be seen as conveying two main messages: (1) "You can't fell the tree without permission, you will be penalised if you do", and (2) "You must replace the tree you have felled". As far as the two messages are concerned, the related TPO (Act 172) provisions are as shown in Table 3.1. The interpretation of terms 'felling a tree', as inserted by Act A933, includes 'cutting down, topping, lopping, uprooting, damaging or destroying a tree'.

In layman's terms, TPO (Act 172) enables LPAs to forbid the felling of trees that are meant to be preserved (later to be referred to as 'TPO trees') and to prosecute the offences with a fine of not more than one hundred thousand ringgit (RM100,000) or a jail term up to 6 months, or both. Felled TPO trees must be replaced according to the time and place determined by the LPAs. Examples of TPO trees are such as

Table 3.1 Provision of TPO (Act 172) related to tree felling prohibition and tree planting

Section	Title	Provision	Sub-section
35A	Tree preservation order	If it appears to the local planning authority that it is expedient in the interest of amenity to preserve any tree, trees or group of trees in its area, it may make a tree preservation order with respect to such tree, trees, or group of trees	35A(1)
		A tree preservation order may, in particular, make provisions— *(a)* for prohibiting the felling of trees except with the written permission of and subject to conditions, if any, imposed by the local planning authority; and *(b)* for securing the planting of trees or the replacement of trees by replanting in such a manner as may be determined by the local planning authority	35A(2)
		Any person who contravenes any provision in the tree preservation order commits an offence and is liable, on conviction, to a fine not exceeding one hundred thousand ringgit or to imprisonment for a term not exceeding six months or to both	35A(4)
35E	Replacement of trees	It shall be the duty of the person who is found guilty under subsection 35A(4) for felling any tree in respect of which a tree preservation order is for the time being in force, in contravention of the tree preservation order, to replace such tree by planting another tree	35E(1)
35H	Prohibition to fell, etc., tree with girth exceeding 0.8 m	No person shall, without the written permission of the local planning authority, fell a tree with a girth exceeding 0.8 m which is not subjected to a tree preservation order unless the felling— *(a)* is in respect of such tree which is dying or dead; *(b)* is for the prevention of an imminent danger; or *(c)* is to comply with any written law	35H(1)

(continued)

Table 3.1 (continued)

Section	Title	Provision	Sub-section
		Any person who contravenes subsection (1) commits an offence and is liable, on conviction, to a fine not exceeding ten thousand ringgit or to imprisonment for a term not exceeding three months or to both	35H(4)

those with specific aesthetic values, historical values, planted by public figures and endangered species (JPBDSM, 2011).

The TPO (Act 172) also states that, by default, all trees with a girth exceeding 0.8 m (to be referred as ">0.8 m trees") are meant to be preserved and are not allowed to be felled unless they are dying, posing an imminent danger or the felling is required to comply with other legal regulations. In all the cases, the felling of TPO trees and >0.8 m trees must have written permission from the LPAs.

Prohibition of Tree Felling in Development Areas

In addition to TPO (Act 172), amendments were also made and imposed as additional conditions to the application for planning permission (PP) which among others, prohibits the applicants from felling the trees and requires them to plant trees in the proposed development areas. The PP application-related provisions can be found in sections 21 and 22 of Act 172, and those concerning tree felling prohibition and tree planting are as described in Table 3.2.

To section 21, a new section was added, namely, 21A which requires PP applicants to submit a development proposal report (DPR) which must also contain a survey of the trees and all forms of vegetation and the layout plans whose details are as specified in section 21B. With regard to trees, section 21B mentions that for development involving land, the layout plans must include measures to be taken by the PP applicant for the preservation and planting of trees on the land.

Moreover, the layout plans must also show the location and species of >0.8 m trees and other vegetation thereon. As required by section 21C, the layout plan of the trees must be prepared only by qualified professionals such as LAr (certified landscape architect) and whenever the health status of the trees is a concern, it has to be endorsed by CA (certified arborist). Furthermore, section 22 elaborates that, in dealing with PP applications, LPA must be perseverant in realising its ultimate intent to prohibit tree felling and to secure tree planting or replanting on the development land before deciding to grant or decline the PP. LPA is also required to ensure that the PP applicants will preserve and protect all TPO trees, if any, throughout the development project phases.

Table 3.2 Planning permission (PP) application related provisions in A933 involving tree felling prohibition and tree planting

Section	Title	Provision	Sub-section
21	Application for planning permission	An application for planning permission in respect of a development shall be made to the local planning authority and shall be in such form and shall contain such particulars and be accompanied by such documents, plans, and fees as may be prescribed	21(1)
21A	Development proposal report	In addition to the documents and plans required to be submitted under subsection 21(1) for planning permission, the applicant shall submit a development proposal report which shall contain the following: *(f)* layout plans, the details of which are specified in section 21B	21A(1)
21B	Layout plans	The layout plans under paragraph 21A(1)*(f)* shall show the proposed development and in particular— *(a)* where the development is in respect of any land— (iv) measures for the preservation and planting of trees thereon (v) the location and species of trees with a girth exceeding 0.8 m and other vegetation thereon	21B(1)
21C	Preparation of plan, etc., by a qualified person	All plans, particulars, layout plans and other documents required to be submitted under this Act shall be prepared by— *(a)* a person whose qualifications are prescribed under paragraph 58(2)*(h)*; or *(b)* a person who is entitled to do so under any other written law	-

(continued)

Table 3.2 (continued)

Section	Title	Provision	Sub-section
22	Treatment of applications	In dealing with an application for planning permission, the local planning authority shall take into consideration such matters as are in its opinion expedient or necessary for proper planning and in particular— *(bb)* the development proposal report;	22(2)
		Conditions imposed under subsection (3) may include any or all of the following conditions, that is to say, conditions— *(f)* prohibiting the felling of trees of a certain size, age, type or species at any particular location, unless it is to comply with any written law; *(g)* for securing the planting or replanting of trees of a certain size, age, type or species at any particular location in such manner as may be determined by the local planning authority;	22(5)
		It shall be the duty of the local planning authority to ensure where planning permission is granted that a tree preservation order, if any, is complied with	22(5A)

TPO Rules and TPO Making Procedures

TPO (Act 172) is a legislation and it can be considered to be implemented by LPAs only if they make a TPO legitimise the enforcement. A set of procedures to make a TPO and to deal with other matters related to it, known as "TPO Rules", was formulated by PLANMalaysia (formerly, JPBDSM) in 1998 for gazettement by the states and to be adopted by LPAs. "Rules", defined as "subsidiary legislation" in the Interpretations Act 1948 & 1967, are part of Malaysian legal sources that supplement the legislative function of the Malaysian legal system (Muhammad Syahlan et al., 2018).

Usually, rules require publishing in the Government Gazette to become legal. Selangor was the first state government to gazette TPO Rules, i.e., in 2001, followed by Perak and Melaka in 2011 and 2017, respectively. Nik Adlin, Zulhabri, Wan Tarmeze, et al. (2020) elaborate the process of TPO Rules gazettement by the three states and put up the flow chart as shown in Fig. 3.1.

The content of TPO Rules document (TRD) is summarised in Table 3.3. It covers many aspects, from the interpretation of the terminologies and TPO-making procedures to tree felling requests and penalties. In the light of this chapter with a title that gives more emphasis on the tree preservation order rather than the tree felling prohibition, the following text will elaborate only on the procedures to be followed by LPA when making a TPO, i.e., as those stated in rule 3 and rule 4 of Part II and as illustrated in Fig. 3.2.

Rule 3 states that prior to making a TPO, LPA should prepare a list of TPO trees in its area. The format for the tree list is as shown in Schedule I of TRD (Fig. 3.3). When it is ready to make a TPO, according to rule 4, the LPA will use the format as suggested in Form A of Schedule II (Fig. 3.4) which must contain (a) a scale map or plan showing the position of the TPO trees, (b) particulars relating to the species and size of the TPO trees, and (c) a declaration that permission for felling of TPO trees may only be obtained from the LPA and subject to conditions.

The TPO may also be accompanied by diagrams, illustrations including photographs and other materials if necessary. Each TPO document must be kept by LPA in its office with a copy readily made available for inspection during office hours and where applicable, a copy must also be sent to the owner or occupant of the land on which the TPO trees grow. Moreover, a notice (as in Form B of Schedule II, shown in Fig. 3.5) stating that a TPO has been made shall be affixed at a suitable spot on the land.

Status of TPO (Act 172) Implementation

Although TPO (Act 172) has been around for more than two decades, there has not been much progress in terms of its implementation and enforcement. This fact can

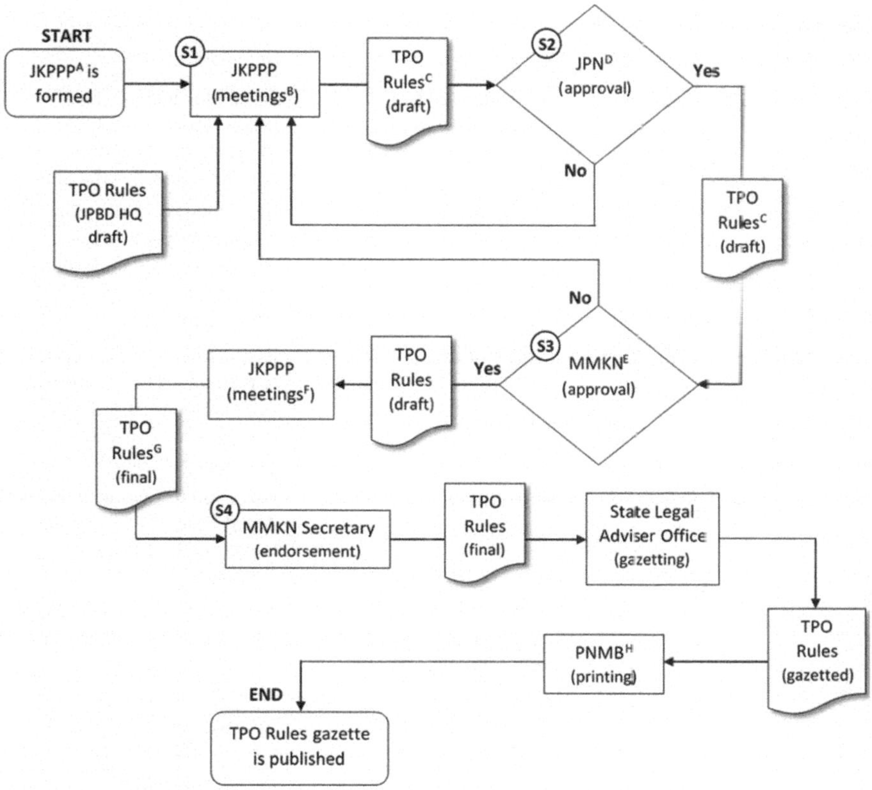

A. JKPPP = *Jawatankuasa Kaedah-kaedah Perintah Pemeliharaan Pokok* (TPO Rules Committee)
B. Discussion/technical meeting/workshop to draft TPO Rules paperwork for approval
C. This TPO Rules paperwork must be vetted by the State Legal Adviser Office representative
D. JPN = *Jawatankuasa Perancang Negeri* (State Planning Committee)
E. MMKN = *Majlis Mesyuarat Kerajaan Negeri* (State Government Council)
F. Discussion/technical meeting/workshop to draft TPO Rules for gazetting
G. This TPO Rules gazette final draft must be vetted by the State Legal Adviser Office representative
H. PNMB = *Percetakan Nasional Malaysia Berhad* (A printing company)

Fig. 3.1 TPO Rules gazettement process flow. *Source* Nik Adlin, Zulhabri, Wan Tarmeze, et al. (2020)

be understood from the findings of the study and remarks by various stakeholders as tabulated in Table 3.4 (extended from Nik Adlin et al., 2017).

TPO (Act 172) and the like, if correctly implemented and enforced, would prohibit illegal felling or damage of trees. However, there have been cases where trees were felled without the LPA consent, for example, those of the Melaka Raya in 2015 and of the Jalan Cochrane in 2016 (Nik Adlin et al., 2017). The cases have raised questions about whether the legislation has effectively been implemented or not. In a more recent case, i.e., in 2018, a very old raintree (believed to be over 100 years) in the Ipoh Methodist Girls' School was topped and attempted to be removed by the school board to make way for a new building. Fortunately, some members of the

Table 3.3 Summary of TPO Rules document (e.g. Government of Selangor Gazette, 2001)

Part	Title	Rule	Summary
I	Preliminary	1. Citation and commencement	To cite the name "Tree Preservation Order Rules", the year, and the date of commencement in the State
		2. Interpretation	Interpretation of the terms "damaging trees and destroying trees", "dead or dying tree", "group of trees", "tree", and "trees"
II	Tree Preservation Order	3. List of trees	Requirement for LPA (prior to making a TPO) to prepare and maintain a list of TPO trees in a format such as shown in Schedule I
		4. Tree preservation order	Form and format (as in Form A and Form B of Schedule II) to be used by LPA when making a TPO
		5. Tree preservation order may be made if the tree not subjected to conditions	Affirmation that a TPO made must not affect a PP given within the same area
		6. Written permission under tree preservation order	Form C—Application to fell TPO trees; Form D—LPA reply to Form C; Form E—Notification of compliance by applicant; Form F—LPA reply to Form E
		7. Certification by a qualified person	Appointment of qualified persons, e.g. CA (certified arborist), to certify that a tree is dead, dying or that causes imminent danger
		8. Compensation	Procedures to make claims for compensation in relation to a TPO: Form G—Claim for compensation Form H—Award of compensation offered by LPA

(continued)

Table 3.3 (continued)

Part	Title	Rule	Summary
		9. Replacement of tree	In dealing with a TPO trees felling offence: Form I—LPA directives to the offender to replace the tree(s); Form J—Application by the offender to be exempted from the tree replacement directives; Form K—LPA reply to Form J; Form L—Application by the offender for extension of time to replace the trees; Form M—LPA reply to Form L In the failure of the offender to replace the tree: Form N—LPA is to notify the offender about its intention to replace the tree; Form O—LPA to require the offender to bear the tree replacement cost
		10. Revocation of tree preservation order	Form P—LPA to revoke or amend a TPO
III	Prohibition to fell etc. tree with girth exceeding 0.8 m	11. Prohibition to fell etc. tree with girth exceeding 0.8 m	Form Q—Application to fell > 0.8 m trees Form R—LPA reply to Form Q
		12. Illustration Schedule I Schedule II Schedule III	Method to measure the girth of a tree from the ground

school alumni intervened by lodging a complaint to the LPA who later exercised the provisions of sections 21 and 22 of Act 172 to prohibit the tree from being felled and to ask it to be relocated.

From time to time, there emerges news about urban trees being felled and causing public outcries. Regrettably, the explanation by the culprits to justify the tree felling has often been unconvincing, especially to the environmental concern public who knew that there are always other options besides tree felling.

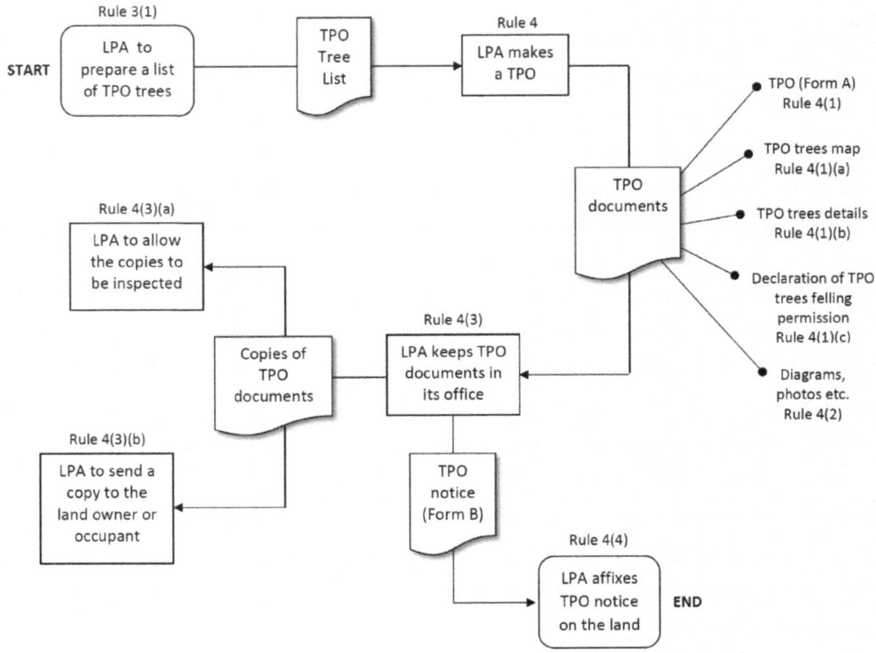

Fig. 3.2 TPO making process flow as in TPO Rules

The Way Forward

Make the Best of a Bad Situation

As a saying goes, "it is never too late to start making things better", and this is also true for TPO (Act 172) implementation. To move forward, we could use the approach of "making the best out of every deprivation situation". To illustrate this thought, let us consider the following circumstances. With the almost non-existence of TPO made by LPAs so far (Nik Adlin, Zulhabri, Wan Tarmeze, et al., 2020), we could conclude that the TPO tree selection activities must have significantly been lacking too. Therefore, new publicity and training must be carried-out continually to instill awareness and maintain the skill of LPA officials in deciding the selection of TPO trees.

With the growing concern on sustainability among urban dwellers, we need to review our current criteria for TPO tree selection to cater to that matter. The criteria to select TPO trees may no longer be confined to the tree's amenity factor obvious to us (immediate and tangible pleasantness the trees give us) but extended to their environmental functions as well. Hence, in addition to the criteria suggested in JPBDSM (2011), tree environmental functions such as carbon sink, air quality improvement, soil conservation, salinity, stormwater management and animal shelters and nesting

TREE PRESERVATION ORDER RULES 2001

SCHEDULE I

Rule 3(1)

LIST OF TREES

For the purpose of the Tree Preservation Order 2001, ... *(name of local planning authority)* .... hereby makes a list of trees in its area with effect from .... *(effective date)* ....

No.	*Species*	*Speciality*	*Remarks*
1.			
2.			
3.			
4.			
5.			

(Authorised Officer)

(Local Planning Authority)

Dated: ....

Fig. 3.3 Tree list format to be used by LPA for the purpose of making TPO. *Source* Government Selangor Gazette (2001)

sites must also be taken into consideration when selecting TPO trees. Fortunately, through the new training, we could effortlessly do the review and introduce the new more appropriate criteria.

New Study by PLANMalaysia

The reputedly lacking in TPO (Act 172) implementation have prompted PLAN-Malaysia to conduct a study, namely, *Kajian Pemeliharaan Pokok di Kawasan Pihak Berkuasa Tempatan Mengikut Peruntukan Bahagian VA, Akta 172* (A Study on Tree Preservation in Local Authority Areas Pursuant to the Provisions of Part VA, Act

SCHEDULE II

Form A

Rule 4(1)

(Name of Local Planning Authority)

TREE PRESERVATION ORDER

Register Reference ....

Whereas it appears to .... *(name of local planning authority)* ....that it is expedient in the interest of amenity to preserve the tree, trees or group of trees stated in the Schedule to this Order on a land situated at ....*(state address)* .... as shown on the map or plan annexed to this order and referred to by the relevant Schedule.

WHEREFORE this Local Planning Authority in exercise of the powers conferred by section 35A(1) of the Town and Country Planning Act 1976.

HEREBY

(a) MAKE this order known as the Tree Preservation Order................. *(give the citation)* ....;

(b) PROHIBIT the felling of any tree comprising the tree, trees or group of trees stated in the Schedule to this order and shown on the map or plan annexed;

EXCEPT with the prior written permission of the Local Planning Authority.

(c) DECLARE

(i) that the permission to fell, if given by the Local Planning Authority, may be subject to conditions;

(ii) that it is in the interest of amenity to preserve the tree, trees or group of trees stated in this Schedule.

TAKE NOTICE THAT—

(a) under section 35A(4) of the Act, any person who contravenes a tree preservation order commits an offence and is liable to a maximum fine of RM50,000.00;

(b) under section 35C of the Act, any person aggrieved by a tree preservation order or any of the provisions of a tree preservation order may appeal to the Appeal Board within one month from the date of the order; and

(c) under section 35D(1) of the Act, the owner of the land to which a tree preservation order relates proves that he has, in complying with the order, suffered damages in the form of depreciation in the value of the land, he may claim compensation from the local planning authority within six months from the date of the order.

By Order

(Authorised officer),

(Local planning authority)

Dated:........................

Fig. 3.4 The format to be used by LPA when making a TPO. *Source* Govt of Selangor Gazette (2001)

TOWN AND COUNTRY PLANNING ACT 1976

TREE PRESERVATION ORDER RULES 2001

SCHEDULE II

Form B

Rule 4(4)

TOWN AND COUNTRY PLANNING ACT 1976
SECTION 35A

NOTIFICATION

*This/These tree/trees/group of trees is/are subjected to a tree preservation order.

Any person who chops, prunes, trims, uproots, damages or destroys this/these *tree, trees or group of trees commits an offence and is liable to a maximum fine of RM50,000.00.

By Order, Register Reference: ....

(Local planning authority)

Dated:..............................

Delete whichever is not applicable.

Fig. 3.5 The format of TPO notice by LPA to be affixed at the site. *Source* Govt of Selangor Gazette (2001)

172). Begun in July 2020, this one-year-long study enables PLANMalaysia to officially review the status quo of the TPO (Act 172) implementation process and TPO-making procedures by LPAs. Its objectives include identifying the related issues and problems faced by LPAs, evaluating the effectiveness and management of TPO in LPAs, laying-out strategies and solutions to deal with the problems, and suggesting viable mechanisms and guidelines of TPO implementation for the LPAs.

The study involving a hundred and more officials and professionals of various urban forestry-related fields such as town planning, landscape architecture, arboriculture, legal, engineering and GIS gives PLANMalaysia a great opportunity to explain TPO (Act 172) so that everybody would have a similar understanding about the legislation from its purposes down to the basic terminologies it uses. This cohesiveness of thinking among the participants is critical to the success of this study and the future efforts to improve TPO (Act 172) implementation.

Table 3.4 Research findings and stakeholder remarks on TPO (Act 172)

No	Publication	Remarks and findings on TPO (Act 172)
1	EPU (2002)	Emphasizing the enforcement of TPO (Act 172) by local planning authorities
2	Syed Abdul Kader (2006)	Mocking the TPO (Act 172) as 'teeth without the bite'
3	CIDB (2007)	Expressing dissatisfaction over TPO (Act 172) enforcement by local planning authorities and urging the State Governments and the ministry to work together to increase the awareness on TPO
4	JLN (2011)	Strategizing the re-evaluation and formulation of landscape related legislation including the TPO (Act 172)
5	Rafiuddin (2011)	The implementation of TPO legislation is still lacking within KL City Hall (DBKL)
6	Abdul Aziz et al. (2011)	The level of knowledge on TPO (Act 172) among the greenspace municipal officers is still questionable
7	Amin and Hashim (2014)	TPO (Act 172) as the disaster risk reduction agenda to be adopted in Development Proposal Report (DPR) when assessing the conditions of vegetation in the existing site
8	Bernama (2014)	Tun Jeanne Abdullah, the chairman of Landskap Malaysia urges the authorities to enhance the implementation of TPO (Act 172) forcing the developers to replace the trees that were cut for development projects
9	Mohd Hashim and Hitchmough (2015)	The low level of understanding on TPO (Act 172) implementation among the decision makers is due to lack of exposure and experience in planning and managing physical development
10	Hasan et al. (2016)	80% of the respondents have considered the TPO (Act 172) was not fully utilized during proposed landscape design
11	Muhammad Aiman (2017)	TPO (Act 172) implementation is still weak and there are no specific guidelines on the types of trees and species to be protected and preserved
12	Nik Adlin et al. (2017)	TPO legislation should be accompanied by efforts to guide the people to understand what are required so that it can be effectively implemented and complied with
13	Ibrahim et al. (2019)	The lack of awareness on TPO legislation among the public has posed challenges for the local authority (DBKL) to implement TPO

(continued)

Table 3.4 (continued)

No	Publication	Remarks and findings on TPO (Act 172)
14	Nik Adlin et al. (2019)	The government can play more active roles in publicising the TPO (Act 172) such as through mass media and awareness campaigns

Source Extension of Nik Adlin et al. (2017)

Effective TPO (Act 172) Implementation Framework

Better or more effective TPO (Act 172) implementation should not only increase the number of preserved trees but also improve the handling of other related matters too, for instance, those regarding tree list preparation and PP application. Nik Adlin, Zulhabri, et al. (2020) proposed a framework for an effective TPO (Act 172) implementation in construction projects. The framework is meant for use by all the stakeholders responsible for the implementation of TPO (Act 172) which includes the departments of town and country planning (of federal and the states) and LPAs, and by those interested in the subject matter among academicians and professionals.

The framework was developed using the fundamentals of legislation implementation that pronounces the necessity to adequately address the key components representing the most important issues in the legislation (WHO, 2003). Next, the framework also embraces the concept of Critical Success Factors (CSFs) which has been widely used by organizations in developing and implementing strategies and projects (Amberg et al., 2005) and is known as a tool for measuring performance in an organization to achieve their mission (Zawawi et al, 2011). The concept of CSF offers a smarter way to identify certain factors (Alzahrani & Emsley, 2013), which when handled well in a project, are likely to make the project successful.

Hence, for TPO (Act 172), as shown in Fig. 3.6, there are 4 key components being proposed by the framework: (1) Publicity, (2) Tree List, (3) PP Standard Operating Procedures (SOPs) and (4) Enforcement Plan. For each of the components, there will be a list of CSFs linking to it. The lists of CSFs can be identified through research activities such as literature and document review, surveys, interviews and focus group discussions. For example, in the component 'Publicity', a literature review on Hasan et al (2016) could identify two potential CSFs: 'the awareness campaign must be a continuous effort, not just a one-off' and 'awareness among the stakeholders must be kept at maximum level'. The CSFs can be assigned as CSF1-01 and CSF1-02, respectively.

Conclusion

This chapter has discussed the importance of trees in ensuring the sustainability of an urban forest. Thus, trees should be preserved and one way to achieve this is

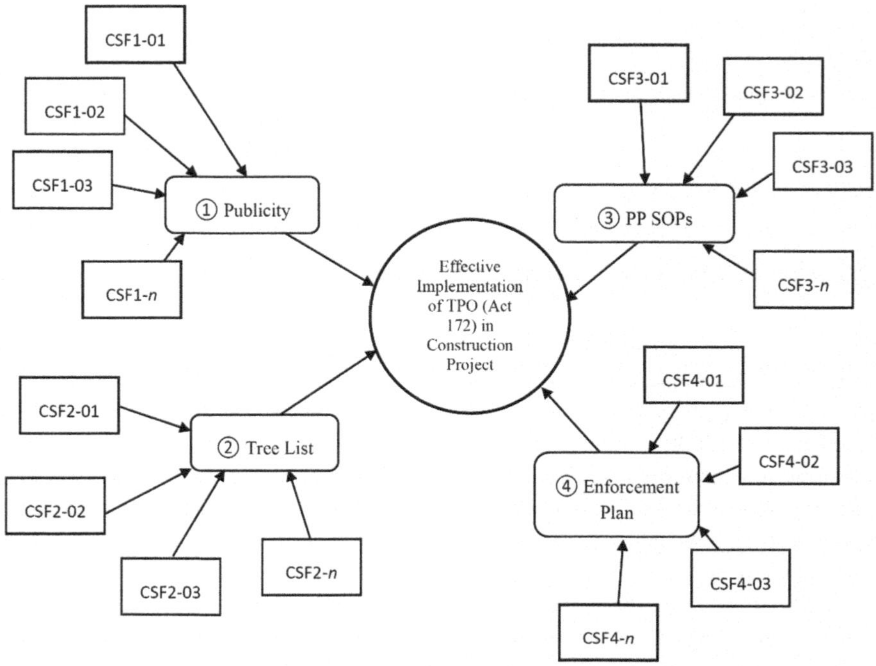

Fig. 3.6 Framework for effective TPO (Act 172) implementation in construction projects. *Source* Nik Adlin, Zulhabri, et al. (2020)

through legislation called Tree Preservation Order or TPO. In Malaysia, the main TPO legislation, referred to as TPO (Act 172) is the Part VA of Town and Country Planning Act 1976 or Act 172. TPO (Act 172) is considered to be implemented once an LPA makes a TPO. The procedure to make a TPO, known as TPO Rules was established by PLANMalaysia in the late 90s and gazetted by the state governments of Selangor, Perak and Melaka for the adoption of their LPAs. Though it had been more than two decades old, TPO (Act 172) was still being considered very unsatisfactory in terms of its implementation, judging by the repeated cases of unconsented urban trees felling. To move forward, we need to have a positive attitude in dealing with this deprivation situation. PLANMalaysia has taken a significant step to conduct a study that looks deeply into the status quo of TPO (Act 172) implementation by LPAs across the country. In the academic sectors, there have been research projects at the postgraduate level taking TPO as the subject matter. One of the studies has developed a framework for the effective implementation of TPO (Act 172). In conclusion, we can say that by introducing TPO (Act 172), Malaysia has already laid a solid foundation for us to overcome all the challenges of making our urban forests sustainable.

References

Abdul Aziz, N. A., Konijnendijk, C. C., Sreetheran, M., & Nilsson, K. (2011). Greenspace planning and management in Klang Valley, Peninsular Malaysia. *Arboriculture & Urban Forestry, 37*(3), 99–107.

Alzahrani, J. I., & Emsley, M. W. (2013). The impact of contractors' attributes on construction project success: A post construction evaluation. *International Journal of Project Management, 31*(2), 313–322. https://doi.org/10.1016/j.ijproman.2012.06.006

Amberg, M., Fischl, F., & Wiener, M. (2005). *Background of critical success factors research* (Working Paper No. 2/2005). University of Erlangen-Nuremberg.

Amin, I. A. M., & Hashim, H. S. (2014). Disaster risk reductions in Malaysian urban planning. *Journal of the Malaysian Institute of Planners, 12*, 35–58.

Bernama. (2014, June 25). *Jeanne*: Ganti Semula Pokok Yang Ditebang (Jeanne: Replace the Felled Trees). *Malaysiakini*. https://www.malaysiakini.com/news/266771

CIDB. (2007). *Strategic recommendations for improving environmental practices in construction industry*. Construction Industry Development Board (Lembaga Pembangunan Industri Pembinaan Malaysia).

EPU. (2002). *Study for the sustainable development of the highlands of Peninsular Malaysia. Final Report Vol II*. Economic Planning Unit (Unit Perancangan Ekonomi). Prime Minister's Department.

Government of Selangor Gazette. (2001). *Kaedah-kaedah Perintah Pemeliharaan Pokok 2001* (Tree Preservation Order Rules 2001). Sel. P.U. 8. Percetakan Nasional Malaysia Berhad.

Hasan, R., Othman, N., & Ahmad, R. (2016). Tree preservation order and its role in enhancing the quality of life. ScienceDirect. *Procedia—Social and Behavioral Sciences, 222*. 493–501.

Ibrahim, P. H., Zahrull Pauzi, H. F., & Mohd Masri, N. N. (2019). The implementation of tree preservation order in urban environment: Public and local authority perception. *Journal of Architecture, Planning & Construction Management, 9*(1), 94–111.

JLN. (2011). *National Landscape Policy*. National Landscape Department (Jabatan Landskap Negara). Ministry of Housing and Local Government of Malaysia.

JPBDSM. (2002). *Ensiklopedia Undang-undang dan Pentadbiran Perancangan Bandar dan Desa* (Town and Country Planning Law and Administration Encyclopedia). Federal Department of Town and Country Planning (Jabatan Perancangan Bandar dan Desa—JPBDSM).

JPBDSM. (2011). *PPA 06—Panduan Pelaksanaan Akta 172: Perintah Pemeliharaan Pokok* (Guidelines of Act 172: Tree Preservation Order). Federal Department of Town and Country Planning (Jabatan Perancang Bandar dan Desa Semenanjung Malaysia—JPBDSM).

Laws of Malaysia. (2014). *Town and Country Planning Act 1976 (Act 172)*. The Commissioner of Law Revision Malaysia. Percetakan Nasional Malaysia Bhd.

Mohd Hashim, N. H., & Hitchmough, J. D. (2015). The comparison of perceptions among landscape professionals' on tree retention and legislation. *International Academic Research Journal of Social Science, 1*(2), 164–176.

Muhammad Aiman, M. R. (2017). *Tree preservation order at Hutan Bandar Public Park, Johor Bahru* [Unpublished Bachelor's thesis]. Universiti Teknologi MARA.

Muhammad Syahlan, S., Mohd Izzat Amsyar, M.A., Hisham, H., & Fareed, M. H. (2018). Subsidiary legislation in Malaysian administrative law: Definition, advantages & grounds to challenge it. *International Journal of Scientific and Research Publications, 8*(10), 292–297. https://doi.org/10.29322/IJSRP.8.10.2018.p8238

Nik Adlin, N. M. S., Noriah, O., & Wan Tarmeze, A. W. (2017). A review on the needs to improve Malaysian Tree Preservation Order (TPO) (Act 172). *Planning Malaysia: Journal of the Malaysian Institute of Planners, 15*(4), 105–114.

Nik Adlin, N. M. S., Wan Tarmeze, W. A., & Noriah, O. (2019). Awareness and knowledge of TPO (Act 172) among construction industry professionals and local planning authority personnels in Klang Valley. *Planning Malaysia: Journal of the Malaysian Institute of Planners, 17*(2), 267–279.

Nik Adlin, N. M. S., Zulhabri, I., & Wan Tarmeze, W. A. (2020). Conceptual framework for developing a model of effective Tree Preservation Order (Act 172) implementation in construction projects. *International Journal of Sustainable Construction Engineering and Technology, 11*(1), 18–30. https://doi.org/10.30880/ijscet.2020.11.01.003

Nik Adlin, N. M. S., Zulhabri, I., Wan Tarmeze, W. A., & Rumaizah, M. N. (2020). Tree Preservation Order (Act 172) adoption process within the national development planning framework. *Planning Malaysia: Journal of the Malaysian Institute of Planners, 18*(4), 271–286.

Rafiuddin, R. (2011). *Urban tree management: Towards best practices and applications* [Master thesis]. Universiti Teknologi MARA.

Syed Abdul Kader, S. Z. (2006, February). Tree preservation orders under Malaysian planning law: Teeth without the bite. *Planners Bulletin. Malaysian Institute of Planners*, 10 & 18.

WHO. (2003). *Mental health legislation and human rights (Mental health policy and service guidance package)*. World Health Organization.

Zawawi, E. M. A., Kamaruzzaman, S. N., Ithnin, Z., & Zulkarnain, S. H. A. (2011). Conceptual framework for describing CSF of building maintenance management. *Procedia Engineering, 20*, 110–117. https://doi.org/10.1016/j.proeng.2011.11.145

Nik Adlin Nik Mohamed Sukri is a Senior Research Officer with Urban Forestry Branch, FRIM. Holding a Degree in Landscape Architecture from the Universiti Teknologi Malaysia (UTM) in 2001, she obtained her MSc in Integrated Construction Project Management from the Universiti Teknologi MARA (UiTM) in 2017 and is currently in the final stage of her Ph.D. on tree protection in construction project management.

Wan Tarmeze Wan Ariffin is a Division Director at the Forest Research Institute Malaysia (FRIM). With his qualifications in Mechanical Engineering (BSc and MSc) and Civil Engineering (Ph.D.), he has also been appointed as a technical advisor in many construction projects conducted by FRIM.

Shahzarimin Salim is a Senior Assistant Director, PLANMalaysia@Negeri Sembilan under the Planning Development Division. Among his tasks is to be involved in the reviews and amendments of Act 172. He has a bachelor's degree in Urban and Regional Planning from the Universiti Teknologi Malaysia (UTM).

Chapter 4
Urban Soil Environment in Malaysia

Jeyanny Vijayanathan

Abstract The urban soil environment is generally heterogeneous and varies greatly within and between sites. This includes soil compaction, impermeable crust, lack of oxygen, soil moisture, altered microbial activities, nutrient availability and presence of artificial materials and contaminants. This chapter discusses the various characteristics of the urban soil environment and challenges in its management strategies. Interventions and proper management of urban soils is a necessity in the changing climate as soil is a finite resource. Despite the abundance of research reported in temperate and subtropical continents, there is a dire need for investigations in tropical countries such as in Malaysia. Although recent advances recommend various plausible solutions in improving soil properties, time tested solutions in the local context need further exploration and findings. Nevertheless, this chapter defines the current scenario of the urban soil environment of Malaysia and prompts future directions in urban soil sustainability and environment.

Keywords Anthropogenic soils · Disturbed · Properties · Amendments · Urbanisation

What Are Urban Soils?

Urban soils are broadly impacted by human activities, urban soil is found generally but not only in urban zones (Morel et al., 2005). They comprise:

1. Soils that are composed of a combination of materials varying from those in adjacent agricultural or forest areas, and that may present a surface layer more prominent than 50 cm, profoundly changed by human actions through mixing, importing, and exporting material, and by contamination;
2. Soils in parks and gardens that are closer to agricultural soils but offer diverse composition, utilization, and management than rural soils; and.

J. Vijayanathan (✉)
Forest Research Institute Malaysia (FRIM), Kepong, Malaysia
e-mail: jeyanny@frim.gov.my

S. Maruthaveeran et al. (eds.), *Urban Forestry and Arboriculture in Malaysia*,
https://doi.org/10.1007/978-981-19-5418-4_4

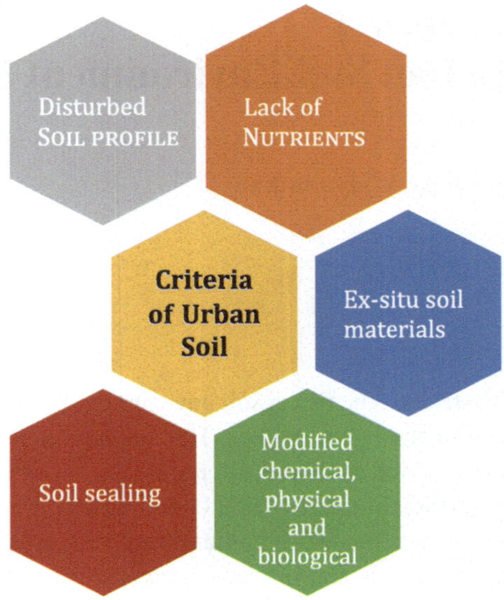

Fig. 4.1 Criteria of urban soil influenced by anthropogenic activities

3. Soils that result from different development activities in urban regions which are regularly sealed.

According to this definition, urban soils are basically beneath strong human impact in urban and rural situations; they may impose strong impacts on human wellbeing, on plants and soil fauna, and on water infiltration.

Many urban soils may have been moved, graded, compacted or contaminated. Soil forming processes in urban soils occur more rapidly, erratically and are governed by human (anthropogenic) impacts and possible natural causes. Some scientists classify it as anthroposols. According to Naeth et al. (2012), anthroposols are normal in renewable resources, industrial, commercial and urban development scenarios and in transportation, fuels and power corridors (Fig. 4.1).

Urban Soil Description

Due to population expansion and aspirations for greater economic growth, the major threats are loss of agricultural soils and reduction in soil efficiency in specific areas due to the urban areas in developing countries. The utilisation of soil or land is inevitable and has taken its toll on urban soils. Alteration of urban soils has made them unsuitable for optimal tree growth and development, in some cases. However, the current sustainable development goals outlined by the United Nations (Tapiador et al., 2021) call for the incorporation of trees within urban settings for various reasons, mainly for disaster risk reduction, mitigating extreme temperatures, urban

Fig. 4.2 Comparison of urban soil (left) and a forest soil profile (right)

gardening for food security and many other functions. Trees growing in most urban soils will need to tolerate the distinctive characteristics detailed below.

Disturbed Soil Profile

Most soil profiles in urban areas do not display the original horizons as in the natural environment. They lack the original organic layer which constitutes high organic matter accumulation and microbial activity. With increasing depths, the layering of the soils differs with respect to textural and structural properties. This was shown in recent studies by Demina et al. (2018) where urban sites showed considerable deterioration of microbial properties in topsoil after the land-use change (i.e., forest to urban) where microbial respiration in urban land in Moscow was reduced 4.4 times compared to the previous land use of forest soils. Similar work was reported by Barrico et al. (2018), when compared to public gardens, the various indices of soil bacteria were significantly greater in remnant forests (Fig. 4.2).

Lack of nutrients

During construction related to urbanisation, the top layer of soil is often scraped off (Howard, 2017) resulting in the loss of important organic matter constituents such as nitrogen (N) and carbon (C). Top layers are scraped off to facilitate cut and fill works on level ground. In Singapore's urban roads, soil C was negatively associated with the percent of commercial and utility land use at the 0–30 cm layer according to a study shown (Ghosh et al., 2016). Trees need nutrients to grow and soils lacking nutrients will give poor tree growth. Many of these situations can be seen visually via stunting, discoloration of leaves and acute attacks by pests and diseases. These adversities are widely known and reported by Jeyanny et al. (2009) [Fig. 4.3] and Zainudin et al. (2003) reported that growth of *Hopea odorata*, commonly used in urban forestry showed significant increments in shoot: root ratio and total heights with slow-release fertiliser applications compared with control. Jeyanny et al. (2009) showed how a tropical forest species, *Khaya senegalensis* which is utilised for landscaping responded negatively to a lack of nutrients. It exhibited severe discolouration for N, P and K deficiencies.

Fig. 4.3 Symptoms of nutrient deficiencies related to nitrogen with pale chlorosis (i), phosphorus with stunted growth (ii), and yellow patches between veins for potassium (iii) displayed by *Khaya senegalensis* seedlings compared to control. (*Source* Jeyanny et al. [2009])

Ex-Situ Soil Materials

Most urban areas are filled with ex-situ (transported from outside) soil materials, which are mainly subsoils, rocks, debris and substrates which are compressed during land preparation. The heterogeneous mixing of soil has an adverse effect due to poor soil moisture, lack of pore spaces for aeration and water holding capacity and inhibits root proliferation due to the presence of pebbles and rocks. Trees planted on shallow soils have a higher tendency to collapse due to reduced root anchorage (see Fig. 4.4). These trees become hazardous, especially during heavy storms and strong winds. Hasan et al. (2017) reported that trees planted on shallow soils have issues with root anchorage and were more prone to fall, and this phenomenon amplified the trees' potential as hazardous trees. Besides uprooting due to depths, conventional tree pits also pose threats such as impeding root systems due to hardscapes, underground utilities and compacted soil (Bartens et al., 2010).

Fig. 4.4 Shallow soils result in the uprooting of a jackfruit tree (*Artocarpus heterophyllus*) in an urban park in Subang Jaya, Selangor

Modified Soil Properties

Soils of the urban environment are usually compacted due to foot and vehicle traffic and the utilisation of heavy machinery during construction (Fig. 4.5) reduces the soil pore space, increases bulk density, and reduces the water holding capacity and affects the tree's physiological traits (Norainiratna et al., 2013; Bartens et al., 2010; Philip & Azlin, 2005). As the original soil in the area is heavily mixed with subsoil and rocks, the soil textural and structural properties differ, causing several problems such as crusting, hard soil peds, poor drainage, flooding, ponding effects and degradation. The conventional impermeable pavements inhibit aeration, water seepage and porosity (Just et al., 2018; Morgenroth et al., 2013). The absence of soil biota decreases beneficial biological activities that enhance soil organic matter, soil aggregation, soil binding and structural development and hinders the nutrient cycling processes (Jim, 2019; Morgenroth & Buchan, 2009; Scharenbroch et al., 2013). Due to the nature of urban soils, chemical properties are highly altered, where possible contamination from alkaline and heavy metal from building materials may occur (Scharenbroch & Catania, 2012; Wei & Yang, 2010). As a result of anthropogenic sources in Seri Kembangan, Selangor, a recent report showed that heavy metals such as elevated Cd, Co, Cr and Cu may pose health hazards to residents in the region (Praveena et al., 2014).

Fig. 4.5 Utilisation of tractors in urban soils increases compaction and reduces aeration and water holding capacity

The proportion of the weight of soil relative to its undisturbed soil volume is known as bulk density. It is commonly measured in units of grams per cubic centimetre (g cm^{-3}) and expressed as a unit of weight per volume. It can indicate the amount of pore space within soil horizons and is inversely proportional to pore space. Bulk density can easily be determined by the core ring method (Fig. 4.6). Soil compaction occurs when there is an increase in bulk density and a decrease in total pore space. The average bulk density of a tropical loam-textured mineral soil is taken to be 1.2 g cm^3. Compacted tropical soils are larger than 1.5 g cm^3. A recent study by Jeyanny et al., 2019b showed that the bulk densities of a football field (1.6 g cm^3) was relatively higher than a virgin jungle (1.1 g cm^3) and a botanical park (1.4 g cm^3) at soil depths up to 50 cm. The soil of finer textures such as clay loams and sandy loams have moderate bulk densities (< 1.4 g cm^3) and do not impede root growth. Besides, these conditions allow for better water holding capacities (Watson et al., 2014). As an example, it was reported that prolonged droughts (November 2009 to March 2010) resulted in significant damage of healthy poplars (*Populus* spp.), elms (*Ulmus* spp.) and plane trees (*Platanus* spp.) in parks and the central business district of Melbourne, Australia. Soil moisture contents recorded were less than 40% during this period (May et al., 2013). There was strong reasoning behind this as trees undergo 'hydraulic failure' due to desiccation or starvation on carbohydrate reserves as photosynthesis fails to occur due to stomatal closure from water stress.

Fig. 4.6 Core rings are inserted into soils to collect soil bulk density in the football field of Forest Research Institute Malaysia (FRIM), Selangor, West Malaysia

Soil Sealing

According to FAO (2016), the permanent covering of the soil surface with impermeable artificial materials (asphalt, concrete, cement) is known as soil sealing as it leads to non-reversible loss of soil and most of its ecosystem amenities. Soil sealing is common in areas with buildings, construction, yards and driveways. The main destructive effects on ecosystem services comprise losses of food and fibre production; decreases in the soil's function as a natural sink and diluter for pollutants, significant declines or total loss of the soil's water retention; loss of soil faunal biodiversity, loss of soil series, reduced neutralisation and purification capacities and decline of carbon sequestration capacity (FAO, 2016; Jim, 2019; Lu et al., 2020; Tobias et al., 2018). Figure 4.7, shows how a native soil series (*Tringkap* soil series) with a prominent spodic horizon that occurs only in the montane forest is lost due to the construction works of Cameron Square in Brinchang, Cameron Highlands. According to a study in Torun, northwest Poland, the soil sealing impeded the carbon and nitrogen content, microbial count and soil respiration compared to non-sealed soils due to adverse effects on soil environment and properties. Soil moisture content and nitrate reductase activity of soils also were affected by sealing (Piotrowska-Długosz & Charzyński, 2015).

Fertilisers and Soil Amendments

Since the processes involved in urban planning and development markedly interrupt the ecological landscape of trees, soil and water, urban soils need special handling

Fig. 4.7 This image shows construction work in progress in the background where the soil surveyor is explaining the unique features of *Tringkap* series with a spodic (bleached horizon) which is lost forever

in terms of sustainable utilisation. Most urban trees in Malaysia grow in constricted pavements, with the absence of topsoil. The bases of trees are cleaned periodically from leaf litter for aesthetic purposes, however, little did we realise that leaf litter is an ideal form of decomposing 'composts' which replenish back nutrients in the soil. Soils have minimal biological makeup and its properties are far from their original conditions. Some of the common practices in urban areas for productive tree growth include the usage of fertilisers, soil amendments and improvement of drainage.

Fertilisers have long sustained the requirements for nutrients of ornamental plants in various urbanscapes. For instance, maple trees *(Acer rubrum)* require multiple fertiliser applications during a growing season. Application of slow-release fertilisers (18:5:12) and irrigated with pond or city water resulted in significant collar diameter increases, greater canopy areas and better root systems (Zhu et al., 2013) on maple trees. Ash trees (*Fraxinus* sp.) in the United States responded positively to the application of a gibberellin inhibitor (paclobutrazol) and compound NPK fertilisers. Results showed that fertilisers significantly increased N levels in leaves (27%) and radial growth (20%) compared to control. There was an increase of 9 to 10% in root:total biomass ratios of trees applied with paclobutrazol compared to control or fertilised trees, respectively (Tanis et al., 2015). Another study looked at how the combination of paclobutrazol and potassium nitrate significantly increased leaf thickness and chlorophyll content of *Xanthostemon chrysanthus* (Roseli et al., 2021). Most nutrient management studies report results from short-term experiments, initial application during planting and nursery trials which are not practical in urban settings. Thus, the use of mulching is quite common in Malaysia to improve soil fertility status over a long period as the substrates decompose naturally (Tasan, 2002).

In the age of sustainable urban forestry and nature-based solutions, the utilisation of composts, biosolids, biochar and microbial-based biofertilizers has increased. Scharenbroch (2009) in his meta-analysis further explored the various studies on organic amendments between 1982 to 2008 such as mulch, and composted materials and their combinations. His work pointed out the biggest improvements were seen in root density, germination, soil porosity, mycorrhizal density, particulate organic matter, and tree diameter. Another recent work by Vidal-Beaudet et al. (2018) demonstrated that high levels of organic composts improved the soil's physical properties (e.g., dry bulk density, aggregate stability and hydraulic conductivity) applied to ornamental trees of *Ostrya carpinifolia* in Anges, France. There was an increase of 12% in trunk diameter for trees applied with soil mixed with green waste compost, and 25% for soil mixed with co-compost of sewage sludge and wood chips. Indirectly, the fine roots of trees modified soil structure and decreased dry bulk density in the top layer of the soil after 5.5 years.

In the urbanscapes, biochar use has been gaining momentum in recent years due to its promising traits in facilitating soil fertility and organic C contents (Ghosh et al., 2015; Somerville et al., 2020). Biochar is produced under partial exclusion of oxygen by burning at 350 to 800 °C through pyrolysis (Lehmann & Joseph, 2015; Scharenbroch et al., 2013). In many studies, biochar has improved soil quality and plant growth, soil surface area, water, and nutrient retention and also increased microbial biomass and activity (Helliwell, 2015; Thomas & Gale, 2015). Scharenbroch et al.

(2013) reported that biochar was able to increase total organic C and, of the trees studied, seems to have the greatest and most significant increases in *Acer saccharum* and *Gleditsia triacanthos* in terms of overall tree biomass and soil properties as compared to conventional composts, and wood chips. On the contrary, some works also reported no significant differences or negative effects on tree growth and soil fertility when tested with *Corymbia maculata* (Somerville et. al., 2020), *Pseudotsuga menziesii* (Sarauer & Coleman, 2018; Sarauer et al., 2019), *Tachigali vulgaris* and exotic *Eucalyptus urophylla x Eucalyptus grandis* trees (de Farias et. al., 2016).

Biofertilizers contain plant growth-promoting microbes such as rhizobium and mycorrhizae. Through diverse mechanisms, these microbes colonise plant roots, conferring several benefits to the plants. Menendez et al. (2016) showed how *Rhizobium* inoculated *Dianthus caryophyllus* (carnation) ornamental species showed positive results related to phosphate solubilization, siderophores production, IAA (Indole Acetic Acid) and precursors biosynthesis with effective root colonizations and root hairs development as compared to control.

Most trees need essential elements to grow well and some are needed in high quantities (macronutrients) such as nitrogen, phosphorus, potassium, calcium and magnesium whereas some are needed in small amounts such as iron, zinc, copper, boron, manganese, silica, molybdenum, sodium, cobalt, chloride (micronutrients). A complete fertiliser is a blend of three main plant nutrients: nitrogen (N), phosphorus (P), and potassium (K), in the forms of potash, phosphoric acid, and nitrogen usually used to provide nutrients for trees. The uses and functions of nutrients are given in Table 4.1.

Factors Affecting Nutrient Uptake

Soil Chemical Properties

Soil pH is an important factor that affects nutrient uptake. The availability of nutrients differs at various soil pH (Fig. 4.8). Acidic soils may have high amounts of iron and aluminium in the soil solution, affecting other nutrient uptakes. Soils that are alkaline such as limestone soils have a high tendency of calcium and magnesium ions. Magnesium has antagonistic effects on potassium (Marschner, 1995). Soil cation exchange capacity also describes the number of ions that the soil particles can hold. Soils with higher CEC have higher capacities to hold ions compared to low CEC soils such as peat and sandy soils. The CEC values are also dependent on the types of clay minerals and their variable charges (Parfitt et al., 1995).

As given above, we see that soil chemical, physical and biological properties are essential in urban soils and play cross-functional roles in keeping both the soils and the trees healthy. In certain urban areas, the dissolution of calcareous materials in the cement and concrete of which most of the paved surfaces in urban areas are made are indications of a high soil pH (Park et al., 2010). Hence, there is a boundary to the

Table 4.1 Mineral elements required by plants

Element	Absorbed form	Major functions
Macronutrients		
Nitrogen (N)	NO_3^- and NH_4^+	In proteins, nucleic acids, etc
Phosphorus (P)	$H_2PO_4^-$ and HPO_4^{2-}	In nucleic acids, ATP, phospholipids, etc
Potassium (K)	K^+	Enzyme activation; water balance; ion balance; stomatal opening
Sulfur (S)	SO_4^{2-}	In proteins and coenzymes
Calcium (Ca)	Ca^{2+}	Affects the cytoskeleton, membranes, and many enzymes; second messenger
Magnesium (Mg)	Mg^{2+}	In chlorophyll; required by many enzymes; stabilizes ribosomes
Micronutrients		
Iron (Fe)	Fe^{2+} and Fe^{3+}	In active site of many redox enzymes and electron carries; chlorophyll synthesis
Chlorine (CI)	CI^-	Photosynthesis; ion balance
Manganese (Mn)	Mn^{2+}	Activation of many enzymes
Boron (B)	$B(OH)_3$	Possibly carbohydrate transport (poorly understood)
Zinc (Zn)	Zn^{2+}	Enzyme activation; auxin synthesis
Copper (Cu)	Cu^{2+}	In active site of many redox enzymes and electron carriers
Nickel (Ni)	Ni^{2+}	Activation of one enzyme
Molybdenum (Mo)	MoO_4^{2-}	Nitrate reduction

solubility and availability of nutrients to plant and soil organisms. Both soil pH and soil biological properties are interrelated as an alteration to soil pH whether more acidic or alkaline indicates the microbial communities function at the given range of soil pH (Joyner et al., 2019). Tree roots in either acidic or alkaline conditions can deteriorate and influence the absorption of nutrients.

Cation exchange capacity (CEC) which is the measure of exchangeable bases are important and values between 10.00 to 13.00 cmol kg^{-1} were considered to be satisfactory in West suburban Chicago (Scharenbroch & Catania, 2012) but values in the tropical context can be far lesser (3–6 cmol/kg) due to soil disturbances (Jeyanny et al., 2019a, 2019b). But this was not a great hindrance to tree growth. Another study by Takahashi et al. (2015) observed that increased exchangeable bases were detected with rapid urbanisation due to the deposition of dust. There was a close positive relationship between exchangeable base concentrations and soil pH in forested parks along the Tamagawa River in metropolitan Tokyo and Kanagawa Prefecture, Japan.

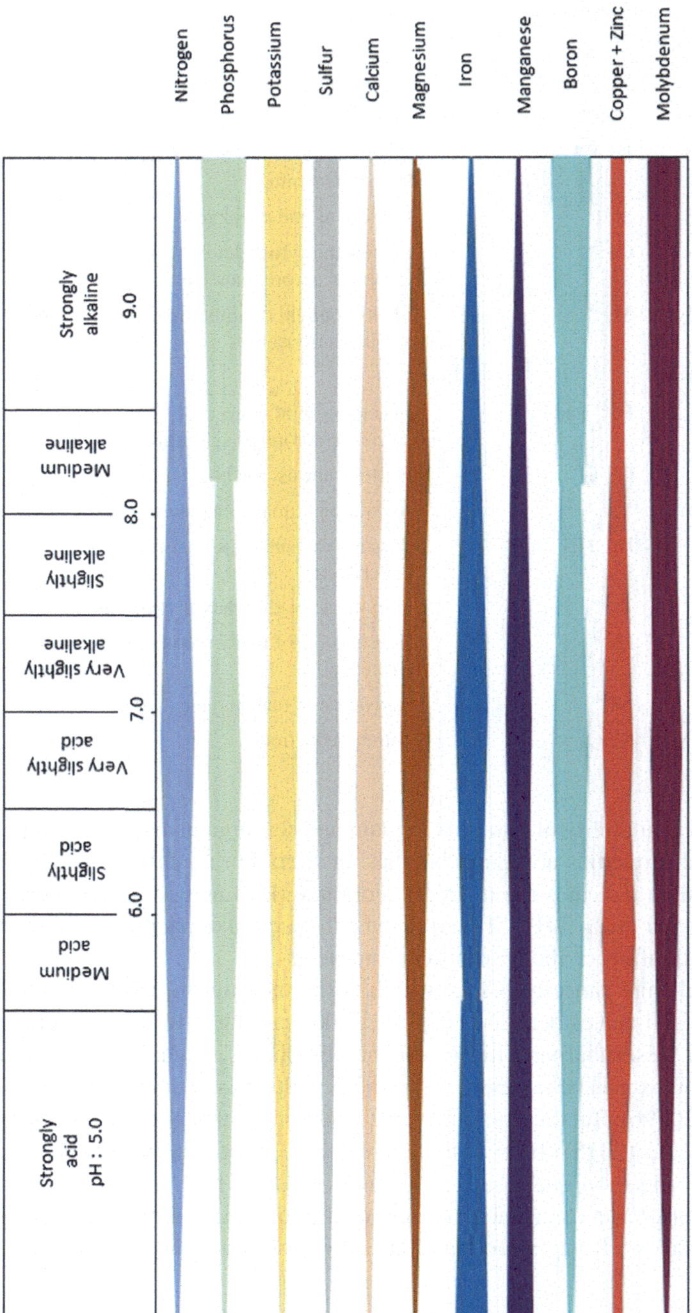

Fig. 4.8 Soil pH controlling selected nutrient availability for plant uptake

Soil Moisture

In order for the ions to move and interchange between the soil solution and plant roots via processes such as diffusion, root interception and mass flow, soil moisture is essential. Water transports the ionic compounds which are positively and negatively charged to the root hairs. However, excessive nutrient application and irrigation will result in leaching, where nutrients are washed away from the surface or to deeper soil depths.

Organic Matter

Organic matter is decomposed and replenishes the nutrient reserves of the soil, besides the natural weathering of parent material that provides essential nutrients. Normally, mulches derived from woodchips, litterfall and pruned cuttings provide an excellent source of organic matter. The soil biodiversity depends on the amount of organic matter that serves as a feedstock for decomposition. Organic matter also binds soil particles into aggregates and improves the water holding capacity of the soil. Most soils contain 2–10 percent organic matter.

Redox Potential

Redox potential (Eh) is the oxidation and reduction reactions that take place in the soil, where it measures the tendency of a chemical species to acquire from or lose electrons to an electrode and thereby be reduced or oxidised. Some studies have reported how Eh influences gas exchange, and other nutrients. Low Eh was reported to inhibit the uptake of N in *Spartina patens* in a wetlands environment and increase the Fe and Mn contents in the soil (Bandyopadhyay et al., 1993). It can be measured using a voltmeter. Soil Eh and soil pH are also inversely related (Husson, 2013). Redox potential particularly becomes important in contaminated urban soils which are inundated when reduced conditions further alleviate mobilisation/immobilisation of certain trace elements (Fe, Mn, As) (Muktawuri et al., 2015) that may have mixed effects on plant growth.

In terms of redox potential, values between 400–700 mV are generally considered well aerated but some specific water-tolerant species (e.g., *Taxodium distichum*) were able to withstand redox potential of 200 mV (Watson et al., 2014).

Microbial Activity

Microbial activity as reported earlier in this chapter plays an important role in nutrient uptake. For instance, mycorrhizas inoculate roots and enhance soil phosphorus absorption in ornamental plants (Crişan et al., 2017) and landscape trees of both *Acacia smallii* and *Fraxinus uhdei* (Stabler et al., 2001). Urban areas have been notoriously reported to affect mycorrhizal fungi detrimentally despite their importance in N uptake and increased plant heights (Tonn & Ibanez, 2016).

Xie (2020) tested two PGPR, and concluded that inoculation of *R. irregularis* and *B. amyloliquefaciens* led to higher shoot biomass and photosynthetic efficiency of urban greens in vegetated building envelopes. Rhizobia enhances N uptake in plants. Studies on inoculated live oak (*Quercus virginiana*), laurel oak (*Q. laurifolia*), and Drake elm (*Ulmus parvifolia*) seedlings with mycorrhizal fungi and rhizobacteria yielded positive results after a year. Root growth, mycorrhizal development and stem callipers were greater for all three species but rhizobacteria were able to increase root and stem the growth of Drake elm only (Rao et al., 2006).

Soil Texture

Clayey soils due to their finer texture, tend to have a higher water holding capacity and higher CEC to hold nutrients. According to Cornell University Cooperative Extension (2007), clayey soils and organic matter particles have negatively charged sites on their sites to adsorb and retain positively charged cations by electrostatic force. Substances such as biochar help to improve soil texture, moisture, CEC and also nitrogen content.

Growth and Morphological Status of Tree

One of the key drivers for nutrient uptake is the tree itself. Trees with good media for rooting and stability will be able to absorb more nutrients depending on their requirements. Plant growth is rapid during the initial stage of cell division and formation of foliage (exponential phase) and moves into a plateau (stationary phase) when it is well established and reaches maturity. Trees facing physiological stresses may also respond differently to nutrient uptake. This may be due to abiotic stresses (e.g. light, temperature, water) or biotic stresses (e.g. pests, diseases).

Conclusion

The reality that we need to face in the age of rapid urbanisation is that urban soils are highly transformed from their natural state, due to anthropogenic advances. The constraints related to physical, chemical and biological properties require careful consideration of interventions, may it be fertilisers, soil amendments such as mulching, composts, biochars, microbes, biofertilizers, etc. Tree growth depends on a multitude of nutrients such as N, P, K, Ca, Mg and trace elements and their inputs very much depend on tree health (foliage and root growth) and the confounding human activities. In conclusion, the management challenge is to provide an urban environment that imitates a natural ecosystem. This can be achieved if all stakeholders from multi-disciplinary backgrounds strategize urban greening designs with sustainability in mind.

References

Bandyopadhyay, B. K., Pezeshki, S. R., DeLaune, R. D., & Lindau, C. W. (1993). Influence of soil oxidation-reduction potential and salinity on nutrition, N-15 uptake and growth of *Spartina patens*. *Wetlands, 13*, 10–15.

Barrico, L., Castro, H., Coutinho, A. P., Gonçalves, M. T., Freitas, H., & Castro, P. (2018). Plant and microbial biodiversity in urban forests and public gardens: Insights for cities' sustainable development. *Urban Forestry & Urban Greening, 29*, 19–27.

Bartens, J., Wiseman, P. E., & Smiley, E. T. (2010). Stability of landscape trees in engineered and conventional urban soil mixes. *Urban Forestry & Urban Greening, 9*, 333–338.

Cornell University Cooperative Extension (CUCE). (2007). Cation Exchange Capacity (CEC). Agronomy Fact Sheet Series # 22. Department of Crop and Soil Sciences, College of Agriculture and Life Sciences, Cornell University.

Crişan, I., Vidican, R., & Stoian, V. (2017). Utilization of arbuscular mycorrhizae in the cultivation of ornamental plants. *Research Journal of Agricultural Science, 49*, 392–397.

de Farias, J., Marimon, B. S., Ramos Silva, L. C., Petter, F. A., Andrade, F. R., Morandi, P. S., & Marimon-Junior, B. H. (2016). Survival and growth of native *Tachigali vulgaris* and exotic *Eucalyptus urophylla* x *Eucalyptus grandis* trees in degraded soils with biochar amendment in southern Amazonia. *Forest Ecology and Management, 368*, 173–182.

Demina, S., Vasenev, V., Ivashchenko, K., Ananyeva, N., Plyushchikov, V., Hajiaghayeva, R., & Dovletyarova, E. (2018). Microbial properties of urban soils with different land-use history in New Moscow. *Soil Science, 183*(4), 132–140.

Food and Agriculture Organization (2016). Soil sealing poster accessed on 23rd August 2019 at http://www.fao.org/3/a-i6470e.pdf

Ghosh, S., Ow, L. F., & Wilson, B. (2015). Influence of biochar and compost on soil properties and tree growth in a tropical urban environment. *International Journal of Environmental Science and Technology, 12*, 1303–1310.

Ghosh, S., Scharenbroch, B. C., & Ow, L. F. (2016). Soil organic carbon distribution in roadside soils of Singapore. *Chemosphere, 165*, 163–172.

Hasan, R., Othman, N., & Ismail, F. (2017). Tree species selection in street planting: It's relationship with issues in urban area. *Environment-Behaviour Proceedings Journal*, 2, 185–194. https://doi.org/10.21834/e-bpj.v2i6.951

Helliwell, R. (2015). Effect of biochar on plant growth. *Arboricultural Journal, 37*, 238–242.

Howard, J. (2017). Anthropogenic landforms & soil parent materials. In Anthropogenic soils. Pp 25–51.
Husson, O. (2013). Redox potential (Eh) and pH as drivers of soil/plant/microorganism systems: A transdisciplinary overview pointing to integrative opportunities for agronomy. *Plant and Soil, 362*, 389–417.
Ibrahim, M. H., Jaafar, H. Z., Karimi, E., & Ghasemzadeh, A. (2013). Impact of organic and inorganic fertilizers application on the phytochemical and antioxidant activity of Kacip Fatimah (*Labisia pumila Benth*). *Molecules, 18*, 10973–10988.
Japakumar, J., Abdullah, R., & Rosli, N. S. M. (2021). Effects of biochar and compost application on soil properties and growth Performance of Amaranthus sp. Grown at Urban Community Garden. *AGRIVITA, Journal of Agricultural Science, 43*(3), 441–453.
Jeyanny, V., Ab Rasip, A. G., Wan Rasidah, K., & Ahmad Zuhaidi, Y. (2009). Effects of macronutrient deficiencies on the growth and vigour of *Khaya ivorensis* seedlings. *Journal of Tropical Forest Science, 21*, 73–80.
Jeyanny, V., Farah, F. M., & A., Syaliny, G., Ahmad Nazarudin, M. R., Norhayati, S. and Syafiqah Nabilah S. B. (2019a). Early growth, biomass and phenolics response of *Labisia pumila* var. *alata* seedlings to fertilisation and soil amendments. *Journal of Tropical Plant Physiology, 11*, 32–39.
Jeyanny, V., Lee, S. S., & Wan Rasidah, K. (2011). Effects of arbuscular mycorrhizal inoculation and fertilisation on the growth of *Acacia mangium* seedlings. *Journal of Tropical Forest Science, 23*, 404–409.
Jeyanny, V., Mohamad Fakhri I., Rozita, A. & Wan Rasidah, K. (2019b). Rich Grounds: FRIM Selangor Forest Park. FRIM Special Publication No.31. 92 pp. ISBN 978–967–2149–45–3.Forest Research Institute Malaysia
Jeyanny, V., Suhaimi, W. C., Wan Rasidah, K., M. Fakhri, I., Hamdan, O., and Ahmad Zuhaidi, Y. (2016). FRIM campus: Carbon conservation in a natural heritage site tropical forest.Pp. 264–266. In: Adzemi Mat Arshad, Ahmed Osumanu Haruna, Rosazlin Abdullah, Che Fauziah Ishak, Christopher The Boon Sung, Jeyanny Vijayanathan, Wan Zaliha Wan Embok (eds.) Proceedings of the SOILS 2016. "Soil Improvement for Sustainable Crop Production".. Malaysian Society of Soil Science, TH Hotel & Convention Centre, Kuala Terengganu,Terengganu 5–7 April 2016. ISBN 978–967–9945–38–6
Jim, C. Y. (2019). Resolving intractable soil constraints in urban forestry through research–practice synergy. *Socio-Ecological Practice Research, 1*(1), 41–53.
Joyner, J. L., Kerwin, J., Deeb, M., Lozefski, G., Prithiviraj, B., Paltseva, A., ... & Muth, T. R. (2019). Green infrastructure design influences communities of urban soil bacteria. *Frontiers in Microbiology, 10*, 1–14.
Just, M. G., Frank, S. D., & Dale, A. G. (2018). Impervious surface thresholds for urban tree site selection. *Urban Forestry & Urban Greening, 34*, 141–146.
Lee, S. S., & Alexander, I. J. (1994). The response of seedlings of two dipterocarp species to nutrient additions and ectomycorhizal infection. *Plant and Soil, 163*, 299–306.
Lehmann, J., & Joseph, S. (Eds.). (2015). *Biochar for environmental management: Science*. Routledge.
Liang, B. J., Lehmann, S. P., Sohi, J. E., Thies, B. O., & Trujillo, L. (2010). Black carbon affects the cycling of non-black carbon in soil. *Organic Geochemistry, 41*, 206–213. https://doi.org/10.1016/j.orggeochem.2009.09.007
Lu, C., Kotze, D. J., & Setälä, H. M. (2020). Soil sealing causes substantial losses in C and N storage in urban soils under cool climate. *Science of the Total Environment, 725*, 138369.
Marschner, H. (1995). Mineral nutrition of higher plants. 2nd. *Edn. Academic Press.*
May, P. B., Livesley, S. J., & Shears, I. (2013). Managing and monitoring tree health and soil water status during extreme drought in Melbourne, Victoria. *Arboriculture & Urban Forestry, 39*, 136–145.
Menéndez E., Escribano-Viana R., Flores-Félix J. D., Mateos P. F., & Rivas, R. (2016). Rhizobial biofertilizers for ornamental plants. In: González-Andrés F., James E. (eds) Biological nitrogen

fixation and beneficial plant-microbe interaction. Springer, Cham. https://doi.org/10.1007/978-3-319-32528-6_2

Morel, J. L., Schwartz, C., Florentin, L. & de Kimpe, C. (2005). Urban soils. Pp 202–208. *Encyclopedia of Soils in the Environment*. https://doi.org/10.1016/B0-12-348530-4/00305-2.

Morgenroth, J., & Buchan, G. D. (2009). Soil moisture and aeration beneath pervious and impervious pavements. *Journal of Arboriculture, 35*, 135.

Morgenroth, J., Buchan, G., & Scharenbroch, B. C. (2013). Belowground effects of porous pavements—Soil moisture and chemical properties. *Ecological Engineering, 51*, 221–228.

Mosse B. 1981. *Vesicular–Arbuscular Mycorrhiza Research for Tropical Agriculture*. Research Bulletin 194. Hawaii Institute of Tropical Agriculture and Human Resources. University Hawaii, Manao

Mukwaturi, M., & Lin, C. (2015). Mobilization of heavy metals from urban contaminated soils under water inundation conditions. *Journal of Hazardous Materials, 285*, 445–452.

Naeth, A. M., Archibald, H. A., Nemirsky, C. L., Leskiw, L. A., Brierley, A. J., Boc, M. D., & Chanasyk, D. S. (2012). Proposed classification for human modified soils in Canada: Anthroposolic order. *Canadian Journal of Soil Science, 92*, 7–18.

Norainiratna, M., Manohar, M., & Mohd. Roslan. (2013). Health of trees in Titiwangsa recreational park, Kuala Lumpur, Malaysia. *Journal of Sustainability Science and Management, 8*, 191–196.

Osman, K. T. 2012. *Soils: Principles, properties and management*. Springer Science & Business Media.

Parfitt, R. L., Giltrap, D. J., & Whitton, J. S. (1995). Contribution of organic matter and clay minerals to the cation exchange capacity of soils. *Communications in Soil Science and Plant Analysis, 26*, 1343–1355.

Park, S. J., Cheng, Z., Yang, H., Morris, E. E., Sutherland, M., Gardener, B. B. M., & Grewal, P. S. (2010). Differences in soil chemical properties with distance to roads and age of development in urban areas. *Urban Ecosystems, 13*, 483–497.

Philip, E., & Azlin, Y. N. (2005). Measurement of soil compaction tolerance of *Lagerstroemia speciosa* (L.) Pers. using chlorophyll fluorescence. *Urban Forestry & Urban Greening, 3*, 203–208.

Piotrowska-Długosz, A., & Charzyński, P. (2015). The impact of the soil sealing degree on microbial biomass, enzymatic activity, and physicochemical properties in the Ekranic Technosols of Toruń (Poland). *Journal of Soils and Sediments, 15*, 47–59.

Praveena, S. M., Yuswir, N. S., Aris, A. Z., & Hashim, Z. (2014). Potential health risk assessment of urban soil on heavy metal content in Seri Kembangan. In *From Sources to Solution* (pp. 77–81). Springer, Singapore.

Rao, B., Marx, D. H., & Jeffers, B. (2006). Response of oaks and elm to soil inoculations with mycorrhizal fungi and rhizobacteria in a nursery. *Arboriculture & Urban Forestry, 32*, 62–66.

Roseli, A. N. M., Ying, T. F., & Osman, N. (2021). Changes in leaf thickness, chlorophyll content, and gas exchange of a landscape tree, Xanthostemon chrysanthus, treated with paclobutrazol and potassium nitrate. *Arboriculture & Urban Forestry, 47*(2).

Sarauer, J. L., & Coleman, M. D. (2018). Converting conventional agriculture to poplar bioenergy crops: Soil greenhouse gas flux. *Scandinavian Journal of Forest Research, 33*, 781–792.

Sarauer, J. L., Page-Dumroese, D. S., & Coleman, M. D. (2019). Soil greenhouse gas, carbon content, and tree growth response to biochar amendment in western United States forests. *GCB Bioenergy, 11*, 660–671.

Scharenbroch, B. C. (2009). A meta-analysis of studies published in Arboriculture & Urban Forestry relating to organic materials and impacts on soil, tree, and environmental properties. *Journal of Arboriculture, 35*, 221–231.

Scharenbroch, B. C., & Catania, M. (2012). Soil quality attributes as indicators of urban tree performance. *Arboriculture and Urban Forestry, 38*(5), 214.

Scharenbroch, B. C., Meza, E. N., Catania, M., & Fite, K. (2013). Biochar and biosolids increase tree growth and improve soil quality for urban landscapes. *Journal of Environmental Quality, 42*(5), 1372–1385.

Singh, S. (1998). Role of mycorrhiza in the tree nurseries. Part 1. Evaluation of mycorrhizal efficiency with and without application of fertilizers. *Mycorrhiza News, 10*, 2–11.

Somerville, P. D., Farrell, C., May, P. B., & Livesley, S. J. (2020). Biochar and compost equally improve urban soil physical and biological properties and tree growth, with no added benefit in combination. *Science of the Total Environment, 706*, 135736.

Stabler, L. B., Martin, C. A., & Stutz, J. C. (2001). Effect of urban expansion on arbuscular mycorrhizal fungal mediation of landscape tree growth. *Journal of Arboriculture, 27*, 193–202.

Takahashi, T., Kanzawa, Y., Kobayashi, T., Zabowski, D., & Harrison, R. (2015). The effects of urbanization on chemical characteristics of forest soil in Tamagawa basin, Japan. *Landscape and Ecological Engineering, 11*, 139–145.

Tanis, S. R., McCullough, D. G., & Cregg, B. M. (2015). Effects of paclobutrazol and fertilizer on the physiology, growth and biomass allocation of three *Fraxinus* species. *Urban Forestry & Urban Greening, 14*, 590–598.

Tapiador, F. J., Navarro, A., Mezo, J., de la Llave, S., & Muñoz, J. (2021). Urban Vegetation Leveraging Actions. *Sustainability, 13*, 4843.

Tasan, J. (2002). Response of three shade tree species to grass and woodchip mulching. PhD thesis. Universiti Putra Malaysia.

Thomas, S. C., & Gale, N. (2015). Biochar and forest restoration: A review and meta-analysis of tree growth responses. *New Forests, 46*, 931–946.

Tobias, S., Conen, F., Duss, A., Wenzel, L. M., Buser, C., & Alewell, C. (2018). Soil sealing and unsealing: State of the art and examples. *Land Degradation & Development, 29*(6), 2015–2024.

Tonn, N., & Ibáñez, I. (2017). Plant-mycorrhizal fungi associations along an urbanization gradient: Implications for tree seedling survival. *Urban Ecosystems, 20*, 823–837.

Verheijen, F., Jeffery, S., Bastos, A. C., Velde, M. V. D., & Diafas, I. (2010). *Biochar application to soils: A critical scientific review of effects on soil properties, processes and functions* (pp. 1–166). Office for the Official Publications of the European Communities.

Vidal-Beaudet, L., Galopin, G., & Grosbellet, C. (2018). Effect of organic amendment for the construction of favourable urban soils for tree growth. *European Journal of Horticultural Science, 83*(3), 173–186.

Watson, G. W., Hewitt, A. M., Custic, M., & Lo, M. (2014). The Management of tree root systems in urban and suburban settings: A review of soil influence on root growth. *Arboriculture & Urban Forestry, 40*, 193–217.

Webb, R. (1998). Urban forestry in Kuala Lumpur. *Malaysia Arboricultural Journal, 22*, 287–296.

Wei, B., & Yang, L. (2010). A review of heavy metal contaminations in urban soils, urban road dusts and agricultural soils from China. *Microchemical Journal, 94*, 99–107.

Xie, L. (2020). Application of plant growth-promoting microbes on urban building vegetated envelopes, from lab to field. *Dissertationes Schola Doctoralis Scientiae Circumiectalis, Alimentariae, Biologicae.*

Zainudin, S. R., Awang, K., & Mohd Hanif, A. H. (2003). Effects of combined nutrient and water stress on the growth of *Hopea odorata* Roxb. and *Mimusops elengi* Linn. seedlings. *Arboriculture & Urban Forestry, 29*, 79.

Zhu, H., Zondag, R. H., Merrick, J., Demaline, T., Jeon, H., Krause, C. R., & Locke, J. C. (2013). Fertilizer applications for container-grown ornamental tree production. *Journal of Environmental Horticulture, 31*, 68–76.

Jeyanny Vijayanathan completed her Ph.D. in Soil Science at the Faculty of Agriculture, Land Management Department, Universiti Putra Malaysia (UPM) in March 2015. She received her MSc. in Soil Fertility and Management at the same faculty. Currently, she is a senior research officer in the Soil Management Branch, Forest Biotechnology Division, Forest Research Institute of Malaysia (FRIM) since November 2005. Jeyanny is actively involved in research related to soil chemistry, plant nutrition, environmental soil science and soil conservation in forestry and agriculture. She is a certified registered chemist with the Chemistry Institute of Malaysia, Past Vice President of the Malaysian Society of Soil Science and a soil expert in the Intergovernmental Technical Panel on Soils (ITPS) for FAO.

Chapter 5
Trees Diseases and Disorders in Urban Forests of Peninsular Malaysia

Mohd Farid Ahmad and Muhammad Syahmi Hishamuddin

Abstract Urban forests play an important role in urban ecosystems. They provide a variety of functions, services and benefits that are essential for urban development to succeed. The health, structural integrity and aesthetics of trees in urban forests are critical factors in determining a city's image and quality of life. Nevertheless, most trees in urban areas encounter various stresses that cause them to deteriorate and become more susceptible to insect and disease attack. This chapter discusses biotic and abiotic problems affecting the health of trees in urban forests, with a focus on Selangor, Putrajaya, and Kuala Lumpur. Plant health problems are diagnosed through field inspections and laboratory analysis of sampled material for identification of the causative organisms. Most plant health problems observed during inspections are caused by environmental factors such as soil compaction, water availability, lightning strikes, nutrient deficiency, limited planting space and mechanical injury. While some fungal infections such as leaf spot, leaf rust, powdery mildew, and sooty mould, are common, they have a negligible effect on tree health. Additionally, it is discovered that *Chrysoporthe* stem canker, *Fusarium* wilt, root rot and basal stem rot, all have significant impacts, and these diseases are destructive, difficult to control and rapidly spread to initiate new infection. Appropriate disease management measures are essential to bring the disease under control.

Keywords Tree disease · Host · Detection · Susceptibility · Mortality · Control measures

Introduction

Urban forests are important components of complex urban ecosystems in which humans, animals and plants coexist and rely on one another. They perform a variety of critical activities, services and benefits that are crucial for the city's long-term sustainability. In urban forests, the health, structural integrity and appearance of

M. F. Ahmad (✉) · M. S. Hishamuddin
Forest Research Institute Malaysia (FRIM), Kepong, Malaysia
e-mail: mohdfarid@frim.gov.my

S. Maruthaveeran et al. (eds.), *Urban Forestry and Arboriculture in Malaysia*,
https://doi.org/10.1007/978-981-19-5418-4_5

trees are all key elements that determine a city's image and the quality of urban life (Solomou et al., 2019). Additionally, trees contribute significantly to the value of the local real estate and to the growth of shopping, retail sales, and tourism activities (Wolf, 2007). Moreover, they bring value to communities by promoting physical and psychological health and instilling a sense of place in residents (Ulrich, 1986; Kaplan & Stephen, 1989). However, trees in the urban areas often face numerous stresses that make them prone to insects and diseases, including limited space for root growth, compacted and low-nutrient soils, and air pollutants (Colombo, 2016). The deterioration in health of urban trees may be exacerbated by climate change where warmer temperatures and drier conditions may enhance urban trees susceptibility to insects and diseases (Dale & Frank, 2017). Thus, proper strategies for preventing and combating pests and diseases that are significant for urban forest management should be implemented.

This chapter discusses the effect of biotic and abiotic stresses on the plant health of urban forests in several states in Peninsular Malaysia. We emphasise diseases that can have a substantial impact on urban trees and vegetation landscapes in Selangor, Putrajaya, and Kuala Lumpur. Additionally, plant diseases are predicted to be detected if field inspection is expanded to other states. This necessitates further studies, particularly on urban forest functions and utilisation in relation to pest and disease management, because this knowledge is crucial in future action planning. The diseases of urban forests discussed in this chapter offer practical guidance to urban forest managers, arborists, horticulturists and landscape architects to improve their knowledge and capability in disease diagnosis and management.

Disease Inspection and Identification of Plant Health Problems

The information on plant health problems discussed in this chapter is based on tree inspections conducted by forest pathologists in urban settings such as golf clubs, residential areas, community parks, and botanical gardens, as well as premises owned by government agencies, private companies, and individuals, from 2002 to 2020. Most of the inspected sites are in Selangor, as well as the Federal Territories of Kuala Lumpur and Putrajaya. In general, tree health problems in these sites are generally associated with abiotic factors, most notably soil compaction (Fig. 5.1). Other factors include lack or an excess of water (Fig. 5.2a, b), girdling roots, lightning strikes (Fig. 5.3), nutrient deficiency, limited growing space and mechanical injury. Nevertheless, most of the symptomatic trees are still alive, but are struggling to survive as indicated by stunted and poor growth, presence of water sprouts, yellowing of foliage, reduced canopy and dieback. Table 5.1 provides a list of the tree disorders observed during the inspections. The management of the stressed trees varies according to the site conditions, cost and clients' demands. Table 5.1 also summarises general recommendations for resolving abiotic-related tree health problems.

Fig. 5.1 *Acacia auriculiformis* trees in a parking area showing symptoms of sparse foliage, pale green to yellowing of leaves and dieback due to soil compaction and narrow growth space (arrows)

Fig. 5.2 Symptoms of a dying *Pteleocarpa lamponga* associated with poor soil water drainage. (**a**) Wilting, yellowing and defoliation, and (**b**) Dark and odorous rotted roots (arrows)

Table 5.1 Abiotic tree health problems observed during tree inspections in Kuala Lumpur, Selangor and Putrajaya, Malaysia

Disorder	Tree species	Symptoms	Locality	Recommendation
Lightning strike	*Acacia auriculiformis*, *Alstonia* sp., *Cyrtophyllum fragrans*	Longitudinal bark splitting, defoliation, dead tree in cluster or individually	Selangor, Kuala Lumpur	Install lightning diverter, remove dead tree
Soil compaction	*Acacia auriculiformis*, *Juniperus* sp., *Khaya ivorensis*	Dieback, stunted growth, water sprouting, yellowing of foliage, sparse canopy	Selangor, Kuala Lumpur	Improve soil by mulching, apply fertiliser, reduce visitor entrance and reduce traffic vehicles
Waterlogged	*Aquilaria* sp., *Hopea odorata*, *Pteleocarpa lamponga*, *Samanea saman*	Dieback, Yellowing and wilting of foliage, defoliation, and dead of tree	Selangor, Kuala Lumpur	Install or improve drainage systems, introduce water tolerant species
Transplant shock	*Syzygium myrtifolium*	Sudden wilting and drying of foliage	Selangor	Provide sufficient watering
Mechanical injury	*Alstonia* sp., *Cyrtophyllum fragrans*, *Juniperus* sp., *Pterocarpus indicus* *Samanea saman*	Bark injury on stem and root collar, decay root surface	Selangor, Kuala Lumpur	Apply fertiliser, practise good care of the tree, provide appropriate space for vehicle movement, install netting around the tree stem
Limited growth space	*Acacia auriculiformis*, *Cycas* sp., *Eucalyptus* sp., *Juniperus* sp., *Khaya ivorensis*, *Pterocarpus indicus*, *Samanea saman*, *Tabebuia* sp.	Stunted growth, dieback	Selangor, Kuala Lumpur, Putrajaya	Apply fertiliser, mulching
Girdling root	*Couroupita guianensis*	Small leaf size	Kuala Lumpur, Selangor	Remove girdling roots
Nutrient deficiency	*Acacia mangium*	Chlorosis	Selangor	Apply fertiliser

Fig. 5.3 Longitudinal splitting of *Cyrtophyllum fragrans* bark due to lightning (arrow)

Diseases discovered during the inspection can be classified into three groups based on plant parts: foliage disease, stem disease and root disease (Table 5.2). Leaf spot, leaf rust, powdery mildew a nd sooty moulds are all common foliage diseases that are caused by a variety of fungal species including. *Fusarium* spp., *Colletotrichum* spp. *Curvularia* sp., *Pestalotiopsis* spp. (leaf spot), *Olivea tectonae* and *Coleosporium plumeriae* (leaf rust), *Oidium heveae* (powdery mildew) and *Cladosporium* sp., *Aureobasidium* sp., *Antennariella* sp., *Limacinula* sp., *Scorias* sp, and *Capnodium* sp. (sooty mould). They usually cause extremely small and inconsequential damage to their hosts, therefore control measures are not required. Some stem diseases have been identified, including *Chrysoporthe* stem canker, basal stem rot (BSR) and *Fusarium* wilt. These stem diseases have caused significant effects to *Eucalypt*, foxtail palm and angsana in urban areas, respectively. Similarly, root rot diseases such as white root, brown root and red root disease are significant tree diseases in urban landscapes. During tree inspections, white root disease was discovered on a *Casuarina* sp., a *Tectona grandis* and a clump of giant bamboo, *Dendrocalamus giganteus*. Brown root disease was found to be harmful to *Elateriospermum tapos*, *Khaya ivorensis* and *Samanea saman* whereas, red root disease was found to be harmful to *Bertholletia excelsa, Dryobalanops oblongifolia, Mesua ferrea, Shorea singkawang* and *Swietenia macrophylla.* These diseases have been observed to affect tree species ranging in size from pole to large timber tree. Some severely infected trees, particularly those with brown root and red root, have uprooted due to rotting anchoring roots, causing damage to neighbouring structures such as houses, electrical lines, and roads, as well as the loss of human life.

In general, the appearance of both stem and root diseases should be taken seriously by the management because they have the potential to kill their hosts, while also posing threats to neighbouring structures, vehicles and the general public.

Diseases with Significant Impacts in Urban Forests

Root Rot Disease

The root rot disease is known to be economically important in tree plantations all over the world. It can stunt plant growth and kill trees regardless of health status or age (Guyot & Flori, 2002; Semangun, 2000). Root rot has been identified as the most devastating disease to many trees in the Tropics, especially in forest plantations. It is the most common cause of failure in the early phase of plantation development (Wingfield, 1999). Similarly, root rot disease is a severe hazard to trees in urban areas. Tree falls caused by this illness can cause major damage to structures, vehicles and roads and tree failure caused by root disease is extremely dangerous and unsafe to the public, as it can result in casualties and loss of life.

In general, three types of root rot disease are common in tropical countries including Malaysia: white root disease, brown root disease and red root disease. White root disease is caused by *Rigidoporus microporus,* while brown root and red root disease are caused by *Phellinus noxius* and *Ganoderma philippi*, respectively. These are pantropical decay fungi with a wide host range that includes many tropical forest tree species, fruit trees, plantation and landscaping trees, and woody shrubs (Ann et al., 2002; Holliday, 1980; Nunez & Ryvarden, 2001). They spread to healthy surrounding trees to initiate new infection via root contact with infected hosts, residual inocula in host debris, infected stumps of forest trees, or prior crops (Chang, 2002). Other than root contact, the infection could also occur through wind-borne basidiospores (Lim, 1977; Hodges & Teneiro, 1984; Page et al., 2020). Infected trees by these types of diseases can only be distinguished by the colour of the fungal tissues present on the infected root surface and fruiting bodies present on the root collar. In contrast, because the disease symptoms are so similar, identifying the disease via observation on the crown is quite difficult. Common root disease symptoms on the crown include pale green foliage, yellowing, wilting and finally defoliation (Fig. 5.4). However, these symptoms typically indicate that the trees are past the point of treatment or recovery and death is certain (Ismail & Azaldin, 1985). The afflicted trees

Table 5.2 Urban forest diseases found during tree inspections conducted in Kuala Lumpur, Selangor and Putrajaya, Malaysia

Disease	Tree species	Fungus	Locality
Leaf spot	*Acacia mangium, A. auriculiformis, Dyera costulata, Endospermum diadenum, Khaya ivorensis*	*Fusarium* spp., *Colletotrichum* spp. *Curvularia* sp., *Pestalotiopsis* spp.	Kuala Lumpur, Selangor
Leaf rust	*Tectona grandis*	*Olivea tectonae*	Selangor
Leaf rust	*Plumeria* sp.	*Coleosporium plumeriae*	Kuala Lumpur, Selangor
Powdery mildew	*Hevea brasiliensis*	*Oidium heveae*	Selangor
Sooty mould	*Acacia mangium, A. auriculiformis, Khaya ivorensis*	*Cladosporium* sp., *Aureobasidium* sp., *Antennariella* sp., *Limacinula* sp., *Scorias* sp.,*Capnodium* sp.	Selangor
Chrysoporthe stem canker	*Eucalyptus* sp.	*Chrysoporthe cubensis*	Kuala Lumpur
Fusarium wilt	*Pterocarpus indicus*	*F. oxysporum* f. sp. *angsanae,*	Kuala Lumpur, Selangor
Basal stem rot (BSR)	*Wodyetia bifurcata*	*Ganoderma boninense*	Kuala Lumpur, Selangor, Putrajaya
White root disease	*Casuarina* sp., *Dendrocalamus giganteus, T. grandis*	*Rigidoporus microporus*	Selangor
Brown root disease	*Elateriospermum tapos, Khaya ivorensis Samanea saman*	*Phellinus noxius*	Selangor
Red root disease	*Bertholletia excelsa, Casuarina* sp. *Dryobalanops oblongifolia, Mesua ferrea, Shorea singkawang, Swietenia macrophylla*	*G. philippii*	Kuala Lumpur, Selangor

are also vulnerable to wind throw at this stage due to a serious rotting of anchoring and lateral roots. If control measures are not implemented, it has the potential to spread and infect healthy nearby trees through root-to-root contact over time.

White root disease caused by *R. microporus* can be identified by the presence of white, rough, thick and flattened mycelia strands (Fig. 5.5a) that grow and adhere

Fig. 5.4 Typical crown symptom of an affected tree from root rot disease infection showing pale green and yellowing of foliage and reduced canopy, which later may turn to heavy defoliation and death if control measures are ignored

strongly to the root surface (Nandris et al., 1987). In the absence of any woody substrate, these rhizomorphs grow rapidly and can reach several metres through the soil. On newly infested trees, the infected root wood is often brown in colour and hard (Fig. 5.5b). When the infection progresses to an advanced level, the wood turns white or cream in colour, soft and friable when crushed with fingers (Fig. 5.5c). Basidiocarps are bracket-shaped, leathery, orange-yellow with zonation (Fig. 5.5d), and often present at the base of trees that have been severely infected by the fungus (Nandris et al., 1987). The presence of basidiocarps is quite common, especially during the rainy season (Nandris et al., 1987).

In the field, brown root disease-infected trees can be identified by root inspection only when basidiocarp is not present. It is distinguished by bark depression or rotted

Fig. 5.5 Signs and symptoms commonly observed on trees infected by *Rigidoporus microporus* white root disease. (**a**) White rhizomorph on the root surface (arrows), (**b**) Brown discoloration of necrotic tissue at root collar, (**c**) Bleach, soft and friable root tissues, and (**d**) Orange-yellow with zonation of *Rigidoporus microporus* basidiocarp on root collar

collar region, dark brown discoloration beneath the depressed zone and the presence of a brown mycelial crust intermingled with soil particles or sand on the root surface (Fig. 5.6a). In its advanced stages, the disease is clearly identified by the development of brown zigzag lines in the wood, forming a honeycomb-like pattern, and the wood becoming friable, light and dry (Fig. 5.6b). It is uncommon to find a fruiting body in the field. If present, it can come in the form of a resupinate (sock-like) or bracket. The resupinate shape of the fungus is characterised by a flat, undulating and finely velvety surface with a hard, crusted and resinous structure when sectioned. It is in pale reddish-brown to umber (dark yellowish-brown) colour in the older regions, and brown towards the margin. The bracket-shaped fruiting body is hard, woody and broadly attached to the substrate (Fig. 5.6c).

Red root disease is usually recognised by the presence of a wrinkled reddish-brown rhizomorphic skin on the affected roots, where sand or soil particles often strongly adhere to the fungal mycelia on the root surface (Fig. 5.7a) (Lee, 2000; Mohammed et al., 2006). When the root bark is ripped open, a white mottling pattern of mycelia with a very characteristic odour can be seen on the underside of the infected bark (Fig. 5.7b) (Lee, 2000). The woody part of the root is physically normal and hard in the early stages of infection. However, in the advanced stages of infection, it turns pale buff and spongy or dry depending on the soil condition (Fig. 5.7c). In the field, the presence of *G. philippii* fruiting bodies on infected trees is rather common. It

Fig. 5.6 Below ground signs and symptoms of brown root disease (**a**) Dark brown discoloration of infected zone (arrow), (**b**) Golden brown pockets of fungal hyphae in wood, and (**c**) Bracket shaped of *Phellinus noxius* basidiocarps on tree root (arrow)

normally grows low on the tree trunk and might be hidden by undergrowth if weeding is neglected (Page et al., 2020). The fruiting body is broadly attached, woody shelf with concentric furrows and warty (Fig. 5.7d). It has a smooth upper surface and is semi-glossy with a dark reddish or purplish-brown colour. The margin of the fruiting body is usually narrow and white. The lower surface is white or brownish with medium to fine pores invisible to the naked eye.

Currently, disease control and prevention strategies are mainly based on planters' experiences in large-scale plantations. However, not all the techniques are applied directly to urban forestry, and certain modifications may be necessary to fit to the site and management policies. The following are the most effective practices for preventing and controlling root disease incidence in urban forestry.

i. Regular and continuous monitoring is the most important programme that should be conducted by the management as a preventative measure to avoid root rot disease-related damages or casualties. This can assist in detecting pest and disease attacks based on visible signs and symptoms. During monitoring, all information related to tree health such as plant species, weather conditions, soil characteristics, locations and previous site history should be thoroughly investigated and analysed. Technical advice from experts may be required to further confirm the pathogen, evaluate tree hazard and risk status, and propose control

Fig. 5.7 Characteristics of tree roots infected by *Ganoderma* red root disease. (**a**) Thin and wrinkled red skin-like mycelial crust intermingled with sand and soil particles on the surface (arrow), (**b**) White mottling pattern of mycelia on the underside of the bark (arrow), (**c**) Light, spongy and friable wood tissue, and (**d**) Dark reddish *Ganoderma* basidiocarps on root collar

measures. Before any proper actions are taken, a warning about a hazardous tree should be made public.

ii. The most common technique of combating root disease is to apply fungicides by soil drenching. Several commercial fungicides have shown promise in the control of root rot fungus, including hexaconazole, tridemorph, propiconazole, tridemefon cyproconazole and penconazole. However, the efficiency of the fungicide treatment decreases as infection levels rise (Ismail & Shamsuri, 1998). Therefore, fungicides should only be applied on newly infected trees or trees with mild infection levels.

iii. Trees that are seriously infected by the disease should be removed immediately either mechanically or manually, especially their root systems. The tree stump, conks present on the root collar and woody debris buried in the soil should be destroyed. The aim of this technique is to reduce the source of potential inoculum in the soil as well as to stop the disease spread via root contact to healthy adjacent trees or to replanting.

iv. Prior to planting, considerable attention should also be given to the management of the selection of appropriate species or resistant varieties against the root pathogens to reduce the risk and mortality of trees due to the fungal infection.

Stem Diseases

Basal Stem Rot of Foxtail Palm, **Wodyetia bifurcata** *Irvine*

Wodyetia bifurcata Irvine or foxtail palm, is among several palm species that are commonly planted in Malaysian urban landscapes. Its unique foliage, fast growth rate and lack of significant pests and diseases are key factors that often attract the attention of many horticulturalists and landscape architects to use the species in their projects. Unfortunately, recent reports revealed that this palm species is vulnerable to basal stem rot disease (BSR) infection (Mohd. Farid et al., 2018). This noxious disease has been found to infect and kill the palm species in several housing areas and golf clubs carried out during tree inspections in Selangor, Kuala Lumpur and Putrajaya. Unlike urban forestry, the incidence of BSR caused by *Ganoderma* spp. is common in oil palm plantations in Malaysia (Idris et al., 2003). Despite the fact that the ecosystem is varied, this information implies that oil palm trees planted in urban areas are equally susceptible to *Ganoderma* BSR infection. It is capable of killing oil palm trees of various ages ranging from one to 25 years old, regardless of size and health status (Flood et al., 2000). Other landscape palm species that are also susceptible to BSR infection includes *Areca* spp. and *Livinstona chinensis* var. *subglobusa* (Chase & Broschat, 1992). In general, many species of *Ganoderma* can cause BSR such as *G. applanatum, G. chalceum, G. lucidum, G. miniatocinctum, G. pseudoferrum, G. zonatum, G. boninense* and *G. tornatum* (Ariffin et al., 2000). For foxtail palm, fungus species that can cause BSR is *G. boninense.* This is a saprophytic soil inhabitant fungus that can be pathogenic under some circumstances, especially when favourable conditions and susceptible hosts are present. Similar to root rot disease, the natural infection of this fungus is also through root contact between healthy roots and diseased tissues remaining in the soil (Turner, 1965). Spores of the pathogen also could cause the airborne spread of the disease.

BSR disease infection on foxtail palm is often noticeable when fronds of the palm species turn to pale green, yellowing, wilting and collapse with only a single spear leaf remaining (Fig. 5.8). The infection becomes more obvious when symptoms such as bark necrosis and depression are present on the basal stem of the infected tree. White primordia and dark brown basidiocarps of the fungal species sometimes can also be seen on the stem base, which indicates a serious infection on the symptomatic plant (Fig. 5.9). Underneath the necrotic bark, the affected tissues are often brown colored, rotten, and moist with the presence of white mottling mycelia, compared to the firm and cream of healthy tissues. It also produces a mushroom-like odour. With time, the tissue becomes dry, friable and disintegrates to form a hollow in the tree stem.

In the field, palm trees exhibiting disease symptoms are advised to be removed immediately, along with their root system and replaced with woody species. In general, woody species are recommended since the majority of Arecaceae are highly susceptible to *G. boninense* but most woody species have a high level of resistance to the pathogen (Moncalvo, 2000). Another method of controlling the transmission of the disease is to construct isolation trenches. It is constructed by excavating the soil around the symptomatic trees to a depth of at least 70 cm and a width of at least 30 cm. This approach is feasible for small areas with few numbers of affected trees due to the labour and expense associated in tranche construction. Similar to root rot disease, fungicides such as hexaconazole, tridemorph, propiconazole, tridemefon, cyproconazole and penconazole also can be applied in controlling foxtail palm BSR (Ismail & Shamsuri, 1998).

Fusarium *Wilt Disease of Angsana*

Fusarium wilt is known as the most destructive disease that affects Angsana, *Pterocarpus indicus* especially in Singapore and Malaysia. It was first reported in Malacca in 1870 and spread to several parts of Peninsular Malaysia down to Singapore (Furtado, 1935). At that time, the cause of the Angsana mortality was still not clear due to the frequent failure of pathogenicity tests. In 1985, a wilt disease epidemic in Singapore had caused approximately 800 roadside trees to be removed to avoid the disease spread (Sanderson, 1992). In the 1990s, the disease became a major threat to Angsana trees on the Peninsula where it spread swiftly over several states and rapidly killed trees. Most of the infected trees were removed and destroyed in order to halt the spread of the disease infection. Consequently, this species is no longer planted as roadside trees in the country at present.

Fusarium oxysporum was identified as the fungus that causes Angsana vascular wilt disease (wilt disease) in Singapore (Sanderson et al., 1997). Later, the real culprit

Fig. 5.8 A dying foxtail palm caused by *Ganoderma* basal stem root disease showing the collapse of fronds with only a spear leaf remaining

of the disease is identified as *F. oxysporum* f. sp. *angsanae* (Crowhurst et al., 1995; Ploetz, 2006). This is a strictly host-specific pathogen (Arie, 2010), and its infection is limited to Angsana only. *F. oxysporum* is an ascomycetous fungus that inhabits various environments, including both plant tissues and the rhizosphere (Inami et al., 2014). It is very dependent on soil and soil water movement to propagate the disease. Apart from soil and water, disease infection was also reported to occur via root contact between the roots of infected trees and healthy neighbouring trees (Anon, 2009). Studies in Indonesia and Thailand found that the pathogen can also be transmitted by ambrosia beetles (Bumrungsri et al., 2008; Tarno et al., 2016).

Initial symptoms of Angsana wilt disease are usually recognized by sudden wilting of the canopy (Fig. 5.10a). During this early stage, the leaves on the infected plant

Fig. 5.9 White primordia (arrows) on the basal root of foxtail palm indicating a serious infection of basal stem rot

part frequently turn pale green and yellow, which later turn brown and dry with time. Defoliation is obvious particularly during strong wind, indicating that the tree is dying (Fig. 5.10b). At these stages, the infected trees may be considered beyond treatment as the pathogen is already well established in the vascular system. Other symptoms that can also be observed on the infected tree include vascular discoloration and necrotic brown tissue on the trunk (Fig. 5.10c). In most cases, the presence of ambrosia beetles on necrotic areas is indicated by tiny holes and fine wood powder of insect frass (Fig. 5.10d). Tree death caused by the disease may occur as early as five weeks after the first signs of yellowing appear (Crowhurst et al., 1995).

A control strategy for *Fusarium* wilt disease has been developed in Singapore (Anon, 2009). However, only some of the measures are feasible in Malaysia as the tree species is no longer of interest to local authorities, city councils and landscapers for planting in urban areas. At present, the management of this disease in the country is to salvage the remaining Angsana trees only. Other control measures as follows could be considered.

i. Similar to root rot disease, it is critical to maintain regular monitoring and inspection of the plant health. These preventive measures are generally meant to detect the early stages of the disease infection before it becomes serious and threatens public safety.

Fig. 5.10 Sign and symptoms of *Fusarium* wilt disease on Angsana. (**a**) Wilting and yellowing of foliage, (**b**) Dying tree, (**c**) Necrotic brown tissue (arrows), and (**d**) White powdery ambrosia beetle frass on a tree trunk

ii. All infected trees by the disease should be removed and destroyed immediately. This approach is vital because those trees have the potential to be an inoculum source of the pathogen. It can spread to healthy neighbouring Angsana trees and initiate new infections.
iii. Chemical fungicides could be applied if the infection is minor and not serious. A field study conducted in Kuantan, Pahang revealed a significant recovery of Angsana trees infected by the disease after the application of thiabendazole (Philip, 1999).
iv. Arboricultural tools such as chain saw, pruning saw, hoe and machete that are used to cut and remove the diseased trees must be surface sterile with disinfectant or bleaching agent (sodium hypochlorite) before and after being used.

Chrysoporthe *Stem Canker of Eucalyptus*

Stem canker is a commonly observed indication of disease infection on trees in urban forestry or in plantations. It can be associated with fungi from various species such as *Ceratocystis fimbriata, Diplodea pinea, Nectria spp., Thyronectria austro-americana, Leucocytospora kunzei, Phomopsis* spp. *Hypoxylon atropunctatum, Botryosphaeria* spp. (Wegulo, 2001). While most canker fungi are unlikely to cause serious damage or death to healthy and vigorous trees, a few of these fungi are capable of causing destruction on certain tree species. In general, stem canker occurs mainly on trees that are already severely weakened or stressed due to environmental factors such as drought, chemical injury, mechanical injury, extreme temperature fluctuation and nutrient deficiency (Wegulo, 2001). Wounds or branch stubs at the stem or stem base are known as the starting point of the canker. It becomes an entry point for pathogens to initiate infection and then develop a canker.

Stem canker caused by *Chrysoporthe cubensis* or previously known as *Cryphonectria cubensis* is regarded as the most important disease of *Eucalyptus* spp. in the tropics and subtropics (Gryzenhout et al., 2009). It is capable of infecting trees of all ages, from young to older stands. However, mild infection of the disease rarely kills its tree hosts, although stunted and distorted growth is common. Nevertheless, a serious infection of the stem canker disease on young *Eucalyptus* could lead to rapid tree death, whereas on older trees, it could weaken the stem and often result in stem breakage.

In the field, most eucalypt trees infected by stem canker disease are usually unnoticeable until the infection becomes severe and the appearance of symptoms is obvious. The first symptoms of *Chrysoporthe* stem canker are usually elongated sunken areas on the infected tree bark, either at the base or up to a metre above the ground (Sharma et al., 1985). At this stage, peeled off the bark reveals brown necrotic sapwood underlying the sunken bark tissue. The bark later splits, shed and callus ridges gradually developed around the infected area whose size increases with time (Fig. 5.11). On certain trees, symptoms of fungus-caused stem canker disease included swollen tree bases surrounded by cracked bark (Ciesla et al., 1996). The presence of gummosis (oozing of kino) that secretes from the cankers is common. This dark gummy exudate usually becomes dry on the bark surface and can be washed down during the rainy season to form a distinct colour on the infected tree bark. With careful investigations, many fruiting structures (perithecia and pycnidia) can easily be seen on the cracked bark.

Until present, *C. cubensis* is known to occur only on trees from the Myrtaceae family particularly *Eucalyptus* spp. including *Syzygium* spp. and the Melastomataceae family such as *Tibouchina* spp., *Miconia* spp., *Rhynchanthera mexicana, Clidemia sericea* and *Melastoma melabathricum* (Chen et al., 2010; Rodas et al., 2005). This pathogen infects its tree hosts through wounds, either naturally occurring or resulting from the pruning of branches. The most common plant parts that

Fig. 5.11 A eucalyptus tree infected by *Chrysoporthe* stem canker showing a serious stem damage that could lead to structural failure and tree mortality

are highly exposed to the fungal infection include tree bases or lower stem of young *Eucalyptus* up to the breast height or higher on the bole. The most common mode of infection is via spores that are dispersed through rainwater splash. Additionally, wind-borne spores also have been reported to cause the disease infection (Anon, 2002).

The application of chemical fungicides may aid in disease control in urban landscapes, particularly when the infection is not severe. Copper fungicide (1% paste) and tridemorph are two fungicides that are efficient against the disease. (0.2% of i.a) (Mohanan, 2014). They can be applied by brushing directly on the infected bark or cutting the surface of the infected stem according to the instructions. Alternatively, disease-tolerant *Eucalyptus* species can be acquired and planted. This method has been widely used to control the disease in many *Eucalyptus* plantations throughout the world. It is more convenient and cost-effective than labour-intensive and expensive fungicide applications. In Brazil and South Africa, the introduction of disease-tolerant *Eucalyptus* hybrids effectively prevented the occurrence and spread of *Chrysoporthe* stem canker in the plantations (Nakabonge et al., 2006).

Conclusion and Recommendations

Most plant health problems observed at the inspected sites are associated with environmental factors, particularly soil compaction, whereas other abiotic factors such as lack and excessive water, lightning strikes, nutrient deficiency, girdling roots and mechanical injuries are considered as minor contributing factors to plant health problems. Except for root rot disease, *Chrysoporthe* stem canker, *Fusarium* wilt, and basal stem rot, which are destructive, difficult to control, and spread easily to initiate new infections if susceptible hosts and favourable conditions are present, most diseases found in the inspected sites are minor and are not very risky or harmful to the healthy neighbouring trees. During inspections, it is abundantly obvious that lack of knowledge about plant health care and pest management are the fundamental causes of all tree health problems. Moreover, plant health monitoring was carried out without proper planning. As a result, early disease diagnosis and identification of factors associated with the tree health problems frequently failed. Most site supervisors and field staff are also not fully aware of the importance of sound planning and adequate knowledge on pest and disease management. Consequently, most of the problematic trees can only be identified at later stages of decline, making treatment to promote plant health and control pests and diseases challenging. If any management approaches can be utilised, strong financial backup, efforts and time are required.

Thus, some additional recommendations are proposed to lessen the chance of tree failure and financial loss in the future due to plant health problems. These are long-term management recommendations that should be considered, such as

i. Implementation of periodic tree risk assessment for the management of hazards associated with abiotic or biotic factors in an identified area. The assessment should be carried out at least once a year by experienced arborists to avoid wrong evaluation and recommendations in improving tree health and risk.

ii. Enhancing and updating field staff and site supervisor's knowledge on pest and disease diagnosis, identification, and management. It can be achieved by participating in hands-on courses, seminars or workshops in related fields.

iii. Establishing a network with related universities and research institutes is highly recommended in order to gain new information, expertise and cutting-edge technologies via dialogues, courses and training.

References

Ann, P. J., Chang, T. T., & Ko, W. H. (2002). *Phellinus noxius* brown rot of fruit and ornamental trees in Taiwan. *Plant Disease, 86*(8), 820–826.

Anonymous, (2002). *Cryphonectria* canker of *Eucalyptus*. Tree Protection Co-operative Programme Pamphlet.

Anonymous, (2009). *Fusarium wilt disease of Angsana.* Technical Notes.

Arie, T. (2010). Phylogeny and phytopathogenicity mechanisms of soilborne *Fusarium oxysporum. The Journal of General Plant Pathology, 76*, 403–405.

Ariffin, D., Idris, A.S., & Singh, G., (2000). Status of *Ganoderma* in oil palm. In J. Flood, P. D.Bridge, & M. Holderness (Eds.), *Ganoderma diseases of perennial crops* (pp. 49–68). CABI Publishing.

Bumrungsri, S., Beaver, R., Phongpaichit, S. & Sittichaya, W., (2008). The infestation by an exotic ambrosia beetle, *Euplatypus parallelus* (F.) (Coleoptera: Curculionidae: Platypodinae) of Angsana trees (*Pterocarpus indicus* Willd.) in southern Thailand. S*ongklanakarin Journal of Science and Technology*, 30(5), 579–582.

Chang, T. T. (2002). The biology, ecology and pathology of *Phellinus noxius* in Taiwan. In R. Watling, J. C. Frankland, A. M. Ainsworth, S. Isaac & C.H. Robinson (Eds.), *Tropical mycology vol. 1, macromycetes* (pp. 87–99). CAB Publishing.

Chase, A. R., & Broschat, T. K. (1992). *Diseases and disorders of ornamentals palms.* American Phytopathological Society.

Chen, S. F., Gryzenhout, M., Roux, J., Xie, Y. J., Wingfield, M. J., & Zhou, X. D. (2010). Identification and pathogenicity of *Chrysoporthe cubensis on Eucalyptus and Syzygium spp. in South China. Plant DisEase, 94*, 1143–1150.

Ciesla, W.M., Diekmann, M. & Putter, C.A.J., (1996). *FAO/IPGRI Technical Guidelines for the Safe Movement of Germplasm. No. 17* (p. 68). *Eucalyptus* spp. Food and Agriculture Organization of the United Nations, Rome/International Plant Genetic Resources Institute, Rome.

Colombo, S.J., (2016). *Canada's urban forests in a changing climate* (Canadian Climate Forum Issue Paper No. 5).

Crowhurst, R. N., King, F. Y., Hawthorne, B. T., Sanderson, F. R., & Choi-Pheng, Y. (1995). RAPD characterization of *Fusarium oxysporum* associated with wilt of angsana (*Pterocarpus indicus*) in Singapore. *Mycological Research, 99*, 4–18.

Dale, A. G., & Frank, S. D. (2017). Warming and drought combine to increase pest insect fitness on urban trees. *PLoS ONE, 12*(3), e0173844. https://doi.org/10.1371/journal.pone.0173844

Farid, A. M., Azrul, W. M., & W.A., Syazwan, S.A., Lee, S.Y. & Mohamed, R. (2018). First record of basal stem rot of foxtail palm *Wodyetia bifurcata* caused by *Ganoderma boninense* in Malaysia. *Plant Disease, 102*(7), 461–1461. https://doi.org/10.1094/pdis-12-17-2027

Flood, J., Bridge, P. D., & Holderness, M. (2000). *Preface.* CABI Publishing.

Furtado, C. X. (1935). A disease of the angsana tree. *Journal of the Malayan Branch of the Royal Asiatic Society, 13*(2), 163–192.

Gryzenhout, M., & Wingfield, B. O. & Wingfield, M. J. (2009). *Taxonomy, phylogeny, and ecology of bark-inhabiting and tree pathogenic fungi in the Cryphonectriaceae*. American Phytopathological Society.

Guyot, J., & Flori, A. (2002). Comparative study for detecting *Rigidoporus lignosus* on rubber trees. *Crop Protection, 21*, 461–466.

Hodges, C. S., & Teneiro, J. A. (1984). Root rot of *Delonix regia* and associated tree species in the Mariana Islands caused by *Phellinus noxius*. *Plant Disease, 68*, 334–345.

Holliday, P. (1980). *Fungus diseases of Tropical crops*. Cambridge University Press.

Idris, A. S., Yamouka, M., Hayakawa, S., Basri, M. W., Noorhasimah, I., & Ariffin, D. (2003). PCR technique for detection of *Ganoderma* (MPOB Information Series No. 202).

Inami, K., Takeshi, K. T., Masato, K. M., Akiko, O. A., Nobuko, I. N., Enrique, R. P., Hozumi, T., Liliana, A. C., Fatima, C. B., Mauricio, J. R., Khalid, A. M., Tobin, L. P., Teraoka, T., Kodama, M., & Arie, T., (2014). The tomato wilt fungus *Fusarium oxysporum* f. sp. *lycopersici* shares common ancestors with nonpathogenic *F. oxysporum* isolated from wild tomatoes in the Peruvian Andes. *Microbes Environment, 29* (2), 200–210. doi:https://doi.org/10.1264/jsme2.ME13184

Ismail, H., & Azaldin, M. Y. (1985). Interaction of sulphur with soil pH and root disease of rubber. *Journal Rubber Research Institute Malaysia, 33*, 5–69.

Ismail, H., & Shamsuri, M. H. (1998, October). Current status of root diseases of rubber. In *CABI workshop on Ganoderma diseases* (pp. 5–8).

Kaplan, R., & Stephen, K. (1989). *The experience of nature: A psychological perspective*. Cambridge University Press.

Lee, S. S. (2000). The current status of root diseases of *Acacia mangium* Willd. In J. Flood, P. D. Bridge, & M. Holderness (Eds.), *Ganoderma diseases of perennial crops* (pp. 71–79). CABI Publishing.

Lim, T. (1977). Production, germination and dispersal of basidiospores of *Ganoderma pseudoferreum* on *Hevea*. *Journal of the Rubber Research Institute Malaysia, 25*, 93–99.

Mohammed, C.L., Barry, K.M. & Irianto, R.S.B., (2006). Heart rot and root rot in *Acacia mangium*: identification and assessment. In K. Potter, A. Rimbawanto & C. Beadle (*Eds.*), *Heart rot and root rot in tropical Acacia plantations* (pp. 26–33). Proceedings of a workshop held in Yogyakarta, Indonesia, 7–9 February 2006. Canberra, ACIAR Proceedings No. 124.

Mohanan, C. (2014). Diseases in *Eucalyptus*: Status and management. In P. P. Bhojvaid, S. Kaushik, Y. P. Singh, D. Kumar, M. Thapliyal, & S. Barthwal (Eds.), *Eucalypts in India* (pp. 280–314). ENVIS Centre on Forestry National Forest Library and Information Centre Forest Research Institute.

Moncalvo, J. M. (2000). Systematics of *Ganoderma*. In J. Flood, P. D. Bridge, & M. Holderness (Eds.), *Ganoderma diseases of perennial crops* (pp. 23–46). CABI Publishing.

Nakabonge, G., Roux, J., Gryzenhout, M., & Wingfield, M. J. (2006). Distribution of *Chrysoporthe* canker pathogens on *Eucalyptus* and *Syzygium* spp. in eastern and southern Africa. *Plant Disease, 90*, 734–740.

Nandris, D., Nicole, M., & Geiger, J. P. (1987). Root rot diseases of rubber tree. *Plant Disease, 71*, 298–306.

Nunez, M., & Ryvarden, L. (2001). *East Asian Polypores* (Vol. 2, p. 522). Polyporaceae s. lato. Synopsis Fungorum 14. Fungiflora.

Page, D.E., Glen, M., Puspitasari, D., Prihatini, I., Gafur, A. & Mohammed, C.L., (2020). *Acacia* plantations in Indonesia facilitate clonal spread of the root pathogen *Ganoderma philippii*. *Plant Pathology* [online]. [Viewed 12 January 2020]. Available from https://doi.org/10.1111/ppa.13153

Philip, E. (1999). Wilt disease of angsana *(Pterocarpus indicus)* in Peninsular Malaysia and its possible control. *Journal of Tropical Forest Science, 11*(3), 519–527.

Ploetz, R. C. (2006). *Fusarium*-induced diseases of tropical, perennial crops. *Phytopathology, 96*, 648–652.

Rodas, C. A., Gryzenhout, M., Myburg, H., Wingfield, B. D., & Wingfield, M. J. (2005). Discovery of the *Eucalyptus* canker pathogen *Chrysoporthe cubensis* on native *Miconia* (Melastomataceae) in Colombia. *Plant Pathology, 54*, 460–470.

Sanderson, F. R. (1992). Angsana Wilt Disease. *Garden Wise, 4*, 9.

Sanderson, F.R., Fong, Y.K., Yik, C.F., Ong, K.H. & Anuar, S., (1997). A *Fusarium* wilt (*Fusarium oxysporum*) of angsana (*Pterocarpus indicus*) in Singapore: Epidemiology and identification of the causal organism. *Arboricultural Journal: The International Journal of Urban Forestry, 21*(3), 18–204. https://doi.org/10.1080/03071375.1997.9747165

Semangun, H. (2000). *Diseases of plantation crops in Indonesia.* Gadjah Mada University Press.

Sharma, J. K., Mohanan, C., & Florence, E. J .M., (1985). *Disease survey in nurseries and plantations of forest tree species grown in Kerala.* (Kerala Forest Research Institute Research Report 36).

Solomou, A. D., Topalidou, E. T., Germani, R., Argirι, A., & Karetsos, G. (2019). Importance, utilization and health of urban forests: A review. *Notulae Botanicae Horti Agrobotanici Cluj, 47*(1), 10–16.

Tarno, H., Erfan Dani, S., & Luqman, Q. A. (2016). Microbial community associated with ambrosia beetle, *Euplatypus parallelus* on sonokembang, *Pterocarpus indicus* in Malang. *AGRIVITA Journal of Agricultural Science, 38*(3), 312–320.

Turner, P. D. (1965). The incidence of *Ganoderma* disease of oil palms in Malaya and its relation to previous crop. *Annual Applied Biology, 55*, 417–423.

Ulrich, R. S. (1986). Human responses to vegetation and landscapes. *Landscape and Urban Planning, 13*, 29–44.

Wegulo, S. (2001). *Fungal stem cankers of trees.* Sustainable urban Landscapes. Iowa State University.

Wingfield, M. J. (1999). Pathogens in exotic plantation forestry. *International Forestry Review, 1*, 163–168.

Wolf, K. L. (2007). The environmental psychology of shopping: Assessing the value of trees. *International Council of Shopping Centers Research Review, 14*(3), 39–43.

Mohd Farid Ahmad is a senior researcher at the Forest Research Institute Malaysia (FRIM). He obtained his bachelor's and master's degree from the Universiti Putra Malaysia (UPM). He also has a Doctor of Philosophy degree from the Universiti Sains Malaysia (USM). The author has served with the research institute since 1998 and his research interests include forest pathology and the development of bio-control agents for pest and disease management. He is currently the Head of the Forest Health and Conservation Programme, Forest Biodiversity Division. He is also a certified arborist (MY-0360A) by the International Society of Arboriculture since 2014.

Muhammad Syahmi Hishamuddin works as a forest pathologist at the Forest Research Institute Malaysia (FRIM). He earned a bachelor's degree with honours from Universiti Malaysia Pahang and a master's degree with honours from Universiti Putra Malaysia. He also has a Doctor of Philosophy degree in Plant Biotechnology from Universiti Putra Malaysia (UPM). Forest biotechnology is one of the author's research interests. He is currently a junior researcher at the Forest Health and Conservation Programme, Forest Biodiversity Division.

Chapter 6
Potential Carbon Storage and Sequestration by Urban Trees in Malaysia

Kasturi Devi Kanniah, Rohayu Abdullah, and Ho Chin Siong

Abstract Urban trees provide a wide range of ecosystem services that can address climate change mitigation and adaptation. In this study, we estimated for the first time the carbon storage potential of 2,245 trees, covering 19 different families and 41 species, planted in parks and roadsides in two cities in southern Peninsular Malaysia. The aboveground, leaf and root biomass of the trees were estimated using allometric equations. The carbon sequestration was obtained using the estimation of trees' radial growth increments. Results show that the highest carbon storage for trees in parks is by *Khaya senegalensis* (2,289 kg C tree^{-1}) and for roadside trees is *Melaleuca cajuputi* (3,644 kg C tree^{-1}). For species with similar sizes, *Mimusop elengi, Syzygium grande* (parks), *Cassia fistula, Pterocarpus indicus* and *Syzygium grande* (roadside) store more carbon than other species. Trees were also compared for their potential capacity to sequester carbon in year x to year x + 1. For park tree species, *Pterocarpus indicus* sequesters more carbon (249.77 kg tree^{-1} year^{-1}) between age 50 and 51 compared to other tree species in the same age group. Results of this study can guide local landscape authorities in planting suitable tree species in urban landscapes to mitigate climate change impacts.

Keywords Urban forest · Urban green space · Climate change · Street trees · Tropical trees

Introduction

Rapid urbanisation combined with increasing population, transportation and industrial activities are responsible for many environmental problems, including global warming and climate change. Cities emit approximately three quarters of global CO_2

K. D. Kanniah (✉) · R. Abdullah
Universiti Teknologi Malaysia, Skudai, Malaysia
e-mail: kasturi@utm.my

H. C. Siong
Centre for Environmental Sustainability and Water Security (IPASA), Research Institute for Sustainable Environment (RISE), Universiti Teknologi, Johor Bahru, Malaysia

S. Maruthaveeran et al. (eds.), *Urban Forestry and Arboriculture in Malaysia*,
https://doi.org/10.1007/978-981-19-5418-4_6

emissions although they cover only 2% of the land surface (Seto et al., 2014). From the Industrial Revolution to the present, the level of atmospheric CO_2 has increased (Fig. 6.1) and currently its concentration has reached 416.93 ppm (Scripps Institution of Oceanography, 2020). The carbon footprints of cities are important as CO_2 emissions are expected to increase to 800 ppm by 2100 (Lashof & Ahuja, 1990; Olivier et al., 2005), which will cause increased air temperature by 1.5–4.5 °C (Houghton et al., 1996). One nature-based solution to reduce CO_2 in the atmosphere is to increase the coverage of vegetation, particularly urban trees and forests as a natural carbon sink (Kanniah & Chin, 2017).

In Malaysia, by 2020, 75% of the population is projected to migrate to urban areas (Ministry of Housing and Local Government, 2010) and CO_2 emissions are expected to increase 69% compared to 2000 level (Safaai et al., 2011). The main sources of CO_2 emissions from Malaysian cities are energy generation, transport, industry, and residential building. Overall, in Malaysia, the energy generation contributes more than 50% of the total emissions followed by the transport, which accounted for 28% (Mustapa & Bekhet, 2016) and the resultant CO_2 emissions in Malaysia were 7.83 metric tons per capita in 2019 (Global Carbon Budget, 2020). Climate change mitigation actions have been taken by several cities/regions in Malaysia by setting emission intensity reduction goals. For example, the Iskandar Malaysia Special Economic Corridor in the southern part of Johor, Peninsular Malaysia has set an emission intensity reduction target of 50% by 2025 relative to 2005 levels (UTM-Low Carbon Asia Research Centre, 2013). Large cities like Kuala Lumpur (the capital of Malaysia) have agreed to reduce 70% of its CO_2 emissions by 2030 (UTM-Low Carbon Asia Research Centre, 2017). Various actions such as promoting green growth, green mobility, sustainable energy systems, low carbon green building, waste management and low carbon lifestyle have been proposed by these cities to reduce CO_2 emissions (UTM-Low Carbon Asia Research Centre, 2017).

Trees have the potential to sequester carbon from the air and accumulate it in stems and branches (above ground) and in roots and soil (below ground). Forested areas

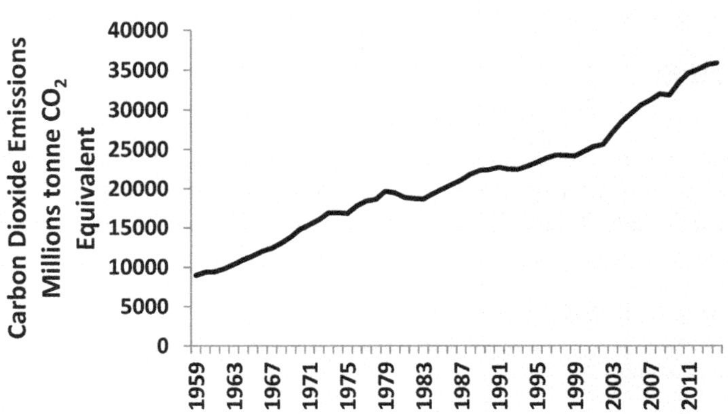

Fig. 6.1 Global CO_2 emissions from 1959 to 2014. *Source* Le Quéré et al. (2015)

and trees in parks can significantly contribute to carbon storage in cities (Fares et al., 2017; Guo et al., 2014; Tang et al., 2016). However, in the last 20 years, research quantifying carbon storage and sequestration by trees found in urban parks and streets has increased in other developed and developing countries because trees are a cost-effective and nature-based solution for moderating the impact of climate change in cities. Some researchers have shown that trees in urban forests are able to stock more carbon than surrounding suburban areas due to CO_2 and nitrogen deposition, and intensive urban tree management (Carreiro & Tripler, 2005; Churkina et al., 2015; Davies et al., 2011; Hutyra et al., 2011; McHale et al., 2009; Woodbury et al., 2007). Large quantities of carbon were found in the above-ground components of trees in Leicester, UK (Davies et al., 2011) and in southwestern Bangladesh (Rahman et al., 2015). All these studies found a significant contribution of urban trees in absorbing CO_2 from the atmosphere. Thus, it is strongly recommended to increase tree cover in Malaysian cities (Kanniah & Chin, 2017). Nevertheless, the effect of urban trees on carbon sink and climate change is not well quantified nor understood in tropical regions of Asia, excluding Singapore (Velasco et al., 2013).

Malaysia is currently developing into an urbanised country. Thus, assessing the carbon storage capacity of urban trees can guide urban planners and local authorities in maintaining existing trees, protecting them from further land conversion, and in planting more trees in cities as a climate change mitigation and adaptation strategy. Therefore, this article aims to calculate the carbon storage and sequestration of trees in urban parks and streets (for the first time) in two cities in southern Peninsular Malaysia. In addition, this study also intends to propose suitable tree species to be planted in Malaysian cities based on their carbon storage and sequestration capacities.

Data and Methodology

Study Area

Two cities, Johor Baharu (JB) and Pasir Gudang (PG), located in the fast-growing economic region of Iskandar Malaysia (IM) in the state of Johor, Malaysia were considered in this study (Fig. 6.2). These cities are located in two of the five 'flagship zones' selected to guide IM's overall development. These cities were selected due to the availability of (i) tree biometric data and (ii) the nature of highly populated, commercial and industrial activities that release more CO_2 compared to other flagship zones in IM. Johor Baharu (1.4927° N, 103.7414° E) is Malaysia's second-largest city and is developing rapidly, with an expected increase in population from 541,508 in 2010 to 1,197,000 in 2025. Johor Bahru (32,721 ha) is a commercial and services centre, while PG (1.4703° N, 103.9030° E), which covers an area of 36,887 ha, is an important port city that handles an edible oil tankage facility, bulk cargo, and petrochemical and agro-processing industries (Ho et al., 2015a). Rapid economic development is anticipated to increase the greenhouse gas emission of these cities

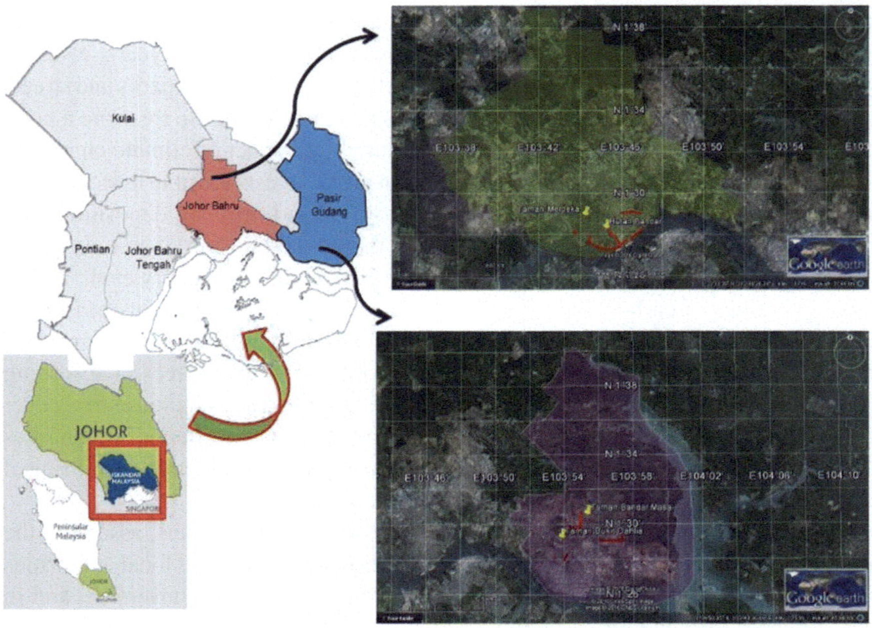

Fig. 6.2 Johor Bahru (top right) and Pasir Gudang (bottom right) municipalities within Iskandar Malaysia region (left panel). Red linear features on Google Earth images (right panel) show the locations of urban and street trees that were used in this study

from 5,298 kt CO_2eq in 2005 to 16,158 kt CO_2eq in 2025 if no mitigation measures are considered (Ho et al., 2015a, 2015b).

The mainland uses in these two cities include urban/settlements, agricultural land use (oil palm), forest, mangrove and vacant land (Kanniah et al., 2015; Kanniah & Chin, 2017). Forest and urban trees cover ~ 8% and 27% of JB and PG, respectively, while impervious surfaces cover 52 and 13% of the total area of these municipalities (Kanniah & Chin,2017) Iskandar Malaysia is characterized by warm weather (temperature ranging between 21 °C and 32 °C) and consistent annual rainfall (2,000 mm to 2,500 mm), suitable for various plants to grow in this region (https://iskandarmalaysia.com.my/).

Data

Carbon storage of trees in parks and roadsides in JB and PG was calculated using tree inventory data obtained from the municipalities (Fig. 6.2). Data on tree stem circumference, height and species of 2,245 trees covering 19 genera and 41 different species (representing approximately 21–27% of the total 150 to 200 tree species found in urban areas in Peninsular Malaysia) from both roadsides and parks were obtained. These data were obtained from 4 parks (2 in each municipality covering areas 6.37–14.65 ha) and 19 streets (5 in PG and 14 from JB) that range from 0.10 to 2.72 km.

Method

Carbon storage of urban trees has largely been estimated using allometric equations that were developed to get aboveground biomass (Nowak & Crane, 2002; Nowak et al., 2013; Weissert et al., 2014). The i-Tree Eco Model (Nowak & Crane, 2002; Nowak et al., 2013) designed for calculating the carbon storage and sequestration of urban trees mainly includes equations for trees in the temperate regions. Since there are no specific allometric equations available to estimate the biomass/carbon storage of tropical urban trees (Weissert et al., 2014), we estimated the above-ground biomass (AGB) of the urban trees with DBH > 26 cm using the allometric equation developed by Chave et al. (2005) (Eq. 6.1), leaf biomass (LB) equation of Chave et al. (2008) (Eq. 6.2) and root biomass (RB) equation of Cairns et al. (1997) (Eq. 6.3). All three equations were developed specifically to quantify the biomass of tropical primary forest trees. The allometric equation of Chave et al. (2005) was developed using data collected at 27 study sites in the tropical forests including one site in Peninsular Malaysia. This equation was previously used by Velasco et al. (2013) for estimating the aboveground biomass of trees in parks and streets in a residential area in Singapore.

$$\mathrm{AGB} = \rho \times \exp\begin{pmatrix} -1.499 + 2.148\ln(D) \\ +0.207(\ln(D))^2 - 0.0281(\ln(D))^3 \end{pmatrix} \tag{6.1}$$

Where:

AGB = above-ground biomass (kg tree^{-1}).
ρ = wood density of trees.
D = DBH = diameter at breast height.

$$\mathrm{LB} = \exp\begin{pmatrix} -5.136 + 2.882\ln(D) \\ -0.156(\ln(D))^2 \end{pmatrix} \tag{6.2}$$

Where:

LB = leaf biomass (kg tree^{-1}).

D = DBH = diameter at breast height.

$$RB = \exp(-1.085 + 0.9256 \ln(AGB)) \quad (6.3)$$

Where:

RB = root biomass (kg $tree^{-1}$).
AGB = above ground biomass.

For trees with smaller size ($\leq$ 25 cm) the equation developed by Breugel et al. (2011) was used (Eq. 6.4).

$$AGB = \exp(-1.130 + 2.267 (\ln(D)) + 1.186 (\ln(p))) \quad (6.4)$$

The roadsides and parks in the study area were found to contain more than 50 species of trees, thus the equation of Chave et al. (2005) and Breugel et al. (2011) are the most appropriate equations as they allow the use of a wood density[1] (ρ) value that differs according to tree species in the AGB calculation. This enables us to estimate AGB on the tree species level. First, we extracted the ρ values of the known tree species in the study area from the database available at http://hdl.handle.net/10255/dryad.234 and calculated the AGB using Eq. 6.1. We used the average ρ values of genera for trees only identified to that taxonomic level. In this study, ρ ranges between 0.29 and 0.88 g cm^{-3} (mean $\pm$ SD of 0.59 $\pm$ 0.10 g cm^{-3}). Meanwhile, LB was obtained using the equation developed by Chave et al. (2008) and only DBH is required in this equation. RB was calculated from the non-linear equation relating AGB and RB as shown in Eq. 6.3.

Trees with the same DBH, but growing in urban environments were found to have less AGB than trees in forests due to pruning, limited availability of resources, etc. Thus, the AGB, LB and RB values of urban trees were reduced by 20% based on studies conducted in temperate (McHale et al., 2009; Nowak, 1994; Nowak et al., 2013) and tropical (Rahman et al., 2015) climates. The dry biomass values were then converted to carbon by considering that the carbon content of tropical forest trees is approximately 50% of the AGB, LB and RB (Brown, 1997). Finally, AGB, LB and RB are added to get the total carbon per tree.

The carbon sequestration of various street and park trees (particularly trees protected under the Tree Preservation Order (TPO) was calculated as the difference in carbon storage between year 1 (y) and year 1 + 1 (y + 1). The trees proposed to be maintained, preserved and gazetted under TPO are to have low risk, high economic

---

[1] Wood density = mass of oven-dried wood ÷ fresh (wet) volume. The ρ values for various tree species are available at http://hdl.handle.net/10255/dryad.234 (Chave et al., 2009; Zanne et al., 2009).

value, have a stem circumference measuring more than 0.8 m and are a species that is relatively rare. Trees protected under TPO have information about their age and health status. This enabled us to adjust the calculation of carbon storage by tree health conditions. Data on tree circumference, height, tree species and health condition of 485 trees covering 35 different species from roadsides and parks were obtained from PG and JB municipality councils. Following Nowak and Crane (2002), we adopted adjusting factors of 1, 1, 0.42, 0.15, and 0, to represent excellent, fair, poor, critical, and dead conditions of the trees respectively. However, the tree inventory data obtained from PG and JB does not cover the age of every species at yearly intervals and no data on the annual growth rate of trees' DBH was available. Following Velasco et al. (2013), we used the metabolic theory of ecology (MTE) model developed by Muller-Landau et al. (2006) to predict the DBH of the trees for year y + 1 (Eq. 6.4). With the new DBH, we could estimate the AGB, RB and LB for the next year and subsequently calculate the carbon sequestration of each tree.

$$\mathrm{Dt} = [\mathrm{D01} - \mathrm{c} + \mathrm{r}\,(1 - \mathrm{c})\,\mathrm{t}]^{1/(1-\mathrm{c})} \tag{6.5}$$

Where;

Dt = expected DBH at time t of a tree.
D01 = initial DBH of a tree

Where;

r = growth intercept and c = growth exponent for a particular forest. These growth scaling parameter values were derived from tropical forests in Malaysia Muller-Landau et al. (2006). Dt at $t = 1$ year was calculated for each tree in the study area using r and c values reported for trees with three different size ranges (trees with D < 20 cm, D > 20 cm and all sizes). The r and c values used in this study are as follows (Table 6.1)

The average Dt from the three sets of scaling parameters was used to estimate the AGB and carbon storage of each tree after one year. The carbon sequestration (annual carbon production) is estimated as the difference between the carbon storage computed with D01 and Dt.

Results

Carbon Storage According to Species

Based on the parks and roadside tree data obtained from PG and JB municipalities, we calculated the carbon storage of each tree and averaged them according to species (Table 6.2). Among the species in the parks, the average carbon storage value was

Table 6.1 The growth intercept and exponent values used to estimate the diameter of urban trees in the study area

Tree size	growth intercept (r)	growth exponent (c)
D < 20 cm	0.0525	0.636
D > 20 cm	0.101	0.554
All size	0.0457	0.677

Source Muller-Landau et al. (2006)

highest for *Khaya senegalensis* accumulating 2,289 ± 2,647 kg tree^{-1} or 8,394 kg CO_2 equivalent over its lifetime in the study area (Table 6.2). This was followed by *Alstonia angustifolia* that showed a carbon storage value of 1,525 ± 946 kg tree^{-1} (5,592 CO_2 equivalent). Meanwhile, for roadside, the average carbon storage value was greatest for *Melaleuca cajuputi* which accumulated 3,644 ± 2,010 kg tree^{-1}, followed by *Pterocarpus indicus* with a carbon storage value of 2,669 ± 1,996 kg tree^{-1} (Table 6.2). The carbon storage values estimated in this study are comparable to the values reported by Velasco et al. (2013). The greatest carbon storage in these species could be owing to the presence of several big and older trees that have high DBH (121 to 124 cm). On the other hand, the lowest amount of carbon was stored by *Shorea leprosula* in parks, and *Alstonia angustifolia* on the roadside. Although these species had the highest ρ value of 0.44 and 0.61 g cm^{-3} (Table 6.2), the reason for the lowest carbon storage could be attributed to their size (lower DBH) and age (younger trees).

The mean carbon storage (kg tree^{-1}) of the trees was further divided into groups based on tree DBH intervals and species as follows: < 5 cm, 6 –10 cm, 11–15 cm, 16–20 cm.... > 130 cm (Table 6.3). Any DBH ranges with fewer than 5 trees were not included in the analysis. The trees were grouped by size to eliminate the effect of tree size on carbon storage. Thus, the tree species that showed the greatest carbon storage for each DBH range was identified. For trees grown in parks, *Mimosup elengi* demonstrated the greatest carbon storage for most of the 5 cm DBH intervals (Fig. 6.3). The difference in carbon storage among tree species in these DBH groups was significant at $p < 0.05$. The species *Alstonia angustifolia*, *Syzygium grande* and *Gymnostoma nobile* in parks also demonstrated the relatively high carbon storage (Fig. 6.3). For roadside trees, *Cassia fistula* showed the greatest carbon storage for most of the 5 cm DBH intervals. *Pterocarpus indicus*, *Syzygium grande*, *Peltophorum pterocarpum*, *Melaleuca cajuputi* and *Alstonia angustiloba* also demonstrated relatively high carbon storage (Fig. 6.3). Again, the difference in carbon storage among different tree species in these DBH groups was significant at $p < 0.05$ except for the DBH interval of less than 5 cm ($p = 0.365$). The higher carbon storage by *Mimusop elengi* and *Cassia fistula* could be due to their relatively higher wood density values (Table 6.2). Trees with higher wood density (ρ) enable maximum photosynthesis and it has been shown that ρ alone could explain 29–45% of the total above-ground biomass (AGB) variation in the Amazon forest (Baker et al., 2004). When comparing trees with similar DBH and h, the tree with higher ρ has more carbon storage compared to the trees with lower ρ (Chave et al., 2005).

Table 6.2 Tree species and their carbon storage and CO_2 equivalent of park and roadside trees

No	Species	Local name	Family	No of Trees	Wood density (g/cm^{-3})	DBH (cm)	Height (m)	Mean carbon (kg $tree^{-1}$)	Standard Deviation	Average CO_2 eq
Park trees										
1	*Khaya senegalensis*	Khaya	Meliaceae	16	0.64	25–121	12–24	2,289.29	2,647.33	8,394.08
2	*Alstonia angustifolia*	Pulai	Apocynaceae	49	0.61	25–95	8–19	1,525.18	946.35	5,592.32
3	*Samanea saman*	Hujan-hujan	Fabaceae	25	0.5	37–146	8–18	1,376.08	2,160.54	5,045.64
4	*Mimusop elengi*	Bunga Tanjung	Sapotaceae	9	0.85	29–70	7–11	1,082.36	905.33	3,968.64
5	*Syzygium grande*	Jambu Laut	Myrtaceae	46	0.71	29–86	9–26	1,008.00	828.17	3,696.01
6	*Hevea brasiliensis*	Getah	Euphorbiaceae	40	0.47	25–80	10–22	524.76	496.76	1,924.13
7	*Peltophorum pterocarpum*	Jemerlang	Fabaceae	19	0.57	25–48	9–14	436.43	215.88	1,600.24
8	*Milletia pinnata*	Mempari	Fabaceae	9	0.59	27–40	1–11	336.91	106.96	1,235.35
9	*Hopea odorata*	Merawan Siput Jantan	Dipterocarpaceae	16	0.63	25–32	9–17	250.54	78.96	918.65
10	*Syzygium polyanthum*	Salam	Myrtaceae	7	0.56	25–37	9–15	246.98	107.40	905.61
11	*Dyera costulata*	Jelutong	Apocynaceae	5	0.34	24–45	15–18	214.33	130.05	785.87
12	*Shorea leprosula*	Meranti Tembaga	Dipterocarpaceae	6	0.44	25–31	14–18	157.53	58.55	577.62

(continued)

Table 6.2 (continued)

No	Species	Local name	Family	No of Trees	Wood density (g/cm^{-3})	DBH (cm)	Height (m)	Mean carbon (kg tree^{-1})	Standard Deviation	Average CO_2 eq
Roadside trees										
1	*Melaleuca cajuputi*	Gelam	Myrtaceae	7	0.7	67–111	9–15	3,644.23	2,009.66	13,362.19
2	*Pterocarpus indicus*	Angsana	Fabaceae	108	0.64	25–124	8–28	2,669.04	1,995.57	9,786.50
3	*Syzygium grande*	Jambu laut	Myrtaceae	7	0.71	14–83	8–15	1,132.64	1,498.48	4,153.03
4	*Samanea saman*	Hujan-hujan	Fabaceae	656	0.5	1–141	5–20	731.44	994.19	731.44
5	*Swietenia macrophylla*	Mahogany	Meliaceae	12	0.52	32–72	6–15	709.58	493.84	2,601.80
6	*Khaya grandifolia*	African Mahogany	Meliaceae	293	0.54	8–127	3–16	448.60	685.55	1,644.86
7	*Tabebuia rosea*	Tecoma	Bignoniaceae	578	0.53	6–64	2–18	339.86	298.60	1,246.14
8	*Cassia fistula*	Rajah Kayu	Fabaceae	122	0.8	8–47	2–9	167.42	172.50	613.86
9	*Hopea odorata*	Merawan Siput Jantan	Dipterocarpaceae	10	0.64	14–32	6–14	125.32	89.51	459.49
10	*Alstonia angustiloba*	Pulai	Apocynaceae	122	0.35	3–45	4–14	90.16	89.82	90.16
11	*Cinnamomum verum*	Kayu Manis	Lauraceae	13	0.5	13–38	4–14	81.05	102.81	297.17
12	*Alstonia angustifolia*	Pulai Penipu Raya	Apocynaceae	11	0.61	6–14	2–4	23.65	8.01	86.71

Table 6.3 Tree size, number of trees and species in each size interval and the mean carbon storage for every 5 cm DBH intervals of (a) park and (b) roadside trees. The difference in carbon storage within each size interval is significant ($p < 0.05$) at 95% confidence level

DBH interval (cm)	No. of trees	No. of tree species	Average carbon (kg tree^{-1})	Standard deviation
Park trees				
21–25	20	9	105.27	27.58
26–30	32	12	245.05	46.39
31–35	45	15	329.84	79.43
36–40	48	14	512.27	87.94
41–45	35	13	598.82	147.48
46–50	18	11	828.11	171.60
51–55	23	8	1,086.77	221.55
56–60	17	6	1,422.42	240.27
61–65	9	4	1,710.52	361.79
66–70	14	5	2,004.40	453.91
71–75	7	4	2,433.67	558.79
76–80	8	4	2,671.14	311.65
81–85	1	1	3,343.95	#
86–90	3	3	3,899.73	523.97
91–95	1	1	4,751.34	#
101–105	2	2	4,634.64	1,573.45
106–110	1	1	6,670.17	#
116–120	1	1	8,339.20	#
121–125	1	1	8,766.65	#
>130	1	1	11,158.47	#
Street Trees				
<5	164	2	0.17	0.14
6–10	13	5	9.99	4.81
11–15	78	8	28.61	11.19
16–20	184	7	52.17	19.82
21–25	238	10	93.15	25.20
26–30	168	7	214.18	39.02
31–35	238	9	310.94	42.69
36–40	176	8	435.23	57.00
41–45	171	7	582.30	62.68
46–50	118	6	739.08	58.54
51–55	131	13	928.15	89.31
56–60	96	6	1,240.25	142.21
61–65	48	5	1,576.78	172.88

(continued)

Table 6.3 (continued)

DBH interval (cm)	No. of trees	No. of tree species	Average carbon (kg tree^{-1})	Standard deviation
66–70	34	5	1,972.11	303.45
71–75	18	5	2,274.35	302.13
76–80	19	4	2,735.18	360.98
81–85	9	5	3,172.90	750.89
86–90	10	3	3,661.17	386.15
91–95	11	2	4,199.32	642.68
96 –100	7	3	5,171.22	370.49
101–105	9	3	5,568.47	681.00
106–110	5	2	6,521.78	594.12
111–115	4	3	6,861.32	908.56
116–120	3	3	7,306.56	764.27
121–125	2	2	8,464.84	1,451.39
126–130	1	1	8,540.23	#
>130	15	3	10,186.54	1,812.17

Carbon Sequestration

The results of carbon sequestration for several park and roadside tree species are shown in Fig. 6.4. Trees were compared for their potential capacity to sequester carbon in year 1 (y) to year 1 + 1 (y + 1). We grouped the trees based on their age. For park tree species (Fig. 6.4a), *Pterocarpus indicus* sequesters more carbon (249.77 kg tree^{-1} year^{-1}) between age 50 and 51 compared to other tree species in the same age group. For the age group 15–16 years old, *Ixonanthes reticulata* Jack sequesters the highest amount of carbon (67.48 kg tree^{-1} year^{-1}) compared to all 15 tree species found in this group. For the age group 20–21 years, *Alstonia angustifolia* sequesters 53.85 kg tree^{-1} year^{-1} compared to 19 other species with the same age. Meanwhile, for roadside tree species (Fig. 6.4b), *Syzygium grande*, *Pterocarpus indicus*, *Samanea saman*, *Delonix regia* and *Peltophorum pterocarpum* sequester the highest carbon for ages 20–21, 30–31, 40–41, 50–51, 55–56, and 60–61 respectively, compared to other tree species found in the same age group.

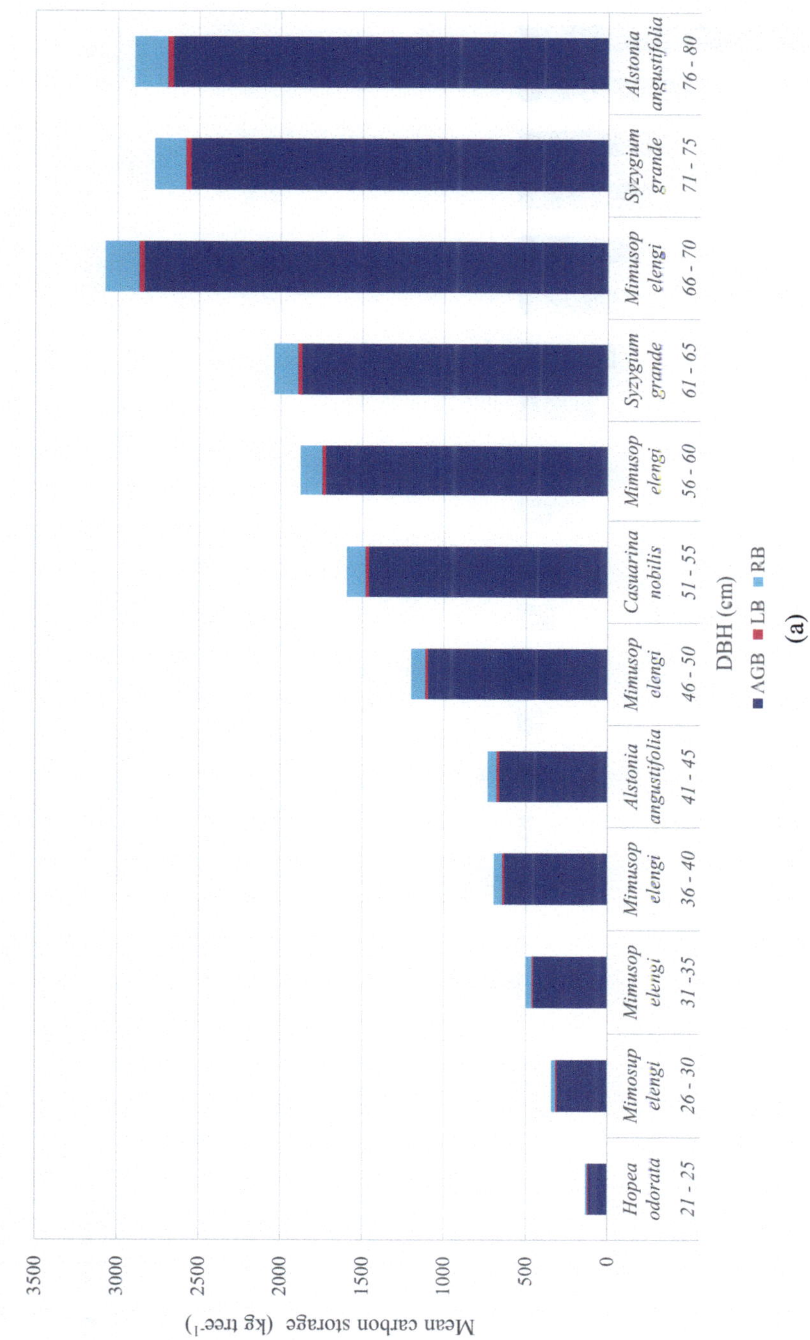

Fig. 6.3 Carbon storage, based on size of trees, of (a) parks and (b) roadside trees in Johor Bahru and Pasir Gudang. (X-axis displays the tree species that stored the largest amount of carbon among the species found within the age group)

Fig. 6.3 (continued)

Discussion

Trees Mitigate CO_2 Emissions

Planting tree species that could sequester more CO_2 in urban centres and industrialised regions, for example, can help to reduce atmospheric pollution in the urban environment. The results of this study show that *Khaya senegalensis*, *Alstonia angustifolia*, *Melaleuca cajuputi* and *Pterocarpus indicus* have a higher capacity than other tree species in this study to store carbon. Meanwhile, *Pterocarpus indicus*, *Khaya senegalensis*, *Delonix regia* and *Samanea saman* have a higher capacity, relative to other tree species in this study to sequester carbon (Fig. 6.4). AGB results obtained in this study were compared with the results of AGB of urban trees in Singapore. Ngo and Lum (2018) destructively sampled 31 street trees in Singapore and obtained their AGB by oven drying the samples. The results obtained using allometric equations that were developed for forest trees (Ngo & Lum, 2018) over-estimated the values obtained using destructive methods with percentage differences ranging between 1 and 11% for trees with >15.8 cm and an over-estimation of 52% for smaller trees. Ngo and Lum (2018) also found that allometric equations developed for tropical forests overestimated street tree AGB by 33–57% for trees with DBH ~ 30 cm and 35–324% for larger trees (60 cm DBH), respectively. Meanwhile, equations for secondary forests slightly underestimated street tree AGB. These differences could be due to variations in tree architecture between trees grown in natural forest and urban environments.

Suitability of Current Park/Roadside Trees in Johor Bahru and Pasir Gudang

Tree species' carbon storage and sequestration capability is not the only factor to consider when selecting trees for urban space. Other ecosystem services and disservices, such as their maintenance, crown size (and hence shading potential), hazard potential (i.e. surface roots and large limbs that could cause obstructions), pollutant removal capability, aesthetic value, and disease resistance also need to be considered. Some tree species such as *Khaya senegalensis* and *Pterocarpus indicus* have high carbon sequestration (Fig. 6.4) and also have large structures, wide crowns, and high vegetative cover (Table 6.4) from big leaves that enable maximum photosynthesis and provide shading and cooling effects. Yet these species may not be suitable for roadsides because *Khaya senegalensis* is not only an exotic species to Malaysia but also a deciduous tree that sheds leaves during Malaysia's hottest months. The large leaves of this species were found to clog drains in the city of Kota Kinabalu in Sabah, Malaysia which was found partly responsible for triggering flash floods in urban areas (Daily Express, 2015). Although *Pterocarpus indicus* produces beautiful flowers twice each year that beautify the roadsides, the tree structure is not sufficient

Fig. 6.4 Carbon sequestration of various tree species in (a) parks and (b) along roadsides in Johor Bahru and Pasir Gudang. Numbers in parentheses next to species names represent the number of trees for each species in each age group. Vertical lines on the bar chart show one standard deviation of the mean

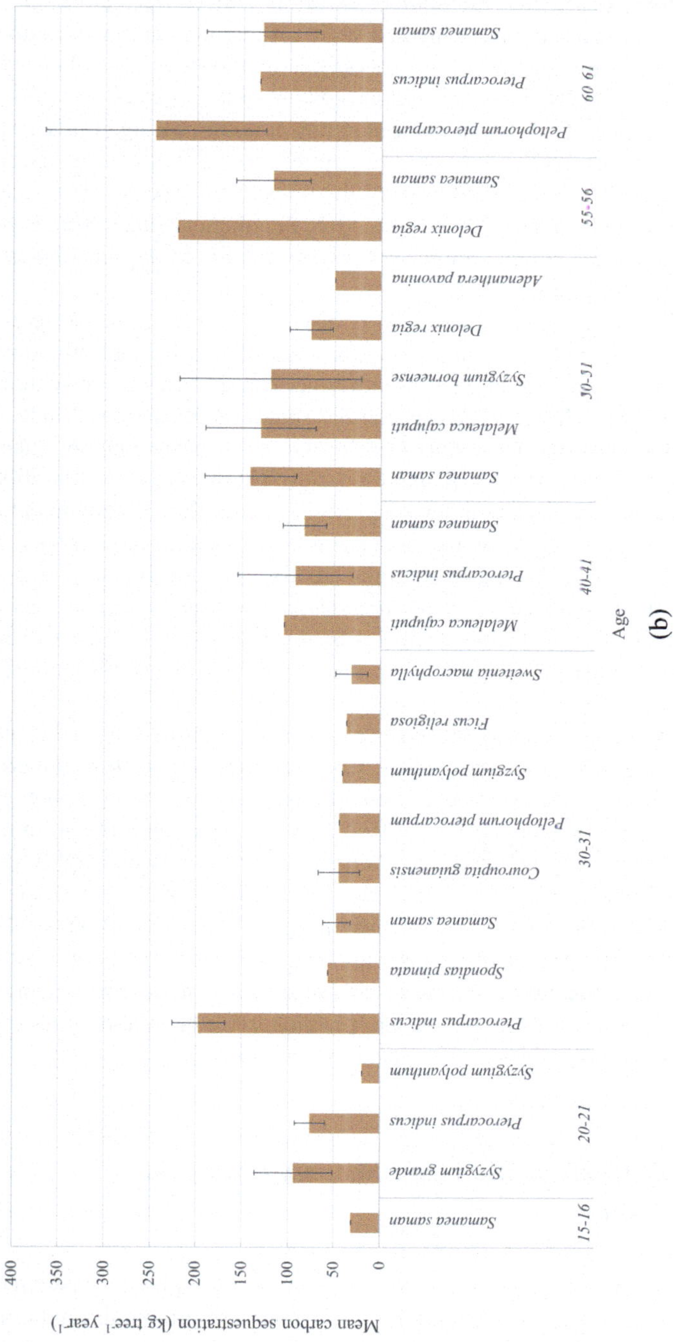

Fig. 6.4 (continued)

to support its heavy crown, making the tree prone to falling and posing a danger to vehicles. Recently, in Kuala Lumpur Pterocarpus indicus trees along the roadsides older than 10 years fell and damaged street restaurants, motorcycles and cars (Sinar Harian, 2014) making the species not only dangerous to property but also life-threatening to citizens. The chronic deciduous nature of this species also prevents it from producing functions of shade and evapotranspiration cooling (Daily Express, 2015). *Pterocarpus indicus* is also very susceptible to the wilt disease caused by *Fusarium oxysporum*, which killed most of the trees of this species that existed in Malaysia and Singapore (Chin, 2003; FAO, 2015; Sreetheran et al., 2006). For these reasons, this species is no longer preferred for street trees in Malaysia although they can be considered for parks.

Similarly, *Samanea saman* (Table 6.4) is also a good carbon sequesterer (Tan et al., 2009) and it is widely planted along the roadsides in IM and other cities in Malaysia (Sreetheran et al., 2011). This South American species has been naturalised throughout the tropics. However, this species has shallow surface roots (Table 4) that thrust above the soil level as they mature, causing damage to pavements and sidewalks as well as building foundations if planted too close (Staples & Elevitch, 2006). In PG, the large limbs of Samanea saman trees growing above the roadway in industrial and port areas are found to create overhead hazards that damage large trucks. Along other major roadsides in IM, the large limbs are also found to block road signage. Although it seems impractical to plant *Samanea saman* along roadsides in certain areas for these reasons, these trees are still a good selection for parks. *Samanea saman* are large trees with umbrella-shaped crowns that can provide plenty of shading (Table 6.4).

Both *Pterocarpus indicus* and *Samanea saman* have beautiful flowers that provide aesthetic value and seeds that attract birds. However, these trees attract the types of birds that make noise in the evenings and create a nuisance as their droppings fall on the cars parked below the trees (Chin, 2003). This is a common problem in resting areas along the roadsides in Malaysia. *Pterocarpus indicus* is not highly disease resistant and is found to rot after some time, causing the branches to break easily in high winds. Consequently, this may cause property damage and physical harm (Zakaria, 2012). This also involves high maintenance costs for municipalities to clear fallen limbs, compensate for damaged property or injury, and replant trees. Therefore, local authorities should consider planting trees that attract types of birds and other fauna that do not cause nuisance or harm to people.

Limitations of Current Work and Recommendations for Future Studies

This study was limited to analysing carbon storage of a sample of urban street and park trees in two cities in Iskandar Malaysia region. The authors did not consider the carbon release to the atmosphere as result of autotrophic respiration by plant

Table 6.4 Tree species found in urban parks and roadsides in Iskandar, Johor, Malaysia and their characteristics

Tree/ Plant	Origin	Physical characteristics i.e. Wood density (g/cm^3), height (m), Size	Type/Crown/Growth rate/Longevity	Aesthetic values	Role in soil improvement	Root system	Other values	Images
P. indicus	Native	0.52, 30 m, Large	Deciduous, Dense, dome-shaped, wide-spreading, Fast, >200 years	Yellow flowers	Nitrogen fixing	Thick roots grow out of its trunk	Best fine-furniture wood in Malaysia	
K. senegalensis	Africa	0.52, Up to 15–30 m, Tall	Semi- deciduous, Wide dense crown, Fast, 70–80 years	Used as street trees	Nitrogen fixing	Deep rooted	Useful and expensive woods	

(continued)

Table 6.4 (continued)

Tree/ Plant	Origin	Physical characteristics i.e. Wood density (g/cm^3), height (m), Size	Type/Crown/Growth rate/Longevity	Aesthetic values	Role in soil improvement	Root system	Other values	Images
P. pterocarpum	Native	0.53, 15–35 m, Tall	Deciduous, Umbrella like crown, Fast, ~20 years	Yellow blossoms	Nitrogen fixing bacteria in its roots	Deep root system	Shade trees	
S. saman	South America	0.45, Up to 60 m tall, Large	Evergreen, Big umbrella-shaped crown, Fast, 80–100 years	Pink-white flowers	Nitrogen fixing	Massive surface root system	Excellent for shading	

(continued)

Table 6.4 (continued)

Tree/ Plant	Origin	Physical characteristics i.e. Wood density (g/cm^3), height (m), Size	Type/Crown/Growth rate/Longevity	Aesthetic values	Role in soil improvement	Root system	Other values	Images
A. indica	Native	0.66, 15 m, Medium to Large	Evergreen, Fairly dense crown is roundish, Fast, >200 years	Large structured tree which suitable for wide city roads		Long, thick, penetrating and can go up to 20 m	High medicinal values	

roots and associated microorganisms, and heterotrophic respiration via microbial decomposition of soil organic matter. It should be noted that different conclusions might have been reached if the complete biogenic component (i.e. trees and soil) were considered. In tropical locations, the warm and humid conditions of the soil may offset the carbon removed by photosynthesis. Similarly, the carbon emissions associated with the maintenance of urban vegetation were not considered in this study. Future studies should make a holistic assessment of the ecological service of urban trees to reduce CO_2 emissions at the local scale for emissions associated with pruning, watering, trimming and debris management.

As described in the method section, allometric models are not destructive and easy to use because they rely on easily collectable tree biometric data such as tree height, stem diameter, species type, age, crown size and health condition. However, allometric techniques incur high tree inventory costs, tedious procedures and data collection consumes tremendous amounts of time especially when the study area is large. Consequently, not many allometric models have been developed specifically for the urban forest. Since developing new allometric models requires a destructive approach to estimating biomass (i.e. cutting down urban trees and drying their components in an oven to determine their dry biomass and carbon content) most previous research, including this study, applied models that were developed for forest trees. We addressed the difference in carbon storage by urban trees compared to the carbon storage estimated using an allometric equation developed for forests by reducing the value by 20% (Nowak, 1994). However, some studies in the temperate region show a difference of 60–427% between the carbon storage of urban and forest trees (McHale et al., 2009; Yoon et al., 2013). When the allometric models developed for tropical forests (Chave et al., 2005) were applied to estimate the AGB of urban street trees in Singapore, Ngo and Lum (2018) found a difference ranging between 33–324%. However, the difference became smaller (25–145%) when they applied a 0.8 correction factor as suggested by Nowak (1994). The 20% reduction applied to the carbon storage values calculated for the urban forest in this study is still likely to be a gross overestimate. It should be noted that the study of Ngo and Lum (2018) only considered 31 trees from 11 species with tree DBH from 3.2 to 71.9 cm. The errors may increase/decrease depending on the number of trees and their characteristics, species types, and the spatial scale (McHale et al., 2009).

In this study, the carbon storage and sequestration analysis were performed using tree inventory data from a limited number of trees (2,245) and species (41) obtained from two municipalities in IM. Urban trees are generally not systematically monitored, and their data are not regularly updated by the city councils in Malaysia (Sreetheran et al., 2006) mainly because field data collection is expensive, labour intensive and time-consuming. Limited availability of the data is usually restricted to selected sites and may not represent the characteristics of urban trees across a larger region. In Peninsular Malaysia, approximately 150–200 different tree species are found (Sreetheran et al., 2011). A greater number of trees that includes more species should be considered by future studies to analyse the carbon storage of many more species. Future studies in the region may consider employing remotely sensed LiDAR data for the fast and frequent acquisition of tree inventory data such as tree height, DBH, crown size etc. over a large spatial extent. This technology also can be used to obtain tree volume that can be converted into biomass and carbon storage (Yoon et al., 2013, Schreyer et al., 2014, Raciti et al., 2014, Saarinen et al., 2014, Zhao & Sander, 2015, McGovern & Pasher, 2016, Alonzo et al., 2016, Lee et al., 2016,Abd Rahman et al., 2017, Tigges et al., 2017; Wilkes et al., 2018; Huang et al., 2019).

Conclusion

Identifying tree species that have a high potential to sequester carbon is important to enable landscape authorities, urban planners and municipal leaders in Malaysia to conserve these species from further extermination and also choose appropriate tree species that can help mitigate climate change impacts on urban landscapes. From the carbon storage and sequestration analysis of roadside trees and trees in parks in JB and PG, this study proposes that *Khaya senegalensis*, *Alstonia angustifolia*, *Melaleuca cajuputi*, *Pterocarpus indicus*, *Delonix regia* and *Samanea saman* be considered for planting in urban parks and roadsides in IM. Other forest species such as *Intsia bijuga*, *Koompassia malaccensis*, *Dipterocarpus fagineus*, *Pentaspadon motleyi* and *Hopea pedicellata* are recommended for urban parks. Tree species like *Azadirachta indica* and *Tamarindus indica* can be considered for roadsides and parks in IM. The long lifespan, evergreen, indigenous nature, disease resistance and high wood density of these species make them the most appropriate trees for Malaysia.

This conclusion is based on the analysis of a limited number of trees (2,245) and species (41) obtained from two municipalities in IM. A greater number of trees that includes more species may change the results, thus future studies should consider analysing the carbon storage of many more species. The current study considered only one component (carbon uptake/storage) of the carbon cycle of urban trees. Future studies should consider carbon dioxide release from trees i.e., both autotrophic and heterotrophic (soil) respiration to understand the net carbon uptake by urban trees. Advanced flux measurement techniques such eddy covariance-based flux towers can be considered for the purpose. Remotely sensed LiDAR data also should be considered for deriving tree attributes that can be used to calculate tree volume and carbon storage. The selection of tree species is also dependent on other ecosystem services and disservices, such as their potential to remove pollutants, provide shading, and the level of maintenance required. These aspects warrant future investigation to produce scientific data to support the selection of the most suitable species for planting in parks and along roadsides in Iskandar Malaysia.

Acknowledgements The authors acknowledge the Ministry of Education Malaysia (through research grant Q.J130000.2452.08G51 and R.J130000.7301.4B145) and Universiti Teknologi Malaysia for providing funding to conduct the study. We thank the Pasir Gudang and Johor Bahru Municipality Councils for providing tree inventory data and their personnel's assistance to this research.

References

Abd Rahman, M. Z., Abu Bakar, M. A., Razak, K. A., Rasib, A. W., Kanniah, K. D., Wan Kadir, W. H., Omar, H., Faidi, A., Kassim, A. R. & Abd Latif, Z., (2017). Non-destructive, laser-based individual tree aboveground biomass estimation in a tropical rainforest. *Forests*, *8*(3), 86.

Alonzo, M., Mcfadden, P. J., Nowak, D. J., & Roberts, D. A. (2016). Mapping urban forest structure and function using hyperspectral imagery and lidar data. *Urban Forestry & Urban Greening, 17*, 135–147.

Baker, T. R., Philips, O. L., Malhi, Y., Almeida, S., Arroyo, L., Di Fiore, A., Erwin, T., Killeen, T. J., Laurance, S. G., Laurance, W. F., Lewis, S. L., Lloyd, J., Monteagudo, A., Neill, D. A., Patino, S., Pitman, N. C. A., Silva, J. N. M., & Martinez, R. V. (2004). Variation in wood density determines spatial patterns in Amazonian forest biomass. *Global Change Biology, 10*, 545–562.

Brack, C. L. (2002). Pollution mitigation and carbon sequestration by an urban forest. *Environment Pollution, 16*, S195–S200.

Breugel, M. V., Ransijn, J., Craven, D., Bongers, F., & Hall, J. S. (2011). Forest Ecology and Management Estimating carbon stock in secondary forests : Decisions and uncertainties associated with allometric biomass models. *Forest Ecology and Management, 262*(8), 1648–1657.

Brown S (1991) Estimating Biomass and Biomass Change of Tropical Forests: a Primer. For the Food and Agriculture Organization of the United Nations. Rome. FAO Forestry 134. ISBN 92–5–103955–0.

Cairns, M. A., Brown, S., Helmer, E., & Baumgardner, G. A. (1997). Root biomass allocation in the world's upland forests. *Oecologia, 111*, 1–11.

Carreiro, M., & Tripler, C. (2005). Forest remnants along urban-rural gradients: Examining their potential for global change research. *Ecosystems, 8*, 568–582.

Chaparro, L., & Terradas, J. (2009). *Ecological services of urban forest in Barcelona*. Aplicacions Forestals, Universitat Autonoma de Barcelona, Bellaterra, Spain.

Chave, J., Andalo, C., Brown, S., Cairns, M. A., Chambers, J. Q., Eamus, D., Folster, H., Fromard, F., Higuchi, N., Kira, T., Lescure, J. P., Nelson, B. W., Ogawa, H., Puig, H., Riera, B., & Yamakura, T. (2005). Tree allometry and improved estimation of carbon stocks and balance in tropical forests. *Oecologia, 145*, 87–99.

Chave, J., Coomes, D., Jansen, S., Lewis, S. L., Swenson, N. G., & Zanne, A. E. (2009). Towards a worldwide wood economics spectrum. *Ecology Letters, 12*, 351–366.

Chave, J., Olivier, J., Bongers, F., Chatelet, P., Forget, P. M., Van Der Meer, P., & Charles-Dominique, P. (2008). Above-ground biomass and productivity in a rain forest of eastern South America. *Journal of Tropical Ecology, 24*(4), 355–366.

Chin, W. Y. (2003). *Tropical Trees and Shrubs: A Selection for Urban Plantings*. Sun Tree Publishing Limited.

Christen, A., Coops, N. C., Crawford, B. R., Kellett, R., Liss, K. N., Olchovski, I., & Voogt, J. A. (2011). Validation of modeled carbon-dioxide emissions from an urban neighborhood with direct eddy-covariance measurements. *Atmospheric Environment, 45*(33), 6057–6069.

Churkina, G., Brown, D., & Keoleian, G. (2010). Carbon stored in human settlements: The conterminous United States. *Global Change Biology, 16*, 135–143.

Churkina, G., Grote, R., Butler, T. M., & Lawrence, M. (2015). Natural selection? Picking the right trees for urban greening. *Environmental Science & Policy, 47*, 12–17.

Coutts, A. M., Beringer, J., & Tapper, N. J. (2007). Characteristics influencing the variability of urban CO2 fluxes in Melbourne, Australia. *Atmospheric Environment, 41*, 51–62.

Daily Express (2015) Taming a hot capital city by Kan Yaw Chong. Daily Express e- newspaper Monday (accessed on 11 May 11 2015).

Davies, Z. G., Edmonson, J. L., Heinemeyer, A., Leake, J. R., & Gaston, K. J. (2011). Mapping an urban ecosystem service: Quantifying above- ground carbon storage at a citywide scale. *Journal of Applied Ecology, 48*, 1125–1134.

Department Of Environment (2010) Malaysia Environmental Quality Report. Ministry of Natural Resources & Environment.

Dorney, J. R., Guntenspergen, G. R., Keough, J. R., & Steams, F. (1984). Composition and structure of an urban woody plant community. *Urban Ecology, 8*, 69–90.

Escobedo, F., Varela, S., Zhao, M., Wagner, J., & Zipperer, W. (2010). The efficacy of subtropical urban forests in offsetting carbon emissions from cities. *Environmental Science and Policy, 13*, 362–372.

FAO (2015) Strengthening national capacity for control of Pterocarpus indicus wilt disease and forest protection, Republic of Seychelles. Available online: http://www.fao.org/forestry/49410/en/syc/ (retrieved 7 June 2015).

FAO (2016) Guidelines on urban and peri-urban forestry; Salbitano F.; Borelli S.; Conigliaro M.; Chen Y.; FAO Forestry, Rome, Paper No. 178.

Fares S, Paoletti E, Calfapietra C, Mikkelsen TN, Samson R, Le TD (2017) Carbon Sequestration by Urban Trees. In The Urban Forest: 31–39.

Federal Department Of Town And Country Planning (2010) National Physical Plan of Malaysia. Ministry of Housing and Local Government. Available online: http://theredddesk.org/countries/plans/national-physical-plan-malaysia (Accessed on April 2018).

Global Carbon Project. (2020). dataset: Friedlingstein et al, Global Carbon Budget 2020. *Earth Syst. Sci. Data, 12*(3269–3340), 2020. https://doi.org/10.5194/essd-12-3269-2020

Guo, Z. D., Hu, H. F., Pan, Y. D., Birdsey, R. A., & Fang, J. Y. (2014). Increasing biomass carbon stocks in trees outside forests in China over the last three decades. *Biogeosciences, 11*, 4115–4122.

Huang, W., Dolan, K., Swatantran, A., Johnson, K., Tang, H., O'Neil-Dunne, J., Dubayah, R., & Hurtt, G. (2019). High-resolution mapping of aboveground biomass for forest carbon monitoring system in the Tri-State region of Maryland, Pennsylvania and Delaware, USA. *Environmental Research Letters, 14*, 095002.

Ho CS, Chau LW, Teh BT, Matsuoka Y, Gomi K, Rohayu A, Nadzirah J, Nur Syazwani S, Muhammad Akmal Hakim H, Lv Y (2015a) Low Carbon Society Action Plan for Johor Bahru 2025: Vibrant World Class Cosmopolis of the South. Johor Bahru: UTM-Low Carbon Asia Research Centre.

Ho CS, Chau LW, Teh BT, Matsuoka Y, Gomi K, Rohayu A, Nadzirah J, Nur Syazwani S, Muhammad Akmal Hakim H, Lv Y (2015b) Low Carbon Society Action Plan for Pasir Gudang 2025: Green & Clean Industrial City. Johor Bahru: UTM-Low Carbon Asia Research Centre.

Houghton JT, Meira Filho LG, Callander BA, Harris N, Kattenberg A, Maskell K (1996) Climate Change 1995: The Science of Climate Change 1996. Contribution of Working Group 1 to the Second Assessment Report of the Intergovernmental Panel on Climate Change. xii 572 pp. Cambridge, New York, Port Chester, Melbourne, Sydney: Cambridge University Press for the Intergovernmental Panel on Climate Change.

Hutyra, L. R., Yoon, B., Hepinstall, C. J., & Alberti, M. (2011). Landscape and Urban Planning Carbon consequences of land cover change and expansion of urban lands : A case study in the Seattle metropolitan region. *Landscape and Urban Planning, 103*(1), 83–93.

IUCN (2015) The IUCN Red List of Threatened Species. Available online: www.iucnredlist.org (retrieved 8th June 2015).

Jamil, A. B. (2014). *Pokok di Taman Awam Malaysia*. Penerbit UTM Press.

Jarvi L, Rannik Ü, Kokkonen TV, Kurppa M, Karppinen A, Rostislav D, Wood CR (2018) Uncertainty of eddy covariance flux measurements over an urban area based on two towers. Atmospheric Measurement Techniques: 1–27.

Johl, H. S., & Bahga, S. S. (2014). *Trees in Urban Habitat*. White Falcon Publishing.

Kanniah, K. D., & Ho, C. (2017). Urban Forest Cover Change and Sustainability of Malaysian Cities. *Chemical Engineering Transactions, 56*, 673–678.

Kanniah, K. D., Muhamad, N., & Kang, C. S. (2014). Remote sensing assessment of carbon storage by urban forest, International Symposium on Digital Earth, 26–29 August 2013, Kuching, Sarawak. *Malaysia. Earth and Environmental Science, 18*, 012151.

Kanniah KD, Sheikhi A, Cracknell AP, Goh HC, Tan KP, Ho CS, Rasli FN (2015) Satellite Images for Monitoring Mangrove Cover Changes in a Fast Growing Economic Region in Southern, Iskandar Malaysia. Remote sensing. 14360–14385.

Kanniah, K. D., Tan, K. P., Cracknell, A. P., Huete, A. R., Idris, N. H., Meng, A., & Ahmad, A. (2018). Assessment of biophysical properties of Royal Belum tropical forest, Malaysia. *Singapore Journal of Tropical Geography, 39*, 90–106.
Kumar, V. S., & Navaratnam, V. (2013). Neem (Azadirachta indica): Prehistory to contemporary medicinal uses to humankind. *Asian Pacific Journal Tropical Biomedicine, 3*, 505–514.
Lashof, D. A., & Ahuja, D. R. (1990). Relative contributions of greenhouse gas emissions to global warming. *Nature, 344*, 529–531.
Le Quéré, C., Moriarty, R., Andrew, R. M., Canadell, J. G., Sitch, S., Korsbakken, J. I., Friedlingstein, P., Peters, G. P., Andres, R. J., Boden, T. A., Houghton, R. A., House, J. I., Keeling, R. F., Tans, P., Arneth, A., Bakker, D. C. E., Barbero, L., Bopp, L., Chang, J., Chevallier, F., ... Zeng, N. (2015). Global Carbon Budget 2015. *Earth System Science Data, 7*, 349–396, https://doi.org/10.5194/essd-7-349-2015
Lee, J. H., Ko, Y., & Mcpherson, E. G. (2016). The feasibility of remotely sensed data to estimate urban tree dimensions and biomass. *Urban Forestry & Urban Greening, 16*, 208–220.
Liu, C. F., & Li, X. M. (2012). Carbon storage and sequestration by urban forests in Shenyang, China. *Urban Forestry and Urban Greening, 11*, 121–128.
Mcgovern, M., & Pasher, J. (2016). Canadian urban tree canopy cover and carbon sequestration status and change 1990–2012. *Urban Forestry & Urban Greening, 20*, 227–232.
Mchale, M. R., Burke, I. C., Lefsky, M. A., Peper, P. J., & Mcpherson, E. G. (2009). Urban forest biomass estimates: Is it important to use allometric relationships developed specifically for urban trees? *Urban Ecosystem, 12*, 95–113.
Mcpherson EG, Van Doorn NS, Peper PJ (2016) Urban Tree Database and Allometric Equations. Gen. Tech. Rep. PSW-GTR-253. Albany, CA: U.S. Department of Agriculture, Forest Service, Pacific Southwest Research Station: pp. 86.
Mcpherson, E. G. (1998). Atmospheric carbon dioxide reduction by Sacramento's urban forest. *Journal of Arboriculture, 24*, 215–223.
Muller-Landau, H. C., Condit, R. S., Chave, J., Thomas, S. C., Bohlman, S. A., Bunyavejchewin, S., & Ashton, P. (2006). Testing metabolic ecology theory for allometric scaling of tree size, growth and mortality in tropical forests. *Ecology Letters, 9*(5), 575–588.
Mustapa, S. I., & Bekhet, H. A. (2016). Analysis of CO2 emissions reduction in the Malaysian transportation sector: An optimisation approach. *Energy Policy, 89*, 171–183.
National Research Council. (1992). *Neem: A Tree for Solving Global Problems*. The National Academies Press.
Km, N., & Lum, S. (2018). Aboveground biomass estimation of tropical street trees. *Journal of Urban Ecology, 8*, 1–6.
Ngo, K.M. & Lum, S., (2018). Aboveground biomass estimation of tropical street trees. *Journal of Urban Ecology, 4*(1).
Norainiratna, B., Manohar, M., & Mohd, R. (2013). Health of Trees in Titiwangsa Recreational Park, Kuala Lumpur Malaysia. *Journal of Sustainability Science and Management, 8*, 191–196.
Nowak, D. J., & Crane, D. E. (2002). Carbon storage and sequestration by urban trees in the USA. *Environmental Pollution, 116*, 381–389.
Nowak, D. J., Greenfield, E. J., Hoehn, R. E., & Lapoint, E. (2013). Carbon storage and sequestration by trees in urban and community areas of the United States. *Environmental Pollution, 178*, 229–236.
Nowak DJ (1994) Atmospheric carbon dioxide reduction by Chicago's urban forest. In: McPherson, E.G., Nowak, D.J., Rowntree, R.A. (Eds.), Chicago's Urban Forest Ecosystem: Results of the Chicago Urban Forest Climate Project. General Technical Report. Northeastern Forest Experiment Station 186: pp. 83–94.
Olivier, J., Van Aardenne, J., Dentener, F., Pagliari, V., Ganzeveld, L., & Peters, J. (2005). Recent trends in global greenhouse gas emissions: Regional trends 1970–2000 and spatial distribution of key sources in 2000. *Environmental Sciences, 2*, 81–99.

Park, J. H., Baek, S. G., Kwon, M. Y., Je, S. M., & Woo, S. Y. (2018). Volumetric equation development and carbon storage estimation of urban forest in Daejeon. *Korea. Forest Science and Technology, 14*(2), 97–104.

PEMANDU (2013) ETP Annual Report 2013, Greater Kuala Lumpur/Klang Valley. Available online:http://etp.pemandu.gov.my/annualreport2013/upload/ENG/03_NKEA01_ENG_GKL.pdf.

Le Quéré, C, R Moriarty, RM Andrew, JG Canadell et al.(2015) Global Carbon Budget 2015 Earth System Science Data. 7: 349–396 doi:https://doi.org/10.5194/essd-7-349-2015

Raciti, S. M., Hutyra, L. R., & Newell, J. D. (2014). Mapping carbon storage in urban trees with multi-source remote sensing data: Relationships between biomass, land use, and demographics in Boston neighborhoods. *Science of the Total Environment, 500–501*, 72–83.

Rahman, M., Kabir, E., & Asmj, U. (2015). High carbon stocks in roadside plantations under participatory management in Bangladesh. *Global Ecology and Conservation, 3*, 412–423.

Rowntree, R. A., & Nowak, D. J. (1991). Quantifying the role of urban forests in removing atmospheric carbon dioxide. *Journal of Arboriculture, 17*, 269–275.

Saarinen, N., Vastaranta, M., Kankare, V., Tanhuanpaa, T., Holopainen, M., Hyyppa, J., & Hyyppa, H. (2014). Urban-Tree-Attribute Update Using Multisource Single-Tree Inventory. *Forests, 5*, 1032–1052.

Safaai, M., Sharliza, N., Zainura, Z. N., Haslenda, H., Zaini, U., & Juhaizah, T. (2011). Projection of CO2 Emissions in Malaysia. *Environmental Progress & Sustainable Energy, 30*, 658–665.

Scripps Institution of Oceanography, 2020, Schreyer J, Tigges J, Lakes T, Churkina G,. (2014S). Using airborne lidar and QuickBird data for modelling urban tree carbon storage and its distribution—a case study of Berlin. *Remote Sensing, 6*, 10636–10655.

Scripps Institution of Oceanography, 2020, https://scripps.ucsd.edu/programs/keelingcurve/ (Accessed 16 May 2020).

Seto KC, Dhakal S, Bigio A, Blanco H, Delgado GC, Dewar D, Huang L, Inaba A, Kansal A, Lwasa S, Mcmahon JE, Muller DB, Murakami J, Nagendra H, Ramaswami A (2014) Human Settlements, Infrastructure and Spatial Planning. Climate Change 2014: Mitigation of Climate Change. Contribution of Working Group III to the Fifth Assessment Report of the Intergovernmental Panel on Climate Change. Cambridge University Press, Cambridge, United Kingdom and New York, NY, USA.

Sinar Harian (2014) Stall, Motorcycles being crushed by tree. Available online: www.sinarharian.com.my/mobile/politik/gerai-motosikal-dihempap-pokok-1.331229 (accessed on 30 June 2017).

Sreetheran, M., Adnan, M., & Khairil Azuar, A. K. (2011). Street Tree Inventory and Tree Risk Assessment of Selected Major Roads in Kuala Lumpur, Malaysia. *Arboriculture & Urban Forestry, 37*, 226–235.

Sreetheran, M., Philip, E., Adnan, M., & Siti Zakiah, M. (2006). A historical perspective of urban tree planting in Malaysia. *Unasylva, 223*(57), 28–33.

Staples GW and Elevitch CR (2006) Samanea saman (rain tree), Species profiles for Pacific Island Agroforestry. Available online: www.traditionaltree.org (retrieved May 25th, 2015).

Strohbach, M. W., Arnold, E., & Haase, D. (2012). The Carbon Footprint of Urban Green space - A Life Cycle Approach. *Landscape and Urban Planning, 104*, 220–229.

Subiakto, A., Rachmat, H. H., & Sakai, C. (2016). Choosing native tree species for establishing man-made forest : A new perspective for sustainable forest management in changing world. *Biodiversitas, 17*(2), 620–625.

Tan PY, Yeo B, Yip WX, Lua HK (2009) Carbon Storage and Sequestration by Urban Trees in Singapore. Publication of Centre for Urban Greenery and Ecology (CUGE), Singapore.

Tang Y, Chen A, Zhao S (2016) Carbon Storage and Sequestration of Urban Street Trees in Beijing, China. Frontiers in Ecology and Evolution: 1–8.

Tigges, J., Churkina, G., & Lakes, T. (2017). Modeling above-ground carbon storage: A remote sensing approach to derive individual tree species information in urban settings. *Urban Ecosystems, 20*, 97–111.

UTM-Low Carbon Asia Research Center (2013) Low Carbon Society Blueprint for Iskandar Malaysia 2025, Summary for Policymakers, second edition. UTM Low Carbon Asia Research Center, Johor.

UTM-Low Carbon Asia Research Centre (2017) Kuala Lumpur Low Carbon Society Blueprint 2030. UTM Low Carbon Asia Research Center, Johor.

Velasco, E., Roth, M., Tan, S. H., Quak, M., Nabarro, S. D. A., & Norford, L. (2013). The role of vegetation in the CO_2 flux from a tropical urban Neighbourhood. *Atmospheric Chemistry and Physics, 13*, 10185–10202.

Wang, Z., Cui, X., Yin, S., Shen, G. R., Han, Y., & Liu, C. J. (2013). Characteristics of carbon storage in Shanghai's urban forest. *Chinese Science Bulletin, 58*, 1130–1138.

Weissert, L. F., Salmond, J. A., & Schwendenmann, L. (2014). A review of the current progress in quantifying the potential of urban forests to mitigate urban CO_2 emissions. *Urban Climate, 8*, 100–125.

Wilkes, P., Disney, M., Boni Vicari, M., Calders, K., & Burt, A. (2018). Estimating urban above ground biomass with multi-scale LiDAR. *Carbon Balance Management, 13*, 10. https://doi.org/10.1186/s13021-018-0098-0

Woodbury, P. B., Smith, J. E., & Health, L. S. (2007). Carbon sequestration in the US forest sector from 1990 to 2010. *Forest Ecology and Management, 241*, 14–27.

World Bank (2017) Available online: http://databank.worldbank.org/data/reports.aspx?source=2&series=EN.ATM.CO2E.PC&country=# (accessed July 2017).

Yoon, T. K., Park, C. W., Lee, S. J., Ko, S., Kim, K. N., & Son, Y. (2013). Allometric equations for estimating the aboveground volume of five common urban street tree species in Daegu, Korea. *Urban Forest & Urban Greening, 12*, 344–349.

Zakaria AZ (2012) Fallen Tree Problems in The Field of Landscape Architecture in Malaysia, 1st International Conference on Innovation and Technology for Sustainable Built Environment 2012 (ICITSBE 2012), Perak, Malaysia.

Zanne AE, Gonzalez LG, Coomes DA, Ilic J, Jansen S, Lewis SL, Miller RB, Swenson NG, Wiemann MC, Chave J (2009) Global wood density database. Available online: http://hdl.handle.net/10255/dryad.234 (retrieved 27 July 2015).

Zhang, X., Zhou, P., Zhang, W., Zhang, W., & Wang, Y. (2013). Selection of Landscape Tree Species of Tolerant to Sulphur Dioxide Pollution in Subtropical China. *Open Journal of Forestry, 3*, 104–108.

Zhao, C., & Sander, H. A. (2015). Quantifying and mapping the supply of and demand for carbon storage and sequestration service from urban trees. *PLoS ONE, 10*, e0136392.

Zhao, M., Kong, Z., Escobedo, F., & Gao, J. (2010). Impacts of urban forests on offsetting carbon emissions from industrial energy consumption for Hangzhou, China. *Journal of Environmental Management, 91*, 807–813.

Zhonglin X, Zhao C, Feng Z, Zhang F, Sher H, Wang C, Peng H, Wang Y, Zhano Y, Wang Y, Peng S, Zheng X (2013) Estimating realized and potential carbon storage benefits from reforestation and afforestation under climate change: a case study of the Qinghai spruce forests in the Qilian Mountains, Northwestern China. Mitigation Adaptation Strategy.

Kasturi Devi Kanniah is a professor in Geoinformation Science at Universiti Teknologi Malaysia (UTM). For 20 years, she has taught courses in remote sensing, GIS, and environmental impact assessment. Kasturi received her PhD in Geography & Environmental Science from Monash University, Australia and MPhil in GIS and Remote Sensing from Cambridge University, UK. Her remote sensing investigations examine changes in terrestrial carbon stock (e.g. mangroves, oil palms tropical forest and urban forest), atmospheric aerosols and low-carbon cities. She is a visiting scholar to MIT, USA, Tsinghua University, China and Dundee University, UK. She serves as the Associate Editor of the International Journal of Remote Sensing.

Rohayu Abdullah received her Bachelor and Master's degrees in Urban and Regional Planning from Universiti Teknologi Malaysia (UTM), Johor Bahru, Malaysia. Her research interests include urban planning and the role of vegetation to achieve low carbon cities.

Ho Chin Siong is a professor in Urban and Regional Planning at Universiti Teknologi Malaysia (UTM). He received his PhD in System and Information Engineering from Toyohashi University of Technology, Japan and Master of Science in Construction Management from Heriot Watt University, Edinburgh, UK. His research expertise is in the area of sustainability and low carbon cities.

Chapter 7
Insect Pests of Tropical Malaysian Urban Trees

Su Ping Ong and Ahmad Said Sajap

Abstract Urban trees provide many benefits to the well-being of communities. While trees can be resilient and adapt to the challenging growing conditions in urban areas, prolonged exposure to multiple stressors can affect their growth and development. Open wounds and injuries on an already stressed tree, from natural events, accidental or intentional damage, could become entry points for insect pests to enter the tree, which might exacerbate tree decline. Wood feeders can cause substantial damage to the vascular tissues of trees, and their attacks are usually not visible in the early stage or when occurring below ground as in the case of subterranean termites. This chapter explains the damage to different parts of urban trees caused by wood-, leaf- and sap-feeders. In each feeding guild, examples of the common insect pests, their associated host plants, and signs of infestations are given. A combination of management strategies including cultural, mechanical and biological controls can be practiced to improve tree health as well as enhance populations of natural enemies in order to reduce the usage of insecticides, particularly broad-spectrum insecticides that can also kill beneficial insects. Additionally, the termite baiting system developed based on the feeding behaviour of termites, targeted the species of *Coptotermes* specifically, can be applied on termite-infested urban trees with minimal risk on non-target organisms. Regular monitoring for tree abnormalities should be emphasized in urban tree care to facilitate the early detection and intervention of any plant disorders.

Keywords Urban trees · Subterranean termites · Tree health · Control · Monitoring

S. P. Ong (✉)
Forest Research Institute Malaysia, 52109 Kepong, Selangor, Malaysia
e-mail: ongsuping@frim.gov.my

A. S. Sajap
No. 20, Jalan SS18/4C, 47500 Subang Jaya, Selangor Darul Ehsan, Malaysia

S. Maruthaveeran et al. (eds.), *Urban Forestry and Arboriculture in Malaysia*,
https://doi.org/10.1007/978-981-19-5418-4_7

Introduction

With rapid urbanization and population growth, green infrastructure has become increasingly important in the development of sustainable cities in line with one of the Sustainable Development Goals that aims to make cities and human settlements inclusive, safe, resilient and sustainable. Urban trees help to improve the environment as well as increase the social, economic and aesthetic values of the area. They can also enhance urban biodiversity by providing refuge and food resources for wildlife such as urban birds (Puan et al., 2019) and beneficial insects (Smith et al., 2006), which helps to control the insect pest populations in the urban environment (Shrewsbury & Leather, 2012).

Insect-feeding damages such as holes, skeletonized or missing parts of leaves are common and usually do not harm the hosts as some herbivorous insects such as moths depend on plants to complete their life cycles. These damages, however, can be severe when pest species are present in high numbers, particularly during an outbreak. Insect pest outbreaks on trees are usually associated with multiple factors that reduce the vigour of a tree, including improper site-species matching, soil compaction, physical and mechanical injuries, construction damages and prolonged drought. The heat island effect in cities can also create favourable conditions for certain pest species to multiply. For example, in the United States, the oak scale insects are locally adapted to the warmer temperature in the city and are more abundant on trees with direct exposure to sunlight (Meineke et al., 2013). Accumulation of these stress factors over time can weaken and further predispose the trees to pests and diseases, which can lead to a mortality spiral.

In this chapter, common insect pests that occur on Malaysia's urban trees are categorized into three groups based on their feeding habits, and pest management strategies including the subterranean termite baiting system are discussed. A summary of the types of pests and signs of their infestations is given in Table 7.1.

Wood-Damaging Pests

Attacks by borers and termites can be more destructive than the injuries caused by other insect pests due to the long-lasting feeding damage on the vascular tissues. Also, an infestation often goes unnoticed until symptoms such as wilting of leaves or branch dieback become visible. The rubber termite, *Coptotermes curvignathus* is capable of infesting and killing living trees such as coniferous trees in the genus *Pinus*, *Araucaria* and *Agathis* (Tho & Kirton, 1998), plantation trees such as *Acacia mangium*, landscaping trees such as *Khaya* and *Caesalpinia* (Sajap, 2009), and plantation crops such as rubber and oil palm. Trees infested with *C. curvignathus* have extensive layers of soil plastering around the tree trunk and the termites feed beneath the soil cover to protect themselves from desiccation and ant attack. Though the Asian subterranean termite, *Coptotermes gestroi*, is frequently reported infesting the

Table 7.1 Types of pests and signs of their infestations on the different parts of a tree

Parts of tree infested	Signs of infestations	Types of pests
Leaves	Parts of leaf missing, leaf riddled with holes	Lepidopteran larvae, grasshoppers, beetles
	Leaf surface scraped off leaving a network of leaf veins	Leaf skeletonisers, young larval stages of moths
	Mining tunnel on leaf surface	Leaf miners
	Two or more leaves tied together with silk web to form a shelter	Lepidopteran larvae
	Gall/swelling of leaf tissues	Gall wasps, psyllids, mites
	Stippling damage	Sap-sucking insects and mites
	White cottony mass on the underside of leaves	Whiteflies, mealybugs
	Sooty mould on leaf surfaces	Honeydew-producing scale insects, mealybugs, aphids, whiteflies
Trunks/stems	Extensive mud patches covering the trunk, mud tubes along trunk	Termites
	Bore holes	Beetles, moth larvae
	Pupa exuvia protruding from the emergence hole	Moths
	Pin-holes, frass tubes protruding from the bore holes	Ambrosia beetles
	Ants present in high abundance	Honeydew-producing scale insects, mealybugs, aphids
Roots	Chewing damage on roots	Beetle larvae, termites
Bark	Frass and silk web tunnel along the trunk, uplifted bark	Moth larvae

Note Mites are not insects but are mentioned here due to their importance as plant pests

wooden structures in buildings in the urban and sub-urban areas (Kirton & Azmi, 2005), it has also been recorded feeding on *Casuarina equisetifolia* trees by the presence of their mud tubes along the tree trunks together with feeding damages on the roots and heartwood of the tree (Ong & Sajap, 2018). Other species of termites such as *Nasutitermes* sp. do not attack the living tissues of trees but feed on decaying wood and tree bark.

The mahogany shoot borer, *Hypsipyla robusta*, bores into the shoots of young mahogany trees particularly *Swietenia macrophylla* and *Khaya ivorensis* (Lim et al., 2008). The loss of apical dominance from shoot dieback causes the formation of multiple leaders and crooked stems. *Squamura disciplaga* is another species of borer found on several urban trees such as *Erythrina fusca* (Coral Tree), *Monoon longifolia* (False Ashoka) and *Hura crepitans* (Sandbox Tree) (Chung et al., 2010; Sajap, 2009). This bark borer builds feeding trails with wood debris, frass and silk webs to conceal its feeding activities on the bark as well as its borehole on the trunk.

The long-horned beetle, *Xystrocera festiva*, has been found boring into the trunks of *Acacia* and *Falcataria moluccana* (Batai)(Sajap, 2009). Signs and symptoms of the beetle attack are the presence of frass and sawdust at the boreholes and tree sap oozing from the injuries. Rhinoceros beetle, *Oryctes rhinoceros*, is a pest of palm trees whereby the adults bore into the spear of young palms. As a result of the attack, the developed fronds have fan-shaped characteristic damage with boreholes at the petioles. *Euplatypus paralellus* is an ambrosia beetle that attacks a wide range of timber trees and logs (Fig. 7.1). This species was found infesting a *Millettia pinnata* (Seashore Mempari) tree in 2016. Its infestation appears to be localized and does not spread to the other trees in the area, as ambrosia beetle attack is usually confined to injured or unhealthy trees. Ambrosia beetles do not feed on the wood but cultivate fungal gardens inside the tunnels as a main food source for the adults and larvae while the fungi utilize nutrients from the plants, eventually causing wilt, dieback and death of the tree.

Fig. 7.1 Frass tubes produced by the ambrosia beetle, *E. parallelus* (arrow) and the tree bark were stained by sap oozing from the boreholes

Leaf-Feeding Pests

Leaf-feeding pests like the lepidopteran larvae, beetles and grasshoppers damage leaves with their chewing mouthparts. The larval stages of the butterflies and moths are the active feeding stages, while the adults are short-lived and feed on nectar or do not feed at all. Bagworms such as *Pagodiella heckmeyeri, Metisa plana, Pteroma pendula* and *Mahasena corbetti* are generalist feeders on crops, forest and urban trees including palms such as *Copernicia alba* (Fig. 7.2). The bagworm larvae build protective cases or bags that resemble dead leaves or twigs for protection from natural enemies. Additionally, the larvae of some families of moths such as Lymantriidae and Lasiocampidae have tufts of spiny, poisonous hairs covering their bodies. The Angsana leaf miner, *Neolithocolletis pentadesma,* attacks *Pterocarpus indicus* (Angsana) trees by burrowing between leaf surfaces and scraping off the chlorophyll layers. The feeding activity causes transparent blotches on the leaves and premature leaf fall. Some butterfly species are pests of ornamental plants. For example, the caterpillar of the cycad blue, *Chilades pandava,* feeds on the shoots of cycads resulting in damage to the young leaves before they mature, thereby reducing the aesthetic appearance of the plant.

The *Erythrina* gall wasp, *Quadrastichus erythrinae,* is a pest on *Erythrina variegata* (Indian Coral Tree) (Chung, 2009). The feeding activity of the larvae on the young leaves and shoots induces the formation of galls, which are then utilized by the galling larvae as protection against natural enemies. Galls are swollen, abnormal

Fig. 7.2 Fronds of *Copernicia alba* (Caranday Palm) turning brown and wilting due to bagworm infestation

Fig. 7.3 Leaf deformation and wilting from severe galling

growths that protrude from leaf surfaces, giving the tree an unsightly appearance. In severe infestation, the attacked leaves turn yellow and fall off (Fig. 7.3). Other pests such as the gold dust weevil, *Hypomeces squamosus* and Malaysian locust, *Valanga nigricornis,* are generalist feeders on various trees and shrubs.

Sap-Sucking Pests

Sap-sucking pests such as the scale insects, mealybugs, whiteflies and mites suck on plant sap with their piercing-sucking mouthparts. Scale insects produce crawlers or first instar stage that are mobile and buoyant in which they can easily disperse by wind or crawl to another host plant. Once the crawlers settle on a suitable part of the plant, they usually lose their legs and remain sessile for their entire life, especially for female scales. Adult males are weak fliers and often not needed in reproduction as females may produce eggs by parthenogenesis.

Most scale insects and mealybugs secrete copious amounts of honeydew, which provides a source of carbohydrates for ants (Fig. 7.4). In return, the ants confer protection to the honeydew-producing hemipterans from natural enemies, regular removal of secreted honeydew and transport to new feeding sites for some mealybugs (Way, 1963). The ant-scale mutualism can lead to an increase in the abundance and outbreaks of scale insects on trees. Sooty mould fungus, which grows on the honeydew secreted by the sap-sucking hemipterans coats the surface of leaves and stems of the infested plant, reducing its aesthetic value and photosynthetic efficiency.

Fig. 7.4 Ants collecting honeydew from yellow lac scale *Tachardina aurantiaca* on a tree branch

The spiraling whitefly, *Aleurodicus dispersus,* resembles tiny flies, which infests various horticultural, agricultural and nursery plants (Fig. 7.5). A common sign of its presence is the spiraling pattern of egg deposits with white waxy secretion on the underside of leaves. When the leaves are disturbed, the adults will fly away and the white powdery wax on their wings and body can be irritating.

Pest Management Strategies

A common practice to treat pest problems on trees is to spray with pesticides. Repeated applications of pesticides are harmful to urban ecosystems in the long term. For instance, chemical pollution of pesticides might be introduced, pests can develop resistance, and non-targeted insects such as the natural enemies of the pests are killed. There are several alternative methods, including cultural, mechanical, and/or biological controls, which can be considered when managing pest problems. Pest control may not be necessary during the rainy seasons as some insects for example the bagworms of *M. plana* and *P. pendula* can be washed away during heavy rainfall (Ho et al., 2011).

Fig. 7.5 Leaves attacked by the spiralling whitefly, *A. dispersus*

Cultural Control

Cultural control involves promoting tree health by improving site conditions so as to reduce the incidence of pest problems. At least 80% of tree health issues are related to root and soil problems (Fields-Johnson & Abbott, 2020). Therefore site-species matching is important to keep trees healthy and less susceptible to pests and diseases. For example, the large crown radius and high maintenance of *Khaya senegalensis* (African Mahogany) make it unsuitable as a street tree (Rohayu et al., 2018). Different species of trees also have different levels of tolerance to waterlogging (Ma et al., 2019), so the selection of the right species will increase tree survival and reduce costly maintenance. Practices that promote soil health such as adding mulch, compost and biochar help enhance the soil properties. Sanitation practices by removing wood debris and pest-infested plant materials are also important to reduce potential breeding sites for pests.

Mechanical Control

Mechanical control involves the removal of infected plant parts by pruning and installing insect traps and/or barriers. Eggs, larvae and pupae of moths and butterflies can be hand-picked if present in low numbers. A pair of forceps can be used to pick larvae of moths with urticating hairs. Pruning can also help to remove galling and mining insects, and scale insects. Insect trapping using pheromone lures has been used on boring beetles, for example, in the monitoring of the rhinoceros beetle population in oil palm plantations in Malaysia (Norman & Mohd. Basri, 2004) and the detection of the invasive Asian long-horned beetle in the urban landscapes in the United States (Nehme et al., 2014). To protect trees from crawling insects such as ants and social caterpillars, sticky barrier bands can be placed around the trunk diameter.

Biological Control

Biological control is an approach that uses natural enemies to control pest populations. There are three types of natural enemies available in the environment, which are predators, parasitoids and pathogens. The population of natural enemies can be enhanced with flowering plants like *Cassia cobanensis* and *Turnera subulata* as they provide nectar and pollen resources for the parasitoids of bagworms (Nor Sarashimatun et al., 2011). However, biological control agents may sometimes be absent for an exotic pest or present in low abundance and inconsistent in pest suppression, especially during outbreaks. Certain species of native parasitoids have been mass-reared in the laboratory and introduced in the field to control crop pests in Malaysia (Mohd. Norowi et al., 2005) as well as exported to other countries such as Christmas Island to control invasive pests (Ong et al., 2017).

Chemical Control

Chemical control involves the use of pesticides when other pest control options are limited or not effective in suppressing a pest population. Pesticides can be grouped based on their target groups, for example, insecticides for insects and miticides for mites. An insecticide can also be differentiated based on the mode of action as a contact, stomach or systemic poison. Contact or stomach poison such as malathion and cypermethrin can be used on leaf-feeding pests such as beetles, lepidopteran larvae and grasshoppers. Systemic insecticides such as fipronil and imidacloprid are absorbed by tree roots and translocated into different parts of the tree, and thus can be targeted on sap-sucking insects, leaf miners, borers and termites.

Another type of chemical control pertains to the utilization of several biorational and organic products, which are environmentally friendly and do not harm non-target organisms. For example, *Bacillus thuringiensis* (Bt) is a soil-borne bacterium that kills lepidopteran larvae upon ingestion. Oil emulsion spray made from a diluted mixture of soap and oil or horticultural oil can be used on soft-bodied pests such as mealybugs, nymphs of scale insects, whiteflies and mites whereby the solution will suffocate the pest by blocking its breathing pores/spiracles. Plant-derived insecticides containing neem have anti-feedant properties on lepidopteran larvae. Insect growth regulators (IGRs) are widely used in the control of subterranean termites. The IGRs are target-specific insecticides that interfere with the physiology of the pests for example, by preventing the formation of the exoskeleton when an immature insect undergoes moulting process, leading to the death of the insect before maturity.

Case Studies on the Management of Coptotermes *Termites* on Trees in Peninsular Malaysia

The subterranean termites, *Coptotermes* spp., are major pests in forestry, agriculture and urban areas in Southeast Asia. *Coptotermes* attack on trees usually starts below ground in the roots, around tree bases or stumps and then makes its way upwards into the heartwood. Signs of infestations include mud tubes or extensive soil plastering aboveground that conceals the feeding activity of the termites on the outer wood of the tree trunk. In Peninsular Malaysia, *C. gestroi* termite has not been reported as a serious pest of trees. However, in Brazil, this termiteinfested 93% of urban trees, including *Caesalpinia* and *Erythrina* (Zorzenon & Campos, 2015) and in Guam, it infested *Casuarina equisetifolia* trees (Park et al., 2020). In Taiwan, *C. gestroi* threatens forest trees and its invasion into the forest areas shows the high adaptability of this species in a new environment (Chiu et al., 2016; Li et al., 2011).

In 2017, we examined *C. equisetifolia* trees planted along a beachfront in Kelantan, on the east coast of Peninsular Malaysia, for termite attack. *C. equisetifolia* has been used in coastal reforestation projects for erosion control, windbreak, and improving the aesthetic value of the area. It was found that 27 of 1743 trees in about 2.5 ha were infested with *C. gestroi*, with mud tubes along the tree trunks. Given the potential of *C. gestroi* to become a major pest, the termite baiting system was tested.

The field trial in Pantai Senok, Kelantan was focused on an area of about 0.1 ha towards the river mouth, with 13 termite-infested trees, and the control plot was 540 m away from the treatment plot, at the other end of the *C. equisetifolia* plantation (Fig. 7.6).

Underground monitoring stations were placed at the base of the infested trees in the treatment and control plots (Fig. 7.7). The monitoring station was a plastic container (17.5 × 12.5 × 10.3 cm) with a rubberwood bundle (6 pieces, 15 × 4 × 2 cm) to attract the termites. To allow the access of termites into the container, several small holes were drilled at the base and side of the container. A moistened tissue paper was used as a connector to the active mud tube at the base of the tree. The 16 monitoring stations (3 stations in the control plot; 13 stations in the treatment plot) were inspected every 2 weeks to check for termite feeding.

Termite baits with 0.1% chlorfluazuron (Requiem®, Ensystex Sdn. Bhd., Malaysia) were placed on active mud tubes on the tree trunks and wrapped with a black plastic (Fig. 7.8). Each packet of the bait matrix weighed 200 g. A total of 600 ml of distilled water was added to form a paste. Baiting was initiated in May 2017, one month after the installation of the monitoring stations. Baits were checked and replaced every 2 weeks when consumption was more than 70%. If consumption

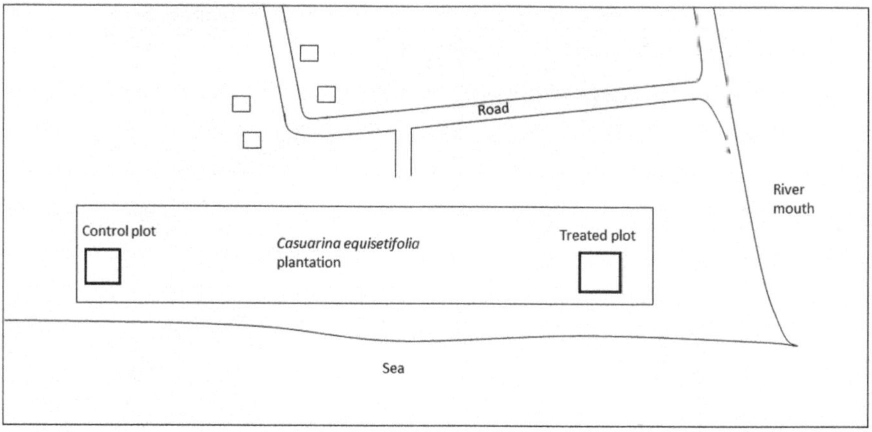

Fig. 7.6 Overview of the experimental plot in Pantai Senok, Kelantan

was less than 70%, the existing bait was replenished with 100 g of bait. The monthly consumption rates of the rubberwood and baits were calculated based on the weight differences before and after consumption (Fig. 7.9).

No termite activities were observed in all underground monitoring stations in the treated plot by July 2017, approximately 10 weeks after baiting was initiated, in comparison to the control plot. While termite activities continued to be observed in the control plot, there was a reduction in the feeding rates probably due to disturbance by human activities. An average of 158.48 g (consumption range: 19.52–268.64 g) of chlorfluazuron was used to eliminate the *C. gestroi* colony from the 0.1 ha treatment plot.

In another study site, severe infestations by *C. curvignathus* termites were observed on three *A. borneensis* (Borneo Kauri) (Fig. 7.10) and two *C. ferrea* (Brazilian Ironwood) trees, which was planted at the sides of the roads in Forest Research Institute Malaysia (FRIM), Selangor Darul Ehsan. These trees are of significant cultural value and thus are labelled as heritage trees. Given this situation, termite treatment was initiated using the same bait matrix, however, without installing the underground monitoring stations. An average of 933.33 g (consumption range: 800–1000 g) and 300 g (consumption range: 200–400 g) of chlorfluazuron were used to control the *C. curvignathus* colonies on the *A. borneensis* and *C. ferrea* trees, respectively.

The baiting system is based on the social behaviour of termites, whereby the bait containing the slow-acting toxicant such as chlorfluazuron is consumed by the worker termite and transferred to the rest of the colony members during trophallaxis (food sharing). Termite baits with different active ingredients have been used in the control of subterranean termites (Rhinotermitidae family) in the urban and forested areas in Malaysia (Lee, 2002; Sajap et al., 2000), Taiwan (Chiu et al., 2016) and Australia (Webb, 2017). Baiting is an environmental-friendly system with low toxicity or not effective on non-rhinotermitids (Lee et al., 2007) such as the higher termite,

Fig. 7.7 An underground monitoring station was placed at the base of an infested tree (left). The termites entered the station and were actively feeding on the rubberwood after one month (right)

Fig. 7.8 The termite bait was placed on the *C. gestroi* mud tube and covered with black plastic (left). Note the soldiers of *C. gestroi* inside the bait matrix (right)

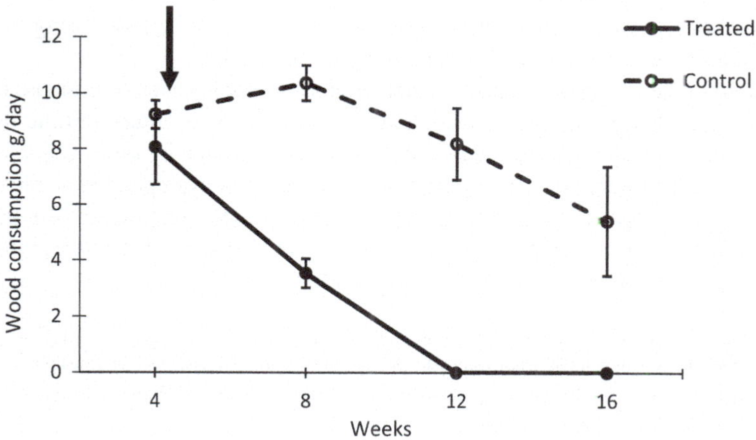

Fig. 7.9 The amount of wood consumed by *C. gestroi* throughout the experiment. Baiting was initiated (arrow) 4 weeks after the termites started feeding in the underground monitoring stations in the control (open circles, $n = 3$ stations) and treated (solid circles, $n = 10$) plots. Monitoring stations with no active feeding were excluded

Fig. 7.10 *Coptotermes curvignathus* feeds under the layers of soil surrounding the trunk of an *A. borneensis* tree (left) and leaves start to wilt as the attack progresses (right)

Microcerotermes crassus and centipedes that were found in the monitoring stations and on the treated trees in our experimental plot in Kelantan.

Although the *C. gestroi* and *C. curvignathus* colonies were eliminated, tree survival depended on the extent of damage caused by the termites and the overall health of the trees. There is also a possibility of re-infestation by other *Coptotermes* colonies and the amount of bait required to eliminate a colony depended on the active ingredient and colony size. Therefore, regular monitoring of tree health is important for early detection of potential problems, given the risks that urban trees pose to public safety and property damage from tree falls.

Acknowledgements We appreciate the contribution of Ensystex (Malaysia) Sdn. Bhd. especially Shahrem Md. Ramli and Mohamad Faiz Mohd Shukry for their assistance and provision of the Requiem® bait matrix in the field experiments. We would also like to thank the Kelantan State Forestry Department for their assistance.

References

Chiu, C.-I., Yeh, H.-T., Tsai, M.-J., & Li, H.-F. (2016). Naturalization and control of *Coptotermes gestroi* (Blattodea: Rhinotermitidae) in a Taiwanese forest. *Journal of Economic Entomology, 109*(3), 1317–1325.

Chung, A. Y. C. (2009). Infestation status of the invasive Erythrina gall wasp *Quadrastichus erythrinae* Kim (Hymenoptera: Eulophidae) on the coral tree *Erythrina variegata* in Sandakan, Sabah, Malaysia. In H. C. Sim (Ed.). *Forest health in a changing world* (vol. 24, pp. 98–100). IUFRO World Series.

Chung, A. Y. C., Ong, R.C., Nilus, R., & Hastie, A. (2010, August 2010). Effects of environmental changes on tree health: A case study on insect infestation (*Squamura disciplaga* Swinhoe) on urban ornamental trees (pp. 23–28). Poster presented in XXIII IUFRO World Congress. Seoul.

Fields-Johnson, C., & Abbott, C. (2020). Applications of biochar for arboriculture. *Arborist News, 29*(1), 12–16.

Ho, C. H., Yusof, I., & Khoo, K. C. (2011). Infestations by the bagworms *Metisa plana* and *Pteroma pendula* for the period 1986–2000 in major oil palm estates managed by Golden Hope Plantation Berhad in Peninsular Malaysia. *Journal of Oil Palm Research, 23*, 1040–1050.

Kirton, L. G., & Azmi, M. (2005). Patterns in the relative incidence of subterranean termite species infesting buildings in Peninsular Malaysia. *Sociobiology, 46*(1), 1–15.

Lee, C. Y. (2002). Control of foraging colonies of subterranean termites, *Coptotermes travians* (Isoptera: Rhinotermitidae) in Malaysia using hexaflumuron baits. *Sociobiology, 39*(3), 411–416.

Lee, C. Y., Vongkaluang, C., & Lenz, M. (2007). Challenges to subterranean termite management in multi-genera faunas in Southeast Asia and Australia. *Sociobiology, 50*(1), 213–221.

Li, H.-F., Yeh, H.-T., Wang, Y.-N., & Tsai, M.-J. (2011). Termite diversity and damage pattern in tropical botanical gardens of Taiwan. *Journal of the Experimental Forest of National Taiwan University, 25*, 139–147.

Lim, G. T., Kirton, L. G., Salom, S. M., Kok, L. T., Fell, R. D., & Pfeiffer, D. G. (2008). Mahogany shoot borer control in Malaysia and prospects for biocontrol using weaver ants. *Journal of Tropical Forest Science, 20*(3), 147–155.

Ma, L., Rao, X., & Chen, X. (2019). Waterlogging tolerance of 57 plant species grown hydroponically. *HortScience, 54*(4), 749–753.

Meineke, E. K., Dunn, R. R., Sexton, J. O., & Frank, S. D. (2013). Urban warming drives insect pest abundance on street trees. *PLoS ONE, 8*(3), e59687. https://doi.org/10.1371/journal.pone.0059687

Mohd Norowi, H., Mohd Roff, M. N., & Lo, N. P. (2005). Country paper—Malaysia. Non-pesticide methods for controlling diseases and insect pests (pp. 100–110). In P. A. C. Ooi (Ed.). Report of the APO Seminar on Non-Pesticide Methods for Controlling Diseases and Insect Pests. 10–17 April 2002, Japan.

Nehme, M. E., Trotter, R. T., Keena, M. A., McFarland, C., Coop, J., Hull-Sanders, H. M., Meng, P., De Moraes, C. M., Mescher, M. C., & Hoover, K. (2014). Development and evaluation of a trapping system for *Anoplophora glabripennis* (Coleoptera: Cerambycidae) in the United States. *Environmental Entomology, 43*(4), 1034–1044.

Nor Sarashimatun, S., Teh, C. L., & Teh, C. C. (2011). Evaluation of beneficial plants as hosts for natural enemies of oil palm bagworms. Poster presented at the International Palm Oil Congress (PIPOC 2011). 14–17 November 2011, Kuala Lumpur.

Norman, K., & Mohd Basri, W. (2004). Immigration and activity of *Oryctes rhinoceros* within a small oil palm replanting area. *Journal of Oil Palm Research, 16*(2), 64–77.

Ong, S.P., O'Dowd, D.J., Detto, T. and Green, P.T. (2017). Introduction of *Tachardiaephagus somervilli*, an encyrtid parasitoid, for the indirect biological control of an invasive ant on Christmas Island (pp. 118–120). In P. G. Mason, D. R. Gillespie & C. Vincent (Eds.), *Proceedings of the 5th International Symposium on Biological Control of Arthropods*. 11–15 September 2017, Langkawi.

Ong, S. P., & Sajap, A. S. (2018). Ekologi, Populasi dan Pengurusan Anai-anai Perosak Pokok Ru di Pesisiran Pantai Semenanjung Malaysia. FRIM Technical Information Handbook No.47.

Park, J.-S., Husseneder, C., & Schlub, R. (2020). Morphological and molecular species identification of termites attacking ironwood trees, *Casuarina equisetifolia* (Fagales: Casuarinaceae). *Guam. Journal of Economic Entomology, 112*(4), 1902–1911.

Puan, C. L., Yeong, K. L., Ong, K. W., Muhd Izzat, A. F., Muhammad Syafiz, Y., & Khoo, S. S. (2019). Influence of landscape matrix on urban bird abundance: Evidence from Malaysian citizen science data. *Journal of Asia-Pacific Biodiversity, 12*, 369–375.

Rohayu, A., Kanniah, K. D., & Ho, C. S. (2018). Identification of suitable trees for urban parks and roadsides in Iskandar Malaysia. *Chemical Engineering Transactions, 63*, 385–390. https://doi.org/10.3303/CET1863065

Sajap, A. S. (2009). *Serangga Perosak Pokok Ameniti*. Jabatan Landskap Negara, pp. 71.

Sajap, A. S., Amit, S., & Welker, J. (2000). Evaluation of hexaflumuron for controlling the subterranean termite *Coptotermes curvignathus* (Isoptera: Rhinotermitidae) in Malaysia. *Journal of Economic Entomology, 93*, 429–433.

Shrewsbury, P. M., & Leather, S. R. (2012). Using biodiversity for pest suppression in urban landscapes. In G. M. Gurr, S. D. Wratten, W. E. Snyder & D. M. Y. Read (Eds.). *Biodiversity and insect pests: Key issues for sustainable management*. John Wiley & Sons, Ltd.

Smith, R. M., Gaston, K. J., Warren, P. H., & Thompson, K. (2006). Urban domestic gardens (VIII): Environmental correlates of invertebrate abundance. *Biodiversity and Conservation, 15*, 2515–2545.

Tho, Y. P., & Kirton, L. G. (1998). A survey of termite attack in Bahau conifer plantation, Peninsular Malaysia. *Journal of Tropical Forest Science, 10*(4), 564–567.

Way, M. J. (1963). Mutualism between ants and honeydew-producing homoptera. *Annual Review of Entomology, 37*, 479–503.

Webb, G. A. (2017). Efficacy of bistrifluron termite bait on *Coptotermes lacteus* (Isoptera: Rhinotermitidae) in Southern Australia. *Journal of Economic Entomology, 110*(4), 1705–1712.

Zorzenon, F. J., & Campos, A. E. C. (2015). Subterranean termites in urban forestry: Tree preference and management. *Neotropical Entomology, 44*, 180–185.

Su Ping Ong is a Forest Entomologist with the Forest Research Institute Malaysia (FRIM) since 2007. She obtained her BSc (Animal Biology) in 2007 and MSc (Applied Entomology) in 2015 from Universiti Sains Malaysia (USM). Her main research interests are forest health protection focussing on the biology and ecology of forest pests including invasive alien species and pest management using biological control agents. She is a Certified Arborist under the International Society of Arboriculture (MY0430A).

Ahmad Said Sajap retired as a Professor from Universiti Putra Malaysia (UPM) in 2007 and served as a Deputy Dean in the Faculty of Forestry before his retirement. He obtained his Diploma in Agriculture from the College of Agriculture Malaya in 1973 and Bachelor of Science (Entomology) from the University of California in 1975. He graduated with an MSc in 1977 and Ph.D. in 1987 in Entomology from Iowa State University. His research interests are insect diversity, biology and ecology of insect pests, and pest management focusing on biological control agents and termite baiting systems. He is a Certified Arborist under the International Society of Arboriculture (MY0229A).

Chapter 8
Common and Potential Insect Pests of Urban Palm Trees in Malaysia

Li Peng Tan, Yew Loong Cheong, Samsuddin Ahmad Syazwan, Wei Chen Lum, and Seng Hua Lee

Abstract Palms belong to one of the largest plant families, Arecaceae. They are among the most extensively cultivated species used for foods and landscaping. This family covers a large variety of species and is widely distributed in warm and humid lowlands of the tropics. Malaysia, which is located in the equatorial region with a typical tropical climate, is home to many palm species. Palms are recommended in landscape planning where the planting space is cramped; they are widely used to create a sense of place and to define the frame of the construction. These evergreen plants could serve as reliable food sources and shelters for insects, and almost all categories of insect pests can be found in palms with various degrees of occurrence. Nevertheless, the relationship between insects and different palm species planted for ornamental functions is still mostly unknown. This chapter summarizes the insect pests associated with the five most common palm species used for the urban landscape in Malaysia viz. *Roystonea regia, Wodyetia bifurcata, Licuala grandis, Livistona*

L. P. Tan (✉)
Department of Paraclinical, Universiti Malaysia Kelantan (UMK), Kota Bharu Pengkalan Chepa, 16100, Kelantan, Malaysia
e-mail: li.peng@umk.edu.my

Y. L. Cheong
Genetic & Agriculture (G&A) Research Centre, Kabupaten Pelalawan PT Musim Mas, Desa Batang Kulim, Kecamatan Pangkalan Kuras,, Riau, Indonesia

S. Ahmad Syazwan
Mycology and Pathology Branch Forest Biodiversity Division, Forest Research Institute Malaysia (FRIM), Kuala Lumpur 52109, Selangor Darul Ehsan, Malaysia

W. C. Lum
Institute for Infrastructure Engineering and Sustainable Management (IIESM), Universiti Teknologi MARA (UiTM), Shah Alam 40450, Selangor Darul Ehsan, Malaysia

S. H. Lee
Institute of Tropical Forestry and Forest Products (INTROP), Universiti Putra Malaysia (UPM), Serdang 43400, Selangor Darul Ehsan, Malaysia

S. Ahmad Syazwan
Department of Forest Science and Biodiversity, Universiti Putra Malaysia (UPM), Serdang 43400, Selangor Darul Ehsan, Malaysia

S. Maruthaveeran et al. (eds.), *Urban Forestry and Arboriculture in Malaysia*,
https://doi.org/10.1007/978-981-19-5418-4_8

chinensis and *Ptychosperma macarthurii*. Similar taxa of insect pests are usually found on different palm species, regardless of their location. Hemiptera, Coleoptera and Lepidoptera are the most documented orders among all insect pests.

Keywords Palm · Insect pest · Ornamental plants · Urban landscape · Malaysia

Introduction

Palms belong to one of the largest plant families, Arecaceae. They are monocotyledonous perennial flowering plants in various forms such as climbers, shrubs, tree-like and stemless plants. Known as palms in general, those having a tree-like architecture are usually called palm trees. Currently, there are around 187 genera with approximately 2,466 species of palms distributed mainly in the tropical and subtropical regions (http://www.theplantlist.org/). Palms can be easily distinguished by their large and distinctive leaves, known as fronds arranged at the top of an unbranched stem. In fact, palms constitute one of the few monocotyledonous plant families with woody-stemmed arborescent species (Howard, 1991) .

Palms inhabit a wide range of habitats from rainforests to deserts. They are among the most extensively cultivated plant families used for foods and also in landscaping. After the grasses and legumes, palms are the third most economically important family of plants. For example, coconut (*Cocos nucifera* L.), oil palm (*Elaeis guineensis* L.) and date palm (*Phoenix dactylifera* L.) are the three most crucial commodity crops in the world (Meerow et al., 2012). Besides, palms are one of the favourite ornamental plants wherever they can grow—in outdoor landscapes or indoor environments, across warm and cold regions. They by all means represent the lure and romance of the tropics, better than any other plant group, and therefore symbolic for the tropics and vacations.

This plant family has a large variety of species and is widely distributed in warm and humid lowlands of the tropics. Malaysia, which is located in the equatorial region with a typical tropical climate, is home to many palm species. Of the 2,600 known palm species, around 443 can be found in Malaysia (Tan, 2014). As an indicator of our rich palm heritage, many Malaysian states and towns are named after palms—Betal Nut Palm (*Areca catechu*), Salak South (*Salacca zalacca*), Nibong Tebal (*Oncosperma tigillarium*), Bertam (*Eugeissona tristis*) and Serdang (*Livistona chinensis*). Although diverse species inhabit this region, palms are endemic with very restricted localities/habitats to survive in. Almost 70% of the palm species found in Malaysia are endemic to this region, their high endemism makes them an important flora in the world of plants in terms of conservation (Tan, 2014).

According to the National Landscape Guideline (2008) published by the National Landscape Department, palm species are recommended in the landscape design where the planting space is cramped. Besides the aesthetic value, landscape trees are also installed for many other benefits, such as shade provision and temperature mitigation. Thus, for unbranched species, like most of the palm trees, planting in a cluster is suggested to achieve the shading effect. Moreover, palms are also planted to create a sense of place and to define the frame of the construction. The recommended palm species for various areas are listed in Table 8.1.

Arecoideae and Coryphoideae are the two subfamilies of palms that have been widely used in the urban landscape design in Malaysia. The Arecoideae is the largest subfamily, which covers more than 50% of the species in a total of 107 genera with pinnate or bi-pinnate fronds (Dransfield et al., 2008). Other morphological characters for this subfamily include highly differentiated primary inflorescence bracts and floral triads. The coryphoid palms have 39 genera, covering all palms with palmate or costa-palmate fronds. The inflorescence in coryphoid palms has a very simple inter-foliar construction (Saw, 2012). According to the report of 'List of Palm Trees around Kuala Lumpur (2011–2020)' by Kuala Lumpur City Hall (DBKL), a total of

Table 8.1 List of palms species recommended in the National Landscape Guideline (2nd Edition, 2008

Scientific name	Local name	Subfamily	Areas
Archontophoenix alexandrae	Palma majestic	Arecoideae	R/B
Areca catechu	Pinang makan	Arecoideae	R/H
Bismarckia nobilis	Bismakia	Coryphoideae	R/I
Borassus flabellifer	Lontar	Coryphoideae	R/B/S
Caryota mitis	Palma fishtail	Coryphoideae	R/I
Chrysalidocarpus lutescens	Pinang kuning	Arecoideae	R/B/H/P/I/T
Cocos nucifera	Kelapa	Arecoideae	R/F/S
Crytostachys lakka	Pinang merah	Arecoideae	R/B/H/I/T
Dictyosperma album	Palma princess	Arecoideae	R/B
Elaeis guineensis	Kelapa sawit	Arecoideae	R
Licuala grandis	Palma kipas	Coryphoideae	R/F/H/T/S
Livistona chinensis	Serdang cina	Coryphoideae	R/B/F/I/T/S
Livistona rotundifolia	Serdang	Coryphoideae	R/F
Neodypsis decaryii	Palma triangula	Arecoideae	R
Nypa fruticans	Nipah	Nypoideae	F
Oncosperma tigillarium	Nibung	Arecoideae	F
Ptychosperma macarthurii	Palma Macarthur	Arecoideae	R/H/P/I
Rhapis excelsa	Palma large lady	Coryphoideae	F/H/P/I
Rhapis humilis	Palma lady	Coryphoideae	F/H/P/S
Roystonea regia	Palma raja	Arecoideae	R/I
Veitchia merrillii	Palma manila	Arecoideae	R/H/T
Wodyetia bifurcata	Palma foxtail	Arecoideae	R

B—Buildings (Public buildings); F—Fronts (Riverfront; Lakeside; Beach); H—Housing area; I—Industrial area; R—Recreational area; S—Suburban area; T—Train station (Railway)
Source National Landscape Guideline (2nd Edition), 2008

29,363 palms have been planted in Kuala Lumpur. The most common species include *Ptychosperma macarthurii* (5,207), *Roystonea regia* (3,891), *Cocos nucifera* (3,053), *Adonidia (Veitchia) merrillii* (2,620), *Saribus (Livistona) chinensis* (2,495), *Licuala grandis* (2,202), *Wodyetia bifurcata* (2,195) and *Elaeis guineensis* (1,495).

The evergreen palms serve as reliable food sources and shelters for insects. Leaves of various ages are present continuously throughout the year, with the size and number of fronds remaining basically the same, once the crown of the palm is fully formed (Howard, 1999). Insect pests found on palms generally show a definite preference for the abaxial surface of the fronds. The broad palmate fronds and the reduplicate roof-like leaflets of pinnate fronds provide incomparable protection from abiotic factors such as heavy rain and hot sun, and the arrangement of the fronds in the horizontal position offers protection from many predatory vertebrates as well (Howard et al., 2001).

In the literature, the insect-palm relationships are very much concentrated only on a few species of palms, for instance, coconut and oil palm with economic importance. Even though palms used as ornamental plants are species relatively easy to propagate, fast-growing, adaptable to urban landscapes or interiorscapes and reasonably resistant to pests and diseases (Howard et al., 2001), various insect pests can still be found in palms with different degrees of occurrence. Ornamental palms have a high aesthetic value, defects that are of minor consequence in a plantation might be a significant problem for the ornamental nursery industry and landscape. A vast array of palm species used as ornamental plants are getting more scholarly attention now. However, a large portion of the insect fauna of the palm genera was still unknown. This chapter summarizes the insect pests associated with the five most common palm species used for the urban landscape planning in Malaysia viz. *Roystonea regia*, *Wodyetia bifurcata*, *Licuala grandis*, *Livistona chinensis* and *Ptychosperma macarthurii*. We categorize the insect pests of palms based on their feeding habits or damage caused viz. leaf defoliators (chewers, skeletonizers and miners), gall-makers, sap-feeders, borers and root-feeders. In general, these insects can cause foliage defects, chlorosis and transmit diseases; in a high population, they can affect the vigour of palms and increase the susceptibility of the palms to some other diseases. Fortunately, in palm-growing regions where most of these pests are native, their populations are usually regulated naturally (Howard, 2001). Human intervention by applying chemical control may exacerbate the problem by delaying the eventual recovery of natural enemies. Therefore, cultural and biological control, the role of semiochemicals in insect pest management, safe pesticides and application techniques, and their incorporation in integrated pest management of palm pests are now the most widely researched areas (Faleiro et al., 2016; Howard, 2001).

Cuban Royal Palm, *Roystonea Regia* (Arecoideae)

Cuban Royal Palm (*Roystonea regia*) is a common exotic palm used for ornamental purposes in Malaysia due to its large and prominent appearance on the landscape

(Fig. 8.1). It is native to the subtropical continents of Central America, Mexico, and Florida, and it is resilient in Malaysia's tropical climates. The fascinating cylindric and smooth trunk is also one of the reasons of why it's always chosen as an ornamental tree in urban areas (Gilbin-Davis & Peña, 1993). Pest infestation on *R. regia* was insufficiently recorded and studied in Malaysia, even though the presence of polyphagous pests has been noticed which might threaten the health of this palm. Red palm weevil (*Rhynchophorus ferrugineus*), an invasive alien species (IAS) was recorded as a minor pest to the most ornamental palm trees including *R. regia* (Malumphy et al., 2016). This ferocious IAS could attack *Cocos nucifera* (a major oil crop) in Malaysia, mostly on the east coast of Peninsular Malaysia (Wahizatul et al., 2013). On the attacked palm, holes on the tree's trunk or crown, crumble of plant fibre, together with brown viscous exudate could be found. The later stage of *R. ferrugineus* infestation is distinctively followed by the withered crown (CABI, 2020a). Another potential borer pest of *R. regia* is the coconut Rhinoceros beetle (*Oryctes rhinoceros*) (Fig. 8.2). It is also recorded as a minor borer pest of *R. regia,* but it mostly attacks oil palm, and coconut palm (John & Kenneth, 2020). Additionally, silky cane weevil (*Metamasius hemipterus*) is recorded as a borer pest for *R. regia,* but it can only be found in South America and West Africa (Gilbin-Davis & Peña, 1993; EPPO, 2016). However, it can potentially attack *R. regia* in Malaysia, as it is able to adapt to and survive in Malaysia's tropical climate.

The infestation of the sap-feeder pests is also poorly recorded in Malaysia. Several sap-feeder pests present in Malaysia are known as major pests of *R. regia* in other

Fig. 8.1 The majestic *Roystonea regia* palms are planted in rows beside the buildings

Fig. 8.2 Rhinoceros beetle (*Oryctes rhinoceros*) on an oil palm frond

countries, including hemipterans such as palm-infesting whitefly (*Aleurotrachelus atratus*), brown soft scale (*Coccus hesperidum*) and pineapple mealybug (*Dysmicoccus brevipes*) (Ben-Dov, 1993; CABI, 2020b; Evans, 2007; Lim, 1972; Martin & Lau, 2011). Additionally, American palm cixiid (*Haplaxius crudus*), and royal palm bug (*Xylastodoris luteolus*) are recorded as a minor sap-feeder pest of *R. regia* at Florida, Central America, and South America (Baranowski, 1958; ; Howard, 2001, 2015; Smith et al., 1997). These hemipterans can potentially become IAS in Malaysia due to similar climates. Palms frond infested by sap-feeder pests usually show necrotic and chlorotic appearance on the upper leaf surfaces. The infesting sap-feeder is commonly enclosed with white wax secretion on the under-leaf surfaces. This will cause sooty moulds that grow from honeydew brought by sap-feeder infestation (CABI, 2020c). Some of the sap-feeder pests might also - become vectors of pathogenic fungus. For example, American palm cixiid is a well-known vector of lethal yellowing disease pathogen, *Candidatus Phytoplasma palmae* which is highly destructive to most of the Arecaceae species (Howard, 2015).

Foxtail Palm, *Wodyetia Bifurcata* (Coryphoideae)

Wodyetia bifurcata is the only species of genus *Wodyetia* and is commonly known as Foxtail Palm. This palm is native to Queensland, Australia, and widely introduced to many countries to be used as an ornamental tree in urban landscapes. A unique feature of this palm is its red–orange fruits which are aesthetically attractive and thus value-added to landscape scenery (Fig. 8.3). The major pest of *W. bifurcate* is sap-feeder insects, palm-infesting whitefly (*A. atratus*). This bizarre sap-feeder is resilient to different temperatures and altitudes, which makes it hard to control (Borowiec et al., 2010). As a typical sap-feeder pest, the by-product of sap-feeding activity, sooty molds growing on the honeydew left by this pest are often interfering with the effectiveness of photosynthetic activity of infested palms. Other sap-feeder pests such as Phantasma scale (*Fiorinia phantasma*) could also cause similar impacts on host palms. In the early stage of infestation, both male and female Phantasma scales are observed to amalgamate on the under-leaf surfaces. The later stages of infestation, this amalgamation continues on the upper-leaf surfaces when the under-leaf surfaces become crowded with a high number of infested sap-feeder pests (Garcia, 2011).

Another type of pest that has been found on Foxtail Palm is Coconut Hispine Beetle (*Brontispa longissimi*), which is a defoliator pest. This beetle was first recorded as native to Indonesia and Papua New Guinea (Nakamura et al., 2006; Singh & Rethinam, 2009). It has been found on various ornamental palm species, and mostly on coconut palm (*Cocos nucifera*). This pest is distributed widely across Southeast Asia (such as Malaysia, Myanmar, Singapore, Thailand, Vietnam, Bangladesh, China, India, Maldives and Sri Lanka) and Australia (Liebregts & Chapman, 2004; Liu et al., 2011). This wide infestation is mainly due to improper quarantine procedures of coconut seeds across the country (Goles, 2003). There are only a limited number of incidences of Coconut Hispine Beetle infestation observed on foxtail

Fig. 8.3 *Wodyetia bifurcata* palms with their prominent orange-red fruit bunch

palm, probably due to lack of economic importance compared to the coconut palm and thus a lack of scholarly attention. In Malaysia, the first incidence of Coconut Hispine Beetle on coconut palm is detected in 2019 (Wan Khairul Anuar et al., 2019). It is believed that the invasion and infestation might have been introduced and have occurred earlier since there was its incidence record in Southern Thailand near the Malaysia border in 2000 (Nakamura et al., 2006). This ravening defoliator feeds on young unopened frond leaflets starting from its larvae stage till adulthood. They nibble off the surface of the leaf in lines that are usually parallel to the midrib. When the front is opened, the nibbled scars may become brown and uneven, leading to scorched leaves when the brown areas curl and wrinkle. The effectiveness of the plant's photosynthetic tissue may have been severely affected when the wide sections of the leaflets break off and leave only partial skeletons. As the leaflets grow, the adults will move to strike younger leaves (CABI, 2019).

Palas Palm, *Licuala Grandis* (Coryphoideae)

Palas Palm (*Licuala grandis*), also called Ruffled Fan Palm or Palas Kipas in Malay, is a medium sized ornamental palm tree (Fig. 8.4). *Licuala grandis* is a minor host of palm-infesting whitefly (*Aleurotrachelus atratus*) and a wild host of red palm mite (*Raoiella indica*) (CABI, 2020a, 2020b, 2020c, 2020d). *Aleurotrachelus atratus* and *R. indica* are common insect pests on coconut palms (*Cocos nucifera*) (Howard et al., 2001). Fortunately, no infestation introduced by these two insects onto Palas Palm has been detected in Malaysia. Only one case has been reported in Sabah. Leaves of Palas Palm at the Forest Research Centre, Sepilok, Sabah were defoliated by green pupae, which were then identified as *Thaumantis klugius* Westwood (Lepidoptera, Satyridae) (Chung, 2008, 2016). It is a butterfly of the family Nymphalidae.

Chinese Fan Palm, *Saribus (Livistona) Chinensis* (Coryphoideae)

Livistona chinensis, also called Chinese Fan Palm or Fountain Palm, is a species of subtropical palm tree in East Asia. It is widely used as ornamental trees as well as in making fans in China (Fig. 8.5). The coconut skipper (*Hidari irava*) is a major pest to coconut palm in Southeast Asia. It also attacks other palm species including *L. chinensis*. In Malaysia, outbreaks of *H. irava* are limited to one generation, apparently controlled by natural enemies (Richards, 1917). *Oryctes rhinoceros* is a borer which can invade a variety of palms, including *L. chinensis*. Meanwhile, *L. chinensis* is also a preferable host to sap feeders such as *Nipaecoccus nipae*, *Ceriantheomorphe brasiliensis* and *Ceroplastes cirripediformis*. *Cerataphis brasiliensis* is an aphid that lives entirely on palms. *N. nipae* infestations are fairly common on mature palms

Fig. 8.4 *Licuala grandis* palms as a landscape tree around the building

of certain species. In Malaysia, *L. chinensis* is a host plant for red palm weevil (*R. ferrugineus*) (Azmi et al., 2017), which is a stem-boring insect pest destructive to a wide range of palm trees (Abbas, 2010).

MacArthur Palm, *Ptychosperma Macarthurii* (Arecoideae)

MacArthur Palm (*Ptychosperma macarthurii*), also called Cluster Palm or Hurricane Palm, is a popular ornamental palm tree in Malaysia (Fig. 8.6). MacArthur palm is native to Australia (Northern Territory and Queensland) and New Guinea (USDA-ARS, 2014). This palm is reportedly resistant to pests and diseases (Gilman & Watson, 1994). It is one of the palm species recommended for landscaping in areas threatened by lethal yellowing owing to its higher resistance against the phytoplasma disease (Howard et al., 2001). However, the infestation of coconut caterpillars (*Brassolis sophorae*) on MacArthur Palm was reported (Howard et al., 2001). This large and brown butterfly has a wide host range of palm species, including MacArthur Palm.

In Malaysia, MacArthur Palm can be severely attacked by bagworms such as *Pteroma pendula*, *Metisa plana*, *Mahasena corbetti* (Fig. 8.7), and *Pagodiella hekmeyeri* (Lee, 2014; Wood, 1968). The bagworm family (Lepidoptera: Psychidae) includes approximately 1000 species, all of which complete larval development

Fig. 8.5 Cluster of *Livistona chinensis* palms by the lake side

Fig. 8.6 Cluster of *Ptychosperma macarthurii* palms planted in front of the building

Fig. 8.7 *Mahasena corbetti* larva which is bigger in size has an irregular-shaped bag compared to *P. pendula*

within a self-enclosing bag (Rhainds et al., 2009). There are about 100 species in Pacific regions (Curran, 1945). At least 10 species of bagworm have been reported as pets in Southeast Asia (Kalshoven, 1981). The larvae of bagworm feed on the upper surface of the leaf, the scraped portion first becomes dried out and then forms a hole. Further damage can be introduced by the removal of leaf pieces to make the case (Hartley, 1967). This leads to discoloration and premature drop of leaves (Ahmad Said, 1983). *Pteroma pendula* was also found infesting some landscape trees in Malaysia such as *Acacia mangium*, *Callerya atropurpurea* and *Cassia fistula* (Lee, 2014). Rusty appearance was commonly spotted on the *P. pendula* infested palm leaves (Lee, 2014). In addition, pupae of bagworms are often observed hanging below the leaves (Fig. 8.8). The damage caused by another bagworm *M. plana* is similar to that caused by *P. pendula* (Fig. 8.9). Resultantly, the infested palm trees lose their aesthetics and shade.

Conclusion

At the end of this chapter, the insect pests that are associated with the five selected ornamental palms and their occurrence in Malaysia are summarized (Table 8.2). Some of the mentioned insects are not reported as pests for these ornamental palms due to various reasons. As mentioned, palms used as ornamental plants are relatively

Fig. 8.8 *Pteroma pendula* pupae hanging under the frond with the rusty appearance caused by this pest

Fig. 8.9 *Metisa plana* pupa (arrow) and larvae (circles) causing damage by scraping the surface of the leaf

resistant to pests and diseases. A prominent number of the insect pests are mainly recorded as pests for the commodity palms like coconut and oil palm. This may be due to their economic importance in comparison with -ornamental palms. And the studies of insect pests on ornamental palms are limited and associated infestations may be under-reported. Gall-makers and root-feeders are not found to be associated with the selected palm species. Nonetheless, a large number of sap-feeders were reported followed by leaf-defoliators and borers. Due to the generalist behaviour of most of these insect pests, the periodical outbreak might occur in some of the selected palms. As a result of climate change and globalization, the introduction of invasive alien species or succession of local pest species to become pests of the ornamental palms potentially therefore should not be neglected.

Table 8.2 List of insect species associated with the selected ornamental palms and their occurrence in Malaysia

Feeding habits	Orders	Species	Record in Malaysia
Leaf-defoliators	Coleoptera	*Brontispa longissimi*	yes
	Lepidoptera	*Brassolis sophorae*	no
		Hidari irava	yes
		Mahasena corbetti	yes
		Metisa plana	yes
		Pagodiella hekmeyeri	yes
		Pteroma pendula	yes
		Thaumantis klugius	yes
Gall-makers		-	
Sap-feeders	Hemiptera	*Aleurotrachelus atratus*	yes
		Cerataphis brasiliensis	yes
		Coccus hesperidum	yes
		Dysmicoccus brevipes	yes
		Fiorinia phantasma	yes
		Xylastodoris luteolus	yes
		Haplaxius crudus	no
		Ceroplastes cirripediformis	no
		Nipaecoccus nipae	no
	Trombidiformes	*Raoiella indica*	yes
Borers	Coleoptera	*Metamasius hemipterus*	yes
		Oryctes rhinoceros	yes
		Rhynchophorus ferrugineus	yes
Root-feeders		-	

References

Abbas, M. S. T. (2010). IPM of the Red Palm Weevil, *Rhynchophorus ferrugineus*. In K. G. Mukerji (Ed.), *Ciancio, Aurelio* (pp. 209–233). Integrated Management of Arthropod Pests and Insect Borne Diseases.

Ahmad Said, S. (1983). Ulat bungkus *Cremastopsyche pendula* Joannis (Lepidoptera: Psychidae) serangga perosak berbagai pokok di taman rekreasi. In Seminar Kebangsaan 1983, Hutan, Taman Negara dan Taman Bandaran untuk Rekreasi, Serdang, Selangor, 26–28 September 1983.

Azmi, A. W., Lian, C. J., Zakeri, H. A., Yusuf, N., Omar, W. B., Wai, Y. K., & Hussain, M. H. (2017). The red palm weevil, *Rhynchophorus ferrugineus*: Current issues and challenges in Malaysia. *Oil Palm Bulletin, 74*, 17–24.

Baranowski, R. M. (1958). Notes on the biology of the royal palm bug, *Xylastodoris luteolus* Barber (Hemiptera, Thaumastocoridae). *Annals of the Entomological Society of America, 51*(6), 547–551.

Ben-Dov, Y. (1993). A systematic catalogue of the soft scale insects of the world (Homoptera: Coccoidea: Coccidae) with data on geographical distribution, host plants, biology and economic importance. Available at https://www.cabi.org/isc/abstract/19951102907

Borowiec, N., Quilici, S., Martin, J., Issimaila, M., & A, Chadhouliati, A.C., Youssoufa, M.A., Beaudoin-Ollivier, L., Delvare, G. & Reynaud, B. (2010). Increasing distribution and damage to palms by the Neotropical whitefly, *Aleurotrachelus atratus* (Hemiptera: Aleyrodidae). *Journal of Applied Entomology, 134*(6), 498–510.

CABI. (2019). *Brontispa longissima* (Coconut Hispine Beetle) Available at https://www.cabi.org/isc/datasheet/10059#F00C2455-0763-48DD-8B47-F3273D80607C.

CABI. (2020a). *Rhynchophorus ferrugineus* (Red palm weevil). Available at www.cabi.org/isc/datasheet/47472.

CABI. (2020b). *Dysmicoccus brevipes* (Pineapple mealybug). Available at www.cabi.org/isc/datasheet/20248.

CABI. (2020c). *Aleurotrachelus atratus* (Palm-Infesting Whitefly). Available at www.cabi.org/isc/datasheet/112108.

CABI. (2020d). *Licuala grandis*. Available at https://www.cabi.org/isc/datasheet/30714.

Christenhusz, M. J. M., & Byng, J. W. (2016). The number of known plant species in the world and its annual increase. *Phytotaxa, 261*(3), 201–217.

Chung, A.Y.C. (2008). Insects associated with some ornamental plants in Sabah, Malaysia. Paper presented at the 7th International Conference on Plant Protection, Kuala Lumpur, 27–29 August, 2008.

Chung, A.Y.C. (2016). An Overview of Insect Pests of Rattans & other Palms in Sabah, Malaysia. Paper presented in Symposium and Training Course on Forest Invasive Pests, Haikou, Hainan, 18–22 October, 2016.

Curran, C. H. (1945). Insects of the Pacific region. *The Infantry Journal, 1*, 317.

DBKL—Senarai pokok palma di sekitar Wilayah Persekutuan Kuala Lumpur (2011–2020).

Dransfield, J., Uhl, N. W., Asmussen, C. B., Baker, W. J., Harley, M. M., & Lewis, C. E. (2008). *Genera palmarum- The evolution and classification of the palms*. Royal Botanic Gardens, Kew, London, UK. https://doi.org/10.34885/92

Edward, F.G. & Dennis, G.W. (1994). *Ptychosperma macarthurii*, Macarthur Palm. Available at https://edis.ifas.ufl.edu/st535

EPPO. (2009). *Mini Data Sheet on Metamasius Hemipterus*. Available at: gd.eppo.int/download/doc/1075_minids_METAHE.pdf.

Evans, G. A. (2007). The Whiteflies (Hemiptera: Aleyrodidae) of The World and Their Host Plants and Natural Enemies. Available at: keys.lucidcentral.org/keys/v3/whitefly/PDF_PwP%20ETC/world-whitefly-catalog-Evans.pdf.

Faleiro, J. R., Jacques, J. A., Carrillo, D., Giblin-Davis, R., Mannion, C.M., Peña-Rojas, E., & Peña, J. E. (2016). Integrated Pest Management (IPM) of Palm Pests In "Integrated Pest Management

in the Tropics". In D. P. Abrol (Ed.), (pp. 439–497, [Part II: Chapter 15]). New India Publishing Agency.
Garcia, J. N. (2011). Fiorinia phantasma Cockerell and Robinson (Hemiptera: Diaspididae), an armored scale pest new to Hawaii. *Proceedings of the Hawaiian Entomological Society, 43*, 59–61.
Gilbin-Davis, R. M., & Peña, J. E. (1993). West Indian sugarcane borer, *Metamasius hemipterus sericeus* (Coleoptera: Curculionidae) An increasing pest problem on field grown ornamental palms. *TropicLine, 6*(4), 1–3.
Gilman, E. F., & Watson, D. G. (1994). *Roystonea spp. Royal palm. Fact Sheet ST-574*, University of Florida. http://hort.ifas.ufl.edu/database/documents/pdf/tree_fact_sheets/roysppa.pdf
Goles, H. G. (2003, May 6–9,). Integrated pest management practices for mango. In Proceedings of 34th PMCP Anniversary and Annual Scientific Conference, Cebu Business Hotel, Cebu City.
Harrison, N. A., & Elliott, M.L. (2015). Lethal yellowing (LY) of palm. Plant Pathology Department, University of Florida (UF)/ Institute of Food and Agricultural Sciences (IFAS) Extension, USA, pp. 222.
Hartley, C. W. S. (1967). Diseases and pests of the oil palm. In Hartley CWA (Ed.), The Oil Palm (*Elaeis guineensis* Jacq.) (pp. 589–590). Longmans, Green and Co. Ltd.
Howard, F. W. (1991). Ecology and control of hemipterous pests of cultivated palms. *American Entomologist, 37*(4), 217–225.
Howard, F. W. (2001). Insect pests of palms and their control. *Pesticide Outlook 12*(6), 240–243. https://doi.org/10.1039/b110547g
Howard, F. W. (1999). An introduction to insect pests of palms. *Acta Horticulturae, 486*, 133–139.
Howard, F. W. (2015). American palm cixiid—*Myndus crudus* Van Duzee. Featured Creatures. University of Florida. Available from http://entnemdept.ufl.edu/creatures/orn/palms/palm_cixiid.htm
Howard, F. W., Giblin-Davis, R., Moore, D., & Abad, R. (2001). *Insects on palms*. CABI publishing.
Hua, L. S., Chen, L., Sajap, A. S., Peng, T., & Ashaari, Z. (2015). Development of Pteroma Pendula Joannis (Lepidoptera: Psychidae) feeding on selected landscape trees In Peninsular Malaysia. *The Malaysian Forester, 78*(1), 87–96.
John, D. V., & Kenneth, A. (2020). A report on the coconut rhinocerous beetle and its menacing life. *Uttar Pradesh Journal of Zoology, 41*(17), 13–21.
Kalshoven, L. G. E. (1981). The pest of crops in Indonesia (p. 701), Ichtiar Baru/Van Hoeve, Jakarta.
Lee, C. Y. (2014). Urban Forest Insect Pests and Their Management in Malaysia. *Formosan Entomologist, 33*, 207–214.
Liebregts, W., & Chapman, K. (2004). Impact and control of the coconut hispine beetle, *Brontispa longissima* Gestro (Coleoptera: Chrysomelidae). Available at http://www.fao org/3/ad522e/ad522e07.htm
Lim, W. H. (1972). Wilting and green spotting of pineapple by the bisexual race of Dysmicoccus brevipes Ckll. in West Malaysia. *Malaysian Pineapple, 2*, 15–21.
Liu, Y. Q., Zhao, C. Y., Yang, L., Zhao, Y. L., & Liu, T. T. (2011). Evaluation of insecticidal activity of podophyllotoxin derivatives against *Brontispa longissima. Pesticide Biochemistry and Physiology, 99*(1), 39–44.
Malumphy, C., Eyre, D., & Anderson, H. (2016). Red Palm Weevil: *Rhynchophorus ferrugineus*. Department for Environment, Food and Rural Affairs.
Martin, J. H., & Lau, C. S. (2011). The Hemiptera-Sternorrhyncha (Insecta) of Hong Kong, China—an annotated inventory citing voucher specimens and published records. *Zootaxa, 2847*(1), 1–122.
Meerow, A. W., Krueger, R. R., Singh, R., Low, E. T., Ithnin, M., & Ooi, L. C. (2012). Coconut, Date, and Oil Palm Genomics. In R. J. Schnell & P. M. Priyadarshan (Eds.), *Genomics of Tree Crops* (pp. 299–351). Springer.
Nakamura, S., Konishi, K., & Takasu, K. (2006, September 18–22). Invasion of the coconut hispine beetle, *Brontispa longissima*: current situation and control measures in Southeast Asia. In *Proceedings of International Workshop on Development of Database (APASD) for Biological Invasion*, Taichung.

Reinert, J. A. (1977). Field biology and control of *Haplaxius crudus* on St. Augustinegrass and Christmas palm. *Journal of Economic Entomology, 70*(1), 54–56.

Rhainds, M., Davis, D. R., & Price, P. W. (2009). Bionomics of bagworms (Lepidoptera : Psychidae). *Annual Review of Entomology, 54*, 209–261.

Richards, R. M. (1917). The diseases and pests of the coconut palm. *Federated Malay States Agricultural Bulletin, 5*, 327–337.

Saw, L. G. (2012). A revision of Licuala (Arecaceae, Coryphoideae) in Borneo. *Kew Bulletin, 67*(4), 577–654.

Selvaraj, K., Sundararaj, R., & Sumalatha, B. V. (2019). Invasion of the palm infesting whitefly, *Aleurotrachelus atratus* Hempel (Hemiptera: Aleyrodidae) in the Oriental region. *Phytoparasitica, 47*(3), 327–332.

Singh, S. P., & Rethinam, P. (2004). Coconut hispine beetle *Brontispa longissima* (Gestro)(Coleoptera: Chrysomelidae). *Cord, 20*(01), 34.

Smith, I. M., McNamara, D. G., Scott, P. R., & Holderness, M. (1997). Palm lethal yellowing phytoplasma. Quarantine Pests for Europe, (pp. 6, 2nd ed.), Wallingford.

Tan, C. L. (2014). Wild about Palms. Available from https://www.thestar.com.my/News/Environment/2014/08/25/Wild-about-palms/

USDA-ARS. (2014). Germplasm Resources Information Network (GRIN). Online Database. Beltsville. National Germplasm Resources Laboratory. Available from https://npgsweb.ars-grin.gov/gringlobal/taxon/taxonomysearch.aspx

Wahizatul, A. A., Zazali, C., Abdul, R., & Nurul'Izzah, A. G. (2013). A new invasive coconut pest in Malaysia: The red palm weevil (Curculionidae: *Rhynchophorus ferrugineus*). *Planter, 89*(1043), 97–110.

Anuar, W. K., W.A., Ahmad Azinuddin, A.R., Nurul Najwa, Z., Mohd Rani. A., et al. (2019). Infestation status of invasive coconut leaf beetle, *Brontispa longissima* (Coleoptera: Chrysomelidae) on coconut palms in three different locations in Malaysia. *Journal of Tropical Agriculture and Food Science, 47*(1), 25–36.

Wood, B. J. (1968). *Pests of oil palms in Malaysia and their control* (p. 204). Incorporated Society of Planters.

Yunus, A., & Ho, T. H. (1980). List of economic pests, host plants, parasites and predators in West Malaysia (1920–1978). *Bulletin Ministry of Agriculture, Malaysia, 153*, 538.

Li Peng Tan is a senior lecturer at the Faculty of Veterinary Medicine, Universiti Malaysia Kelantan (UMK). She obtained her Ph.D. from the Faculty of Forestry and Environment, Universiti Putra Malaysia (UPM). Her research interests include pests and diseases surveillance and biological control of insect pests.

Yew Loong Cheong is an entomologist and research officer at Crop Protection Unit, R&D Musim Mas Plantation, Indonesia. His research interests include oil palm and plant protection.

Samsuddin Ahmad Syazwan is a research officer in the Mycology & Pathology Branch, Forest Biodiversity Division, Forest Research Institute Malaysia. He obtained his MSc from the Faculty of Forestry and Environment, Universiti Putra Malaysia (UPM). His research interests include biological control and also pest and disease.

Wei Chen Lum is a post-doctoral researcher at the Institute for Infrastructure Engineering and Sustainable Management (IIESM), Universiti Teknologi MARA (UiTM), Shah Alam. He obtained his Ph.D. from the Faculty of Forestry and Environment, Universiti Putra Malaysia (UPM). His research interests include forestry and wood science.

Seng Hua Lee is a research fellow in the Laboratory of Biopolymer and Derivatives (BADs), Institute of Tropical Forestry and Forest Products, Universiti Putra Malaysia (UPM). He obtained his Ph.D. from the Faculty of Forestry and Environment, UPM. His research interests include forestry and wood science.

Chapter 9
Tree Vandalism in Malaysia: Criteria for Urban Forest Monitoring

Helmi Hamzah

Abstract Tree vandalism is an unpredictable activity that generally results from a variety of human behaviours or practises, which including planned, spontaneous, or unintentional events that result in tree damage and failure. These failures can be a significant liability that threatens the environment and quality of life. For urban forest managers, liability is a major concern which is strategies in the urban forest monitoring programmes. Unfortunately, the core elements in any urban forest monitoring programmes which can present the subject of tree vandalism incidence are lacking. This chapter focuses on studies related to tree vandalism factors and their significant criteria for urban forest monitoring. The objectives are to understand tree vandalism and its related issues, assessing tree vandalism incidence, and determining tree vandalism criteria for urban forest monitoring. This chapter presented the tree vandalism data from the selected Malaysian local authorities' tree inventory and the experts' consensus related to the tree vandalism subjects. The findings discuss tree vandalism criterion towards improving the urban forest monitoring programme.

Keywords Arborist · Tree damage · Tree care · Tree monitoring · Tree vandalism

Introduction

Transforms are one of the characteristics of urban forests, which are complex and dynamic entities. A variety of biotic and abiotic factors establishes the relative transformations. Age, genetics, water, light, pest and disease concerns, and nutrient availability are just a several of the physiological and external factors to consider. As urban forest transforms and evolves, they need a conducive monitoring effort to ensure the continuity of the ecosystem services they can provide, especially faced with the challenges of urban situations. Monitoring programmes are important for evaluating the structural, health, and hazard status of urban forests. However, until to date, the

H. Hamzah (✉)
Centre of Studies for Landscape Architecture, Department of Built Environment Studies and Technology, Universiti Teknologi MARA (UiTM), Perak Perak Branch, Seri Iskandar Campus,, Seri Iskandar 32610, Malaysia
e-mail: helmi692@uitm.edu.my

S. Maruthaveeran et al. (eds.), *Urban Forestry and Arboriculture in Malaysia*,
https://doi.org/10.1007/978-981-19-5418-4_9

studies that can specifically explain the tree vandalism issues in Malaysia are very limited. Thus, there is a lack of specific approaches for addressing the tree vandalism incidences, that could guide urban forest managers in monitoring programmes.

Vandalism causes mechanical injuries to the tree's structure, which then leads to tree deterioration and eventual tree failure. These unpredictable activities are frequently the result of a variety of human behaviours or practises, including planned, spontaneous, and unplanned events. However, the understanding of vandalism within urban forest monitoring programmes is still vague. Uncertainty in assessing the right attributes in tree vandalism incidents affected the credibility of rectification and improvement recommendations due to inaccurate data during tree care monitoring programmes. Therefore, there is a need to focus on the understanding of urban forest monitoring related to the raising issues of vandalism incidence. It encompasses studies on the tree vandalism factors and their significant criteria for urban forest monitoring programmes. The concern is to understand vandalism activities and their related issues toward urban forests and to identify tree vandalism criteria based on the expert's consensus.

The synopsis of these vandalism issues has been analysed from the tree inventory data from the six Malaysian local authorities as a case study. The dataset in the form of a tree damaged situation extracted from local authorities' tree inventory showed a 27 tree vandalism activities relationship with three tree's structure defects. These findings have subsequently led to a determining a tree vandalism criterion for the urban forest monitoring programme. Consequently, an in-depth understanding of these issues should aid decision-makers and urban forest managers in combating tree vandalism and ensuring the long-term viability of urban forest planting programmes.

Defining the Term "Tree Vandalism"

The tree and the act of vandalism are two ancient themes that deserve a new perspective. Vandalism is a term used to define a kind of misconduct in human behaviour that destroys any public or individual property (Vorobyeva et al., 2015). It is a continual liability for cities that involve stakeholders in terms of cost, time, and energy. Statistics show vandalism causes the state of Selangor, Malaysia to lose RM10 million a year. It involves the cost of repairing various damage to public property such as public toilets, playground equipment, street lighting, and urban trees (Abdul Malek & Mariapan, 2009). Urban trees could render disservices and become a liability due to structural defects affected by the vandalism act. Hence, addressing vandalism incidents is important to ensure the city's liability mitigation and enhance the benefits that should be obtained such as ecosystem services provided by urban forests.

Vandalism in urban forests is widespread and common in most places across the world. (Mullaney et al., 2015). Hong Kong experienced a 10%–15% damaged urban tree related to the vandalism act (Jim, 1987). Furthermore, 30% of urban trees in European cities, and 42% in Eastern Cape, South Africa were affected by the vandalism act (Pauleit et al., 2002; Richardson & Shackleton, 2014). Malaysia's urban areas have been reported to be highly potential for vandalism incidence, and it has an impact on the urban forest as well (Hashim et al., 2012). Although a lack of statistics shows the tree vandalism incidence in Malaysia, some studies reported that tree vandalism incidence defines as a significant factor contributing to the hazardous urban trees in most of the Malaysian local authority areas (Hamzah et al., 2018; Hasan et al., 2018; Sreetheran et al., 2011). This indicates that the vandalism act affects tree performance which deteriorates the safety and security of the urban environment.

Vandalism causes harmful consequences that affect mechanical injuries to a tree's structure such as stem or branch wounds due to nailing, carving, and bark or branch stripping (Richardson & Shackleton, 2014; Travelia & Arifin, 2018). These defects and disorders then become the catalyst for tree damage, potentially increasing tree hazards and failure risks (Moore, 2013). More seriously, vandalism is the magnitude of the damaged tree that can be the extent of urban tree loss. This is to say that the properties of tree vandalism could reduce a comfortable environment, lower tree appraisal value, and increase environmental degradation. These phenomena are tree disservices that affect physical damage to the urban infrastructure, unexpected economic costs and create a sense of fear or inconvenience due to poor health tree conditions (Lyytimäki, 2017; Sreetheran, 2017). Therefore, tree vandalization is defined as an act that could threaten the urban forests toward bringing deterioration effects on the society, economy and environment.

In a depth understanding of the tree vandalism issues, the tree damaged situations extracted from the six Malaysian local authorities, tree inventory data have been analysed. The analysis found that the tree vandalism incidence occurs within the 27 activities (Ahmad, 2010; Hamzah, 2011; Ipoh City Council, 2010; Kuala Kangsar Municipal Council, 2010; Subang Jaya Municipal Council, 2017; Taiping Municipal Council, 2010). There are 11 cases involving the roots, 12 cases involving the stem, and four cases involving the tree canopy (see Table 9.1). These findings indicate that the roots and trunk of the tree are more exposed to a structural defect within various vandalism acts. Hence, addressing the tree vandalism incidence in urban forest monitoring programmes need to be considered these tree vandalism attributes.

Addressing Tree Vandalism Activities

The tree vandalism incident stems from numerous activities. Many social and environmental factors influence it, although the signs and symptoms are often similar (Lu et al., 2010). In many instances, vandalism activities left marks or signs that the trees have been vandalised. The injured tree structure–activity such as carving, nailing, snapping, bark stripping, girdling or ring-barking was determined as the intentional tree vandalism activity (Miller & Miller, 1991; Moore, 2013; Richardson & Shackleton, 2014; Travelia & Arifin, 2018). Besides that, the activities involve inappropriate practices such as tree stem painting and lighting installation classified as unintentional tree vandalism activity (Hernández Zaragoza et al., 2015). On a different note, the carelessness and poor skill work in maintenance practices can also be categorized

Table 9.1 The tree damaged causes by the vandalism act

Tree Defect	Vandalism Act
1. Root zone compaction	1) Carpentry work under the canopy of trees 2) Storage of maintenance equipment beneath the canopy of trees 3) Construction of religious structures in the root zone 4) Constructing a store structure beneath a canopy of trees 5) Surplus earthworks are being dumped beneath the canopy of trees 6) Road premix layering beneath the canopy of trees 7) Constructing concrete constructions beneath the canopy of trees 8) Crusher-run layering beneath the canopy of trees
2. Wounding tree roots	9) At the roots of trees, nails are used to reinforce the structures 10) Water piping work involves root zone excavation 11) Road widening in the root zone
3. Wounding tree stem	12) Garbage burning at the base of a tree 13) Garbage dumping and burning into the cavity of the tree stem 14) At the tree's stem, a signage is attached 15) Ring-barking at tree stem 16) LED light is wrapped around the tree's stem 17) Construction machinery contact with tree stem 18) Slashing tree stem for directions 19) An electric cable is nailed to the tree's stem 20) The commercial space is lighted by a fluorescent lamp attached to the tree stem 21) Contact of a tree stem with a vehicle 22) Attachment of a spotlight to the tree stem 23) Affixing a flag to a tree's stem
4. Tree canopy loss	24) Tree canopy removal 25) Branches that are blocking the walkway are being broken 26) Topping a tree 27) Tree branches that are obstructing road signage are being broken

Source From Local Authorities Tree Inventory (Kajang Municipal Council, Klang Municipal Council, Subang Jaya Municipal Council, Kuala Kangsar Municipal Council, Taiping Municipal Council and Ipoh City Council)

as unintentional tree vandalism activity else well (i.e., wounding on trees by lawn maintenance equipment and improper pruning practices as tree vandalism activities) (Morgenroth et al., 2015). Based on various activities explained above, two types of tree vandalism activities that occur for a deliberate or accidental reasons. Thus, addressing the tree vandalism activities makes sense with understanding the specific factors for deliberate or accidental circumstances in dynamic urban situations. Table 9.2 presents the types of tree vandalism analysed from the tree damaged situations extracted from the six Malaysian local authorities' tree inventory, determined by the Malaysian tree care experts (ISA Certified Arborists).

The term dynamic urban situations synonym with the views of the urban stress thought. Due to risks from numerous deliberate, accidental, and other anthropogenic activities such as construction work, land clearance, and commercial or industrial activities, urban forests are constrained in their ability to grow and perform (Osakabe

Table 9.2 Types of tree vandalism determinations

Types of tree Vandalism	Vandalism Act
1. Intentional	1) Ring-barking at tree stem 2) Garbage dumping and burning into the cavity of the tree stem 3) Tree canopy removal 4) Garbage burning at the base of a tree 5) Slashing tree stem for directions 6) At the tree's stem, a signage is attached 7) An electric cable is nailed to the tree's stem 8) LED light is wrapped around the tree's stem 9) Carpentry work under the canopy of trees 10) The commercial space is lighted by a fluorescent lamp attached to the tree stem 11) Construction machinery contact with tree stem 12) Affixing a flag to a tree's stem 13) Construction of religious structures in the root zone 14) Topping a tree 15) Road premix layering beneath the canopy of trees 16) Contact of a tree stem with a vehicle 17) Water piping work involves root zone excavation 18) Attachment of a spotlight to the tree stem 19) Tree branches that are obstructing road signage are being broken 20) Road widening in the root zone 21) Branches that are blocking the walkway are being broken
2. Unintentional	1) Surplus earthworks are being dumped beneath the canopy of trees 2) Constructing a store structure beneath a canopy of trees 3) At the roots of trees, nails are used to reinforce the structures 4) Storage of maintenance equipment beneath the canopy of trees 5) Crusher-run layering beneath the canopy of trees 6) Constructing concrete constructions beneath the canopy of trees

Source From Local Authorities Tree Inventory (Kajang Municipal Council, Klang Municipal Council, Subang Jaya Municipal Council, Kuala Kangsar Municipal Council, Taiping Municipal Council and Ipoh City Council)

et al., 2012; Vogt et al., 2015). These are the factors that influence unintended tree vandalism incidents due to the conflict interaction between societal reactions and environmental stressors. Furthermore, these have an impact on 'human misconduct' (e.g., bad work skills or approaches, and a lack of knowledge about urban forests), which leads to intentional tree vandalism elsewhere. As a result of the dynamic nature of urban characters, urban woods are susceptible to urban stress, resulting in intentional or unintentional vandalism incidents.

Human misconduct resulted in a widespread occurrence of vandalised urban forests throughout urban areas. Individual differences in attitudes about the urban forest are key human misconduct factors in the tree vandalism act. Pessimistic attitudes, such as boredom, a passion for crime, and a lack of appreciation for trees, contribute to a higher prevalence of tree vandalism than optimistic ones (Long & Burke, 2015; Richardson & Shackleton, 2014). Without a doubt, the character of vandals, which includes their attitudes, knowledge, and awareness, has an impact on the essential factors of intentional tree vandalism incidents.

Poor tree conditions and disservice impacts on urban forest populations are exacerbated by urban stressors, affecting public preferences (Delshammar, Östberg & Öxell, 2015; Lyytimäki, 2017). People are logical in their rejection of troublesome trees that cause them problems in their living environments. This is due to the fear of risk or feeling uncomfortable with poor tree conditions (e.g. defects or hazardous trees, overgrown trees, or poisonous species) that the public felt about their safety inflicted by that trees (Abd Hamzah et al., 2017; Kadir & Othman, 2012). Furthermore, the conflict between urban trees and overhead power wires, as well as aggressive tree roots damaging sidewalks or paved areas, are all contributing to the inconvenience of city residents (Abd Kadir & Othman, 2012). Those were among the factors contributing to the inconvenience situations that could invite to vandalism.

In summary, tree vandalism is caused by three motivating factors that can be divided into three categories: human, environmental, and tree condition variables. Human misconduct factors are a human category, and the tree condition category refers to the characteristics of urban tree factors. Whereas the environmental category refers to anthropogenic factors. Hence, referring to those tree vandalism factors which can explore the tree vandalism criterion is of utmost importance could embrace the subjects of tree vandalism incidence for urban forest monitoring programmes. Figure 9.1 presents the urban tree vandalism incidence related to their motivating factors.

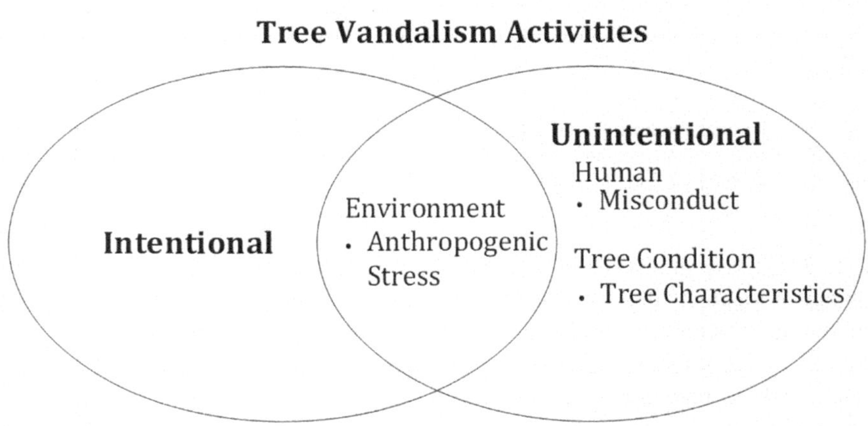

Fig. 9.1 The factors for tree vandalism activities

The Tree Vandalism Criteria

The tree monitoring programme is essential for presenting urban forest performance statuses such as health and hazard conditions. This entails assessing quantitative or qualitative features such as the number of trees, their structure, and their condition on a regular basis in order to observe and describe changes in urban forest populations (Morgenroth et al., 2015). Thus, understanding the particular criteria for tree vandalism could help decision-makers and urban forest managers develop effective tree care monitoring programmes.

Thirty-two criteria for tree vandalism incidents were defined under the three categories based on a consensus among Malaysian tree care specialists (certified arborist); tree conditions (10 criteria), human misconduct (11 criteria), and anthropogenic stress (11 criteria). The 10 criteria in the tree condition category, in particular, demonstrate poor tree condition as a triggering factor for tree vandalism incidents. An example of the triggering factor is the hazardous or failure history of the tree that leads to the vandalization or rejection of the tree (Camacho-Cervantes et al., 2014). The eleven criteria in the human misconduct category represent negative personal attitudes or unappropriated practices that influence tree vandalism incidents; for example, tree damage caused by vandalism due to a lack of knowledge about the benefits of trees or poor maintenance work, such as improper pruning technique (Fickri & Siregar, 2018). The eleven criteria in the anthropogenic stress category, on the other hand, highlight urban stressors as causes of tree vandalism, such as tree damage caused by vandalism acts in conflict with urban infrastructure or utility elements (Gwedla & Shackleton, 2015). The tree vandalism criteria that should be considered in the urban forest monitoring programme are detailed in Table 9.3.

The criteria determined were found to be an appropriate and feasible source of information that guided the care of urban forest monitoring practices, especially in the Malaysian context. These tree vandalism attributes may help to improve the Malaysian local authorities' current tree inventory system. Rather than focusing on physical attributes for monitoring urban forest performance, such as mortality status and hazard rating, the social attributes were important in presenting human misconduct and urban stressors related to tree performance, such as poor maintenance or coordination approaches, and other well-known design failures.

Conclusion

To conclude, this chapter established three outcomes in relation to the chapter's objectives. (1) addressing the tree vandalism incidence for urban forest monitoring programmes need to be considered on the specific 27 tree vandalism act that affected the structure of the tree's roots, stem, and canopy. (2) the chapter addressed tree vandalism incidence stems from the three motivating factors which are human misconduct, environmental stressors, and tree's characteristic factors. (3) the chapter

Table 9.3 The tree vandalism criteria for urban forest monitoring programmes

Tree conditions	Human misconduct	Anthropogenic stress
1. Species of tree	1. Religious and cultural beliefs	1. Tree for structure attachment
2. Age of tree	2. Level of knowledge	2. Trees cause interference or obstruction
3. Size of tree	3. Socioeconomic status	3. Interactions with other activities
4. Location of tree	4. Rule and regulations	4. Prioritization of space utilisation
5. The owner of tree	5. Layout and design	5. Repurposing tree parts
6. Tree characteristic	6. Methods for tree care	6. No protection structure for the tree
7. Tree health condition	7. Tree maintenance status	7. Infrastructure improvement and expansion/urbanisation and development
8. Tree growth rates	8. Collaboration and coordination	8. Occasion and event
9. Tree debris	9. Demographic (age)	9. Human population rate
10. Tree value	10. Monitoring of tree	10. Display of memories
	11. Tree-related information	11. A tree serves as a protective structure

Source The author

has determined 32 tree vandalism criteria within three categories that could guide the decision-maker and urban forest managers in urban forest monitoring programmes.

References

Abd Kadir, M. A., & Othman, N. (2012). Towards a better tomorrow: Street trees and their values in urban areas. *Procedia—Social and Behavioral Sciences, 35*, 267–274. https://doi.org/10.1016/j.sbspro.2012.02.088

Abdul Malek, N., & Mariapan, M. (2009). Visitors' perception on vandalism and safety issues in a Malaysian urban park. *Theoretical and Empirical Researches in Urban Management, 4*(13), 93–107. Retrieved from http://www.um.ase.ro/No13/6.pdf

Ahmad, T. (2010). *Tree inventory and tree risk assessment for Klang Municipal Council*. Klang.

Camacho-Cervantes, M., Schondube, J. E., Castillo, A., & MacGregor-Fors, I. (2014). How do people perceive urban trees: Assessing likes and dislikes in relation to the trees of a city. *Urban Ecosystems, 17*, 761–773. https://doi.org/10.1007/s11252-014-0343-6

Council, S. J. M. (2017). *Tree inventory & tree risk assessment*. Subang Jaya.

Delshammar, T., Östberg, J., & Öxell, C. (2015). Urban trees and ecosystem disservices—A pilot study using complaints records from three Swedish cities. *Arboriculture & Urban Forestry, 41*(4), 187–193. Retrieved from https://www.researchgate.net/publication/281993776_Urban_trees_and_ecosystem_disservices_-_A_pilot_study_using_complaints_records_from_three_Swedish_cities

Fickri, S., & Siregar, I. Z. (2018). Early identification of the quality of tree pruning cuts in a Bogor campus area. *IOP Conference Series: Earth and Environmental Science, 203*, 1–7. https://doi.org/10.1088/1755-1315/203/1/012024

Gwedla, N., & Shackleton, C. M. (2015). The development visions and attitudes towards urban forestry of officials responsible for greening in South African towns. *Land Use Policy, 42*, 17–26. https://doi.org/10.1016/j.landusepol.2014.07.004

Hamzah, H. (2011). *Tree inventory and tree risk assessment for Kajang Municipal Council*. Kajang.

Hamzah, H., Othman, N., & Mohd Hussain, N. H. (2017). Tree removal application by urban dwellers: A case study of Kajang local authority. In C. Dewi & E. N. Rauzi (Eds.), *International conference on architecture 2017 (ICRP-AVAN), Unsyiah (Banda Aceh) and UiTM (Perak)* (pp. 1–5). Banda Aceh, Jurusan Arsitektur Fakultas Teknik Universitas Syiah Kuala. Retrieved from http://icrp-avan.unsyiah.ac.id/open/Cover&Contents.pdf

Hamzah, H., Othman, N., Mohd Hussain, N. H., & Simis, M. (2018). The criteria of urban trees regarding the issues of tree vandalism. *IOP Conference Series: Earth and Environmental Science, 203*, 1–6. https://doi.org/10.1088/1755-1315/203/1/012023

Hasan, R., Othman, N., & Ismail, F. (2018). Social factors influencing hazardous street trees at selected city council in Malaysia. *Asian Journal of Quality of Life, 2*(5), 77–85. https://doi.org/10.21834/ajqol.v3i12.144

Hashim, A. E., Samikon, S. A., Mat Nasir, N., & Ismail, N. (2012). Assessing factors influencing performance of Malaysian low-cost public housing in sustainable environment. *Procedia - Social and Behavioral Sciences, 50*, 920–927. https://doi.org/10.1016/j.sbspro.2012.08.093

Hernández Zaragoza, A. Y., Cetina Alcalá, V. M., López López, M. Á., Hilú, A. C., De La Isla De Bauer, M. D. L., Rosales, D. A., & Rosas, H. G. (2015). Identification of tree damages of three parks of Mexico City. *Mexican Journal of Forest Sciences, 6*(32), 63–82. Retrieved from https://pdfs.semanticscholar.org/b593/36ec05b743e3efa45226a302f9c068c6e0ab.pdf

Ipoh City Council. (2010). *Tree inventory & tree risk assessment*. Ipoh.

Jim, C. Y. (1987). The status and prospects of urban trees in Hong Kong. *Landscape and Urban Planning, 14*, 1–20. https://doi.org/10.1016/0169-2046(87)90002-8

Council, K. K. M. (2010). *Tree inventory & tree risk assessment*. Kuala Kangsar

Long, M., & Burke, R. H. (2015). *Vandalism and anti-social behaviour*. In P. Davies, P. Francis, & T. Wyatt (Eds.), *Critical criminological perspectives*. Palgrave Macmillan. https://doi.org/10.1057/9781137519269

Lu, J. W. T., Svendsen, E. S., Campbell, L. K., Greenfeld, J., Branden, J., King, K. L., & Falxa Raymond, N. (2010). Biological, social, and urban design factors affecting young street tree mortality in New York City. *Cities and the Environment, 3*(1). Retrieved from http://digitalcommons.lmu.edu/cate/vol3/iss1/5/

Lyytimäki, J. (2017). Disservices of urban trees. In F. Francesco, C. C. Van Den Bosch, & F. Alessio (Eds.), *Routledge Handbook of Urban forestry* (pp. 164–176). Routledge. https://doi.org/10.4324/9781315627106

Miller, R. H., & Miller, R. W. (1991). Planting survival of selected street tree taxa. *Journal of Arboriculture, 17*(7), 185–191. https://doi.org/10.1016/0006-3207(92)90963-N

Moore, G. M. (2013). Ring-barking and girdling : How much vascular connection do you need between roots and crown. In N. Wojcik (Ed.), *The 14th National street tree Symposium* (pp. 87–96). Adelaide, University of Melbourne. Retrieved from https://pdfs.semanticscholar.org/21d8/3db275574be4aedb9805071741eddf1afb6a.pdf?_ga=2.146494001.481915270.1570351625-2133761217.1557440011

Morgenroth, J., Santos, B., & Cadwallader, B. (2015). Conflicts between landscape trees and lawn maintenance equipment - The first look at an urban epidemic. *Urban Forestry and Urban Greening, 14*, 1054–1058. https://doi.org/10.1016/j.ufug.2015.10.002

Mullaney, J., Lucke, T., & Trueman, S. J. (2015). A review of benefits and challenges in growing street trees in paved urban environments. *Landscape and Urban Planning, 134*, 157–166. https://doi.org/10.1016/j.landurbplan.2014.10.013

Osakabe, Y., Kawaoka, A., Nishikubo, N., & Osakabe, K. (2012). Responses to environmental stresses in woody plants: Key to survive and longevity. *Journal of Plant Research, 125*, 1–10. https://doi.org/10.1007/s10265-011-0446-6

Pauleit, S., Jones, N., Garcia-Martin, G., Garcia-Valdecantos, J. L., Rivière, L. M., Vidal-Beaudet, L., & Randrup, T. B. (2002). Tree establishment practice in towns and cities—Results from a European survey. *Urban Forestry and Urban Greening, 1*, 83–96. https://doi.org/10.1078/1618-8667-00009

Richardson, E., & Shackleton, C. M. (2014). The extent and perceptions of vandalism as a cause of street tree damage in small towns in the Eastern Cape, South Africa. *Urban Forestry and Urban Greening, 13*, 425–432. https://doi.org/10.1016/j.ufug.2014.04.003

Sreetheran, M., Adnan, M., & Khairil Azuar, A. K. (2011). Street tree inventory and tree risk assessment of selected major roads in Kuala Lumpur. *Arboriculture and Urban Forestry, 37*(5), 226–235. Retrieved from https://www.researchgate.net/publication/236270945_Street_Tree_Inventory_and_Tree_Risk_Assessment_of_Selected_Major_Roads_in_Kuala_Lumpur_Malaysia

Sreetheran, M. (2017). Exploring the urban park use, preference and behaviours among the residents of Kuala Lumpur. *Malaysia. Urban Forestry and Urban Greening, 25*(November), 85–93. https://doi.org/10.1016/j.ufug.2017.05.003

Taiping Municipal Council. (2010). *Tree inventory & tree risk assessment*. Taiping.

Travelia, I., & Arifin, H. S. (2018). Tree management at Boulevard landscape of Alam Sutera Serpong, a new city area in Tangerang. *IOP Conference Series: Earth and Environmental Science, 203*, 1–9. https://doi.org/10.1088/1755-1315/203/1/012022

Vogt, J. M., Watkins, S. L., Mincey, S. K., Patterson, M. S., & Fischer, B. C. (2015). Explaining planted-tree survival and growth in urban neighborhoods: A social-ecological approach to studying recently-planted trees in Indianapolis. *Landscape and Urban Planning, 136*, 130–143. https://doi.org/10.1016/j.landurbplan.2014.11.021

Vorobyeva, I. V, Kruzhkova, O. V, & Krivoshchekova, M. S. (2015). The genesis of vandalism: From childhood to adolescence. *Psychology in Russia: State of the Art*, *8*(1), 139–156. https://doi.org/10.11621/pir.2015.0112

Helmi Hamzah is a lecturer at the Centre of Studies for Landscape Architecture, Department of Built Environment Studies and Technology, Faculty of Architecture, Planning and Surveying, Universiti Teknologi MARA (UiTM) Perak Branch, and has been lecturing at UiTM since 2015. He holds a Diploma in Agriculture from Universiti Pertanian Malaysia (UPM), Bachelor of Landscape Architecture (Hons) from Universiti Teknologi Malaysia (UTM) and Master of Landscape Architecture from the Universiti Putra Malaysia (UPM). He obtained his Ph.D. in Design and Built Environment from UiTM. Helmi Hamzah is an active researcher in the subject of environmental behaviour studies, having presented papers at several conferences and contributing to a number of peer-reviewed journals. He has over 20 years of experience in landscape management and maintenance with several Malaysian local authorities. He is a certified arborist (CA) from the International Society of Arboriculture (ISA) and a corporate member of the Institute of Landscape Architecture Malaysia (ILAM).

Chapter 10
Urban Forestry for Human Health and Well-being in the Tropics

Nor Akmar Abdul Aziz

Abstract The state of peoples' physiology, mental health and well-being are affected when people actively use the forest or urban greenspaces. Studies have been conducted to prove that trees and forests are beneficial for reducing stress, decreasing depression, fighting against obesity and improving physical health. With the continuous usage of forest and urban greenspaces, from a small scale escalating to a big one, passive or active, all will gain the benefits. Accessibility, motivation and a good sense of time consumption to use the forest and urban greenspace wisely is also a chance for people to enhance their immune functions, increase physical activity and social connectedness. Many studies have used a few types of questionnaires, to see the effect of greening areas on human health. Apart from that, applied science measurements are also used in some studies to gain further scientific evidence. Results indicate that people living near and surrounding green spaces showed more health, energy, and happiness. Thus, results also show urban greening has positive benefits physiologically and psychologically compared to the urban environment.

Keywords Affected · Community · Physiology · Psychology · Wellbeing

Benefits of Urban Forest

The benefits of urban forests have been known by people for centuries, and the awareness of the importance of urban forests has increased among the public. The importance of urban forest and trees include social, environmental, economic, as well as physical and psychological (mental) health. In terms of social advantage, urban greening provides a framework for improved community interaction and social activities (Braubach et al., 2017). Public urban green space has also been shown to facilitate social networking and promote social integration among children and youth (Ward Thompson et al., 2016). Quantity and quality of green spaces have been associated with improved social cohesion at the neighbourhood level (de Vries

N. A. Abdul Aziz (✉)
Department of Recreation and Ecotourism, Faculty of Forestry and Environment, Universiti Putra Malaysia (UPM), Serdang, Malaysia
e-mail: norakmar@upm.edu.my

S. Maruthaveeran et al. (eds.), *Urban Forestry and Arboriculture in Malaysia*,
https://doi.org/10.1007/978-981-19-5418-4_10

et al., 2013). Aziz et al. (2018) also found that urban greening makes family and friends spend their time together. Acknowledgement of the benefits of urban forestry for the environment cannot be denied as cleaning the air, cooling the temperature on hot days, conserving heat at night, offering shade, offer watershed protection and preventing soil erosion have been revealed by some researchers (Hartig et al., 2014; Nowak & Greenfield, 2018; Préndez et al., 2019; Zelenáková et al., 2017). Urban greening is also a valuable environmental resource that is often associated with positive impacts on quality of life and property values in many countries, such as the United Kingdom, China, Hong Kong, Malaysia, and also the United States (Chen & Jim, 2008; Sander et al., 2010; Tan, 2011; McCord et al., 2014). Apart from that, trees and landscapes also support tourism (Nesbitt et al., 2017). Aesthetic value in urban greening and landscape plays an important role in influencing visitation to places (Othman et al., 2015).

Health Benefits of Urban Forests

Urban forests are extensive and have numerous tree species and vegetation. Spending time in urban woodlands is increasingly recognised as important for the individual and social health and well-being of children, young people, adults and older people (Fig. 10.1). Since the early 1980s, studies in environmental psychology have focused on the health effects of viewing and spending time in natural landscapes. For many people, solitude and tranquilly are the most important qualities of urban forests, especially in terms of recovery from stress and attention fatigue.

Contact with nature, such as in urban woodlands, can have a positive impact on our psychological, physiological and physical wellbeing. Throughout history, humans have lived near a green environment. City dwellers have tried to combat stress by finding other perspectives on city life and looking for a way to relieve their stress. The suitability of nature for stress relief through pleasure exercises has yet to be deductively and thoroughly evaluated. Many researchers have found that interaction with urban forests, nature, or green spaces positively affects mental health, reduces stress, improves the quality of life, and is positively associated with health outcomes. (Altavilla et al., 2018; Mokhtar, Abdul Aziz & Mariapan, 2018; Aziz et al., 2021; Kabisch et al., 2021; Raman et al., 2021).

Many studies from different parts of the world show that green spaces provide tremendous psychological and physical benefits (Harris et al., 2018; Simkin et al., 2020; Tsai et al., 2018; Wood et al., 2018). According to Barton and Rogerson (2017) from the UK, even simple contact with the natural environment can trigger the process of psychological restoration, which positively affects individuals' feelings and ability to reflect on life problems. On the other hand, green spaces in school landscapes also have an impact on students' academic performance. In Li and Sullivan's (2016) study, students are able to concentrate better when they are in a classroom with a green view. In addition, students' test scores improve when they are in a natural environment (Matsuoka, 2010).

Fig. 10.1 Urban parks and forests offer a variety of activities for people's health and wellbeing

In the city of Belgrade, Vujcic and Tomicevic-Dublijevic (2018) found that forest environments have the potential to support restorative experiences and daily habits in a younger population. Furthermore, in southern Sweden, a rich diversity of green local environments can promote adult well-being (Weimann et al., 2019). A study by Rathmann et al. (2020) in Augsburg, Germany, mentioned that green structures can also have positive effects on human physiological parameters (e.g., lowering heart rate). Moreover, reforestation in New York City can surprisingly improve the health of young children (Jones & Goodkind, 2019).

In the Asian region, only countries such as Korea, China, and Japan are actively exploring the relationship between human health and well-being with forests and nature. Shinrin-yoku or 'forest bathing' is popular in Japan and was developed in the 1980s. A study by Zhou et al. (2019) showed that urban forests can alleviate anxiety among university students in Guiyang City. In addition to the physical structure of urban green spaces, the relationship of urban green spaces within a community is an important component to improve the mental health of older urban residents (Lee & Lee, 2019). Nowadays, not only forests can make people healthy, but also cities can become healthier and fairer for people by having an urban forest or nature in their living space, such as urban gardens and green roofs.

Issues and Challenges

In 2018, about 55% of the world's population will live in urban areas, and by 2050, this number will increase to 68%, with 54% of the world's urban population living in Asia in 2018. In Malaysia, 75% of the Malaysian population is expected to live in an urban area by 2020 (United Nations Population Division, 2018). Life in the twenty-first century is more complex than in previous years. Sensational problems in our country are related to lifestyle, work and study environment which contributes to the unhealthy lifestyle not only for Malaysians but the whole world. Many Malaysians suffered from mental depression and anxiety and this number is expected to increase further due to the increasing challenges in life, 38,815 in 2014 and up to 38,956 in 2015 (Ministry of Health, 2017).

In addition, those living in urbanized areas such as Kuala Lumpur, Selangor, and Sabah tend to be highly stressed (see Fig. 10.2). Traffic congestion, high cost of living, relationship problems and pressure at work are among the causes of anxiety, stress and depression. In Malaysia, 1 in 5 people are depressed (17.7% women, 18.9% men), 2 in 5 people are anxious (42.3% women, 37.1% men) and 1 in 10 people are stressed (10.3% women, 8.9% men). In terms of race, Indians have the highest percentage of stressed people (15%), Bumiputera Sabah and Indians have the highest percentage of anxious people (47%) and Indians also have the highest percentage of depressed people (33%) (see Table 10.1).

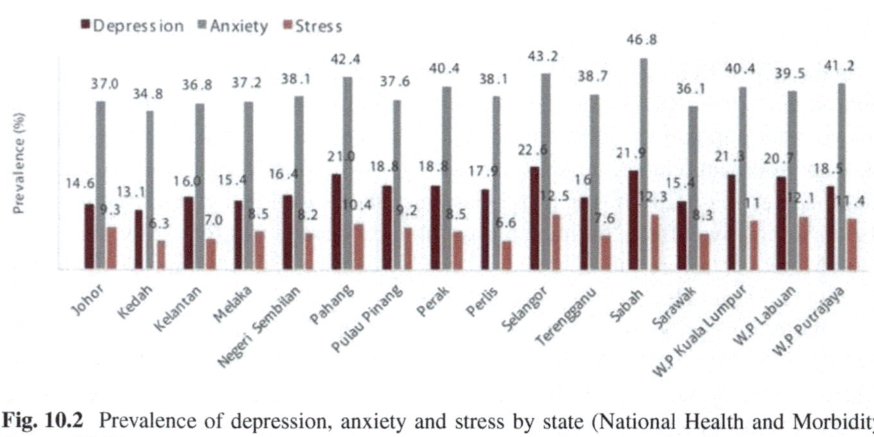

Fig. 10.2 Prevalence of depression, anxiety and stress by state (National Health and Morbidity Survey, 2017)

Table 10.1 Distribution of stress, anxiety and depression among ethnics in percentage (%)

Ethnics	Percentage (%)		
	Stress	Depression	Anxiety
Malay	9	16	39
Chinese	10	21	36
Indian	23	33	47
Bumiputera Sabah	13	22	47
Bumiputera Sarawak	10	17	40
Others	11	22	44

Source Adolescent Mental Health (DASS-21) (National Health and Morbidity Survey, 2017)

Psychological and Physiological Health of Urban Forest in Malaysia

Recently, with the help of advanced technologies, many methods of study and experiments can be done for this study through social sciences and technical fields. The studies conducted by self-reported health, the Profile of Mood States (POMS) questionnaire, and measurement of a general health questionnaire (GHQ) are common studies in social sciences. For applied science, experiments like taking biomarkers such as saliva cortisol, blood and urine, using electroencephalograms (EEG) and blood pressure (BP) are among methods to prove the role of the urban environment in promoting lifelong health and wellbeing. However, the cost, the ethical approval involving human subjects and the hygiene to conduct the sample, are among the restrictions in this research in Malaysia. Also, the combinations of social sciences and technical research have a disagreement.

There is growing evidence and proof of a positive relationship between the amount of green space in the home environment and people's health and well-being. In Malaysia, there are a growing number of studies on urban forests that are associated with human health and well-being (Mohd Saad et al., 2019; Mokhtar et al., 2018; Raman et al., 2021). Trees in streets, squares, city squares or piazzas can actually be one of the health promotion measures to combat non-communicable diseases (NCDs) and promote healthy lifestyles among citizens. Recreational and physical activities such as sitting, walking and jogging can help improve people's health and positively influence wellbeing. In addition, various green infrastructures have helped urban communities to trigger relaxation and well-being.

Some case studies in Kuala Lumpur, e.g. by Abdul Aziz et al. (2012) using self-report health questionnaires, show that visiting urban green spaces can lead to positive and better mood change, while people living near urban parks are reported to be in good health. In addition, a study on visitors' awareness of the importance of the pocket park and the elements of the restorative environment provided by the pocket park in Laman Standard Chartered, Kuala Lumpur, found that the green space provides restorative experiences, using qualitative and quantitative methods (Hashim et al.,

2019). Next, Nath et al. (2018) found that physical and recreational activities in parks have a positive impact on stress reduction, mindfulness, physical fitness, body weight maintenance, body fat burning, and overall well-being.

Mokhtar et al. (2018), conducted a comparison of walking activity between urban parks and city centres in Kuala Lumpur using salivary cortisol concentration, Profile of Mood States (POMS), and Positive and Negative Affect Schedule (PANAS), found that the results indicated that salivary cortisol concentration significantly increased in the city, while no significant change was found in the city park. Diastolic blood pressure decreased significantly after walking in the city park. As for psychological responses, Total Mood Disturbance (TMD) was significantly lower among participants in the city park than in the city. The PANAS (Positive and Negative Affect Schedule) showed that positive affect significantly increased after a walk in the city park, while participants' positive effect significantly decreased after a walk in the city. These results suggest that urban green spaces have positive psychological effects compared to urban environments.

Moreover, a study in Penang using the General Health Questionnaire (GHQ-12) in Penang Youth Park among 400 people who engage in physical activities such as jogging in the urban park was able to reduce their psychological stress levels (Mohd Saad et al., 2019). This proves that urban forest supports more outdoor activities and influences healthy behaviours. Rajoo et al. (2019) conducted a multidisciplinary study to develop a forest therapy program. It showed that systolic (SBP) and diastolic (DBP) blood pressure were reduced, confirming that a forest therapy programme can reduce stress levels.

It can be explained that the frequency of contact with nature or urban forestry with the scale; small, spatial or large, will have an impact on people. More frequent people in contact nature, no matter the size scale; small or large, people will feel better in physical or emotional condition directly or indirectly (Fig. 10.3). These activities need to keep going or be a routine and be part of people's life to make sure they are kept in fitness, good shape and in a better health condition.

The key to reducing stress and improving the state of health is to be involved in positive activities that includes greeneries, even in the small scale such as gardening, or large scale such as wilderness adventure, or passive activities such as window viewing of nature, or up to active actions such as jogging, hiking and cycling (Fig. 10.4). Other than that, a resident that is surrounded by trees and plants can make an individual feel fresh and free from stress. People who live near green trees and plants can also receive the same benefits. This gives them the chance and opportunity to have a walk, jogging and cycling in an improved environment which helps improve their physical and as well as mental health.

It is an undeniable fact that trees help in improving human health; mentally or physically. The increasing awareness among Malaysians in raising their state of health, makes them use urban forestry as a practising place for healthy activities. There are no specific urban forestry designs and species of trees that could help in determining the reduction of stress, anxiety or depression or improving human

Fig. 10.3 Students sitting, viewing and experiencing nature during the therapy program

health physically. Beautification or high aesthetics may not be necessary for a stress-reducing effect but could be helpful in attracting people to the green space and encouraging them to stay longer.

Different densities or qualities of vegetation or greens may lead to different preferences and in addition conditions such as safety, cleanliness and maintenance are also very important factors in order to attract and motivate people to attach to that place. Besides that, it is also important for the green areas to have a quiet, calm and peaceful environment to help increase the stress releasing effect; most people will tend to choose a forest to experience this.

Green spaces encourage exercise and provide a better restorative environment compared with indoor exercise. Various characteristics of urban green space can provide different types of enjoyment and recreation activity. Condition and diversity will also lead to urban green spaces with attractive amenities. Different activities

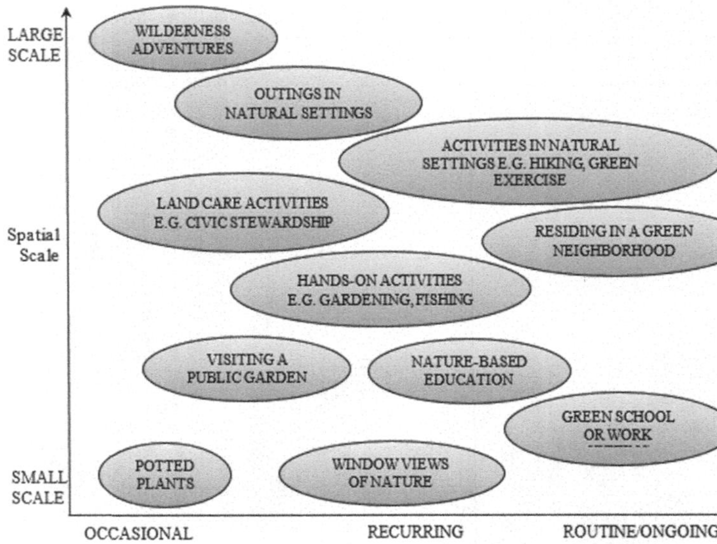

Fig. 10.4 A spectrum of forms of nature contact (Frumkin et al., 2017)

also make the reduction of depression and improvement of physical health vary. For example, jogging is more effective in improving our fitness compared to walking. Meanwhile, sitting and looking at a water fountain helps with relieving stress rather than being in a packed and crowded place. This proves that the diversity of activities done influences the improvement of our health physically, mentally and emotionally.

The frequency and consistency in the use of urban greenspaces also make people feel happy, energetic and lively. However, due to the people of Malaysia living a busy lifestyle, most people get to go to these places only during the weekend or public holidays. Forest or urban greenspaces provide the setting for many social activities, this will influence an individual or a group to visit these places for the purpose of meditation, having fun or even sight-seeing. So, indirectly they are able to relieve stress both emotionally and mentally. At the same time, they are also able to improve their physical health and increase their level of fitness.

Currently, most schools have started gardening as an activity that helps with therapy for the special education needs (SEN). This is proof that when they perform this activity, being surrounded by green trees helps them to relax and stabilize them emotionally. Yet, research shows that children who are exposed to green trees and plants from an early age have a brain that functions more actively compared to other children. Furthermore, there is some evidence that using green space for physical activity will reduce obesity among children. On the other hand, green areas in school landscapes also have an impact on students' academic performance.

Conclusion

In this chapter we have looked at the effects of urban forests on human health. There are many ways and opportunities for people to improve and maintain their health, whether psychological or physiological. The challenge is to make urban green spaces accessible to all people, even when viewing them, regardless of where they are located, live or work. The urban forest should be part of citizens' lives, regardless of gender, age group and different ethnicities or religions.

Future research should examine the relationships between urban greening and psychological and physiological measures, as well as their links to other ecological and biological features of green spaces such as sound, landscape, and air quality, and the effects of exposure to nature on different groups of people (e.g., young people, seniors, and those suffering from sleep-deprived people), apart from the long-term effects of nature. In addition, the use of biomarkers such as hair, blood, and salivary cortisol, as well as electroencephalograms (EEGs) and blood pressure, may provide more definitive cognitive control.

References

Abdul Aziz, N. A., Konijnendijk, C. C., Stigsdotter, U. K., & Nilson, K. (2012). Malaysian case studies on the relation between the use of green space and health promotion. *ALAM CIPTA, International Journal of Sustainable Tropical Design Research and Practice*, *5*(1).

Altavilla, G., D'Elia, F., & Raiola, G. (2018). A brief review of the effects of physical activity in subjects with cardiovascular disease: An interpretative key. *Sport Mont, 16*, 103–106.

Aziz, N. A. A., Shian, L. Y., Mokhtar, M. D. M., Raman, T. L., Saikim, F. H., & Nordin, N. M. (2021). Effectiveness of urban green space on undergraduates' stress relief in tropical city: A field experiment in Kuala Lumpur. *Urban Forestry and Urban Greening*, 127236.

Aziz, N. A. A., van den Bosch, K., & Nillson, K. (2018). Recreational use of urban green space in Malaysian cities. *International Journal of Business and Society, 19*.

Barton, J., & Rogerson, M. (2017). The importance of greenspace for mental health. *Bjpsych International, 14*(4), 79–81.

Braubach, M., Egorov, A., Mudu, P., Wolf, T., Thompson, C. W., & Martuzzi, M. (2017). Effects of urban green space on environmental health, equity and resilience. In *Nature-based solutions to climate change adaptation in urban areas* (pp. 187–205). Springer.

De Vries, S., Van Dillen, S. M. E., Groenewegen, P. P., & Spreeuwenberg, P. (2013). Streetscape greenery and health: Stress, social cohesion and physical activity as mediators. *Social Science and Medicine, 94*, 26–33.

Frumkin, H., Bratman, G. N., Breslow, S. J., Cochran, B., Kahn, P. H., Jr., Lawler, J. J., Levin, P. S., Tandon, P. S., Varanasi, U., Wolf, K. L., & Wood, S. A. (2017). Nature contact and human health environment. A research agenda. *Environmental Help Perspective, 125*(7), 075001. https://doi.org/10.1289/EHP1663

Harris, B., Larson, L., & Ogletree, S. (2018). Different views from the 606: Examining the impacts of an urban greenway on crime in Chicago. *Environmental Behaviour, 50*, 56–85.

Hartig, T., Mitchell, R., de Vries, S., & Frumkin, H. (2014). Nature and health. *Annual Review of Public Health, 35*, 207–228.

Hashim, N. I., Yusof, N. H. S., Anuar, A. N. A., & Sulaiman, F. C. (2019). The restorative environment offered by Pocket Park at Laman Standard Chartered Kuala Lumpur. *Journal of Hotel and Business Management*, *8*(194), 2169–0286.

Jones, B. A., & Goodkind, A. L. (2019). Urban afforestation and infant health: Evidence from MillionTreesNYC. *Journal of Environmental Economics and Management, 95*, 26–44.

Kabisch, N., Püffel, C., Masztalerz, O., Hemmerling, J., & Kraemer, R. (2021). Physiological and psychological effects of visits to different urban green and street environments in older people: A field experiment in a dense inner-city area. *Landscape Urban Planning, 207*, 103998.

Lee, H. J., & Lee, D. K. (2019). Do sociodemographic factors and urban green space affect mental health outcomes among the urban elderly population? *International Journal of Environmental Research and Public Health, 16*(5), 789.

Li, D., & Sullivan, W. C. (2016). Impact of views to school landscapes on recovery from stress and mental fatigue. *Landscape and Urban Planning, 148*, 149–158.

Matsuoka, R. H. (2010). Student performance and high school landscapes: Examining the links. *Landscape and Urban Planning, 97*(4), 273–282.

McCord, J., McCord, M., McCluskey, W., Davis, P. T., McIlhatton, D., & Haran, M. (2014). Effect of public green space on residential property values in Belfast metropolitan area. *Journal of Financial Management of Property and Construction.*

Ministry of Health. (2017). Malaysian mental healthcare performance technical report for 2016. In *Malaysian Healthcare Performance Unit, National Institute of Health Psychiatrist Group*, Ministry of Health, Malaysia.

Mohd Saad, M. R., Abdul Aziz, N. A., & Mariapan, M. (2019). Benefits of forest-related recreational activities towards psychological stress of visitors. *The Malaysian Forester, 82*(2), 361–367.

Mokhtar, D., Abdul Aziz, N. A., & Mariapan, M. (2018). Physiological and psychological health benefits of urban green space in Kuala Lumpur: a comparison between Taman Botani Perdana and Jalan Bukit Bintang. *Pertanika Journal of Social Sciences and Humanities*, *26*(3).

Nath, T. K., Han, S. S. Z., & Lechner, A. M. (2018). Urban green space and well-being in Kuala Lumpur, Malaysia. *Urban Forestry and Urban Greening, 36*, 34–41.

National Health Morbidity Survey. (2017). Adolescent Health Survey. Kuala Lumpur: Institute for Public Health, Ministry of Health, Malaysia, 2017.

Nesbitt, L., Hotte, N., Barron, S., Cowan, J., & Sheppard, S. R. J. (2017). The social and economic value of cultural ecosystem services provided by urban forests in North America: A review and suggestions for future research. *Urban Forestry and Urban Greening, 25*, 103–111.

Nowak, D. J., & Greenfield, E. J. (2018). US urban forest statistics, values, and projections. *Journal of Forestry, 116*(2), 164–177.

Othman, N., Mohamed, N., & Ariffin, M. H. (2015). Landscape aesthetic values and visiting performance in a natural outdoor environment. *Procedia-Social and Behavioral Sciences, 202*, 330–339.

Préndez, M., Araya, M., Criollo, C., Egas, C., Farías, I., Fuentealba, R., & González, E. (2019). Urban trees and their relationship with air pollution by particulate matter and ozone in Santiago, Chile. In *Urban Climates in Latin America* (pp. 167–206). Springer.

Raman, T. L., Abdul Aziz, N. A., & Yaakob, S. S. N. (2021). The effects of different natural environment influences on health and psychological well-being of people: A case study in Selangor. *Sustainability, 13*(15), 8597.

Rathmann, J., Beck, C., Flutura, S., Seiderer, A., Aslan, I., & André, E. (2020). Towards quantifying forest recreation: Exploring outdoor thermal physiology and human well-being along exemplary pathways in a central European urban forest (Augsburg, SE-Germany). *Urban Forestry and Urban Greening, 49*, 126622.

Rajoo, K. S., Karam, D. S., & Aziz, N. A. A. (2019). Developing an effective forest therapy program to manage academic stress in conservative societies: A multi-disciplinary approach. *Urban Forestry and Urban Greening, 43*, 126353.

Sander, H., Polasky, S., & Haight, R. G. (2010). The value of urban tree cover: A hedonic property price model in Ramsey and Dakota Counties, Minnesota, USA. *Ecological Economics, 2010*(69), 1646–1656.

Simkin, J., Ojala, A. M., & Tyrväinen, L. (2020). Restorative effects of mature and young commercial forests, pristine old-growth forest and urban recreation forest—A field experiment. *Urban Forestry and Urban Greening, 48*, 126567.

Tan, T. H. (2011). Neighborhood preferences of house buyers: The case of Klang Valley, Malaysia. *International Journal of Housing Markets and Analysis, 4*(1), 58–69.

Tsai, W. -L., McHale, M. R., Jennings, V., Marquet, O., Hipp, J. A., Leung, Y. F., & Floyd, M. F. (2018). Relationships between characteristics of urban green land cover and mental health in U.S. metropolitan areas. *International Journal of Environmental Research and Public Health, 15*(2), *340*.

United Nations Department of Economic and Social Affairs and Population Division. "World Urbanization Prospects: The 2018 Revision. Retrieved from https://www.population.un.org/2018

Vujcic, M., & Tomicevic-Dubljevic, J. (2018). Urban forest benefits to the younger population: The case study of the city of Belgrade, Serbia. *Forest Policy and Economics, 96*, 54–62.

Ward Thompson, C., Aspinall, P., Roe, J., Robertson, L., & Miller, D. (2016). Mitigating stress and supporting health in deprived urban communities: The importance of green space and the social environment. *International Journal of Environmental Research and Public Health, 13*, 440.

Wood, E., Harsant, A., Dallimer, M., de Chavez, A. C., McEachan, R. R. C., & Hassall, C. (2018). Not all green space is created equal: Biodiversity predicts psychological restorative benefits from urban green space. *Frontiers in Psychology, 9*, 2320.

Weimann, H., Björk, J., & Håkansson, C. (2019). Experiences of the urban green local environment as a factor for well-being among adults: An Exploratory Qualitative Study in Southern Sweden. *International Journal of Environmental Research and Public Health, 16*(14), 2464.

Zelenáková, M. D., Diaconu, D. C., & Haarstad, K. (2017). Urban water retention measures. *Procedia Engineering, 190*, 419–426.

Zhou, C., Yan, L., Yu, L., Wei, H., Guan, H., Shang, C., Chen, F., & Bao, J. (2019). Effect of short-term forest bathing in urban parks on perceived anxiety of young-adults: A pilot study in Guiyang, southwest China. *Chinese Geographical Science, 29*(1), 139–150.

Nor Akmar Abdul Aziz is a Senior Lecturer at the Department of Recreation and Ecotourism, Faculty of Forestry and Environment, Universiti Putra Malaysia, Serdang. Her research has included studies of green space use management, urban forestry, and recreation health and well-being.

Chapter 11
Tree Climbing: From Recreational to Tree Workers in Malaysia

Saifful Pathil

Abstract Tree climbing activity is not new in Malaysia. In the ‘80s, the Malaysia Forestry Department sent their staff to Germany to attend a series of training related to tree climbing and tree work. Since then, tree climbing has been used as a method for trees not accessible by Mobile Elevation Work Platform (MEWP) such as cranes, skylift and boom lifts. Most of the tree climbers in Malaysia are involved with tree maintenance, academic research, installation work and recreational activities. An association for recreational tree climbing was initiated by the Malaysian tree climber community in order to promote tree climbing as a recreational activity in 2009. The Malaysian Society of Arborist (PArM) took the initiative to develop more tree climbers with knowledge and skills related to tree climbing and pruning activities to ensure tree climbers are able to work and comply with international safety requirements in arboricultural operations. Training provider for tree workers is needed in Malaysia to cope with the industry demand and need for skilled workers. Tree climbing Personal Protective Equipment (PPE) is very dynamic with new innovations in line with the evolution of tree climbing techniques from Moving Ropes System (MRS) to Stationary Rope System (SRS). Step by step tree climbing needs to be mastered by tree workers before performing any work on trees and by using the proper equipment for the correct work to ensure the safety of tree climbers.

Keywords Tree climbing · Tree worker · Arborist · Arboriculture · Recreational tree climbing · Pruning · PPE

Introduction

Tree climbing has been introduced by the Malaysian Forestry Department for tree worker purposes either for tree maintenance or seed collection activity in Malaysian tropical forests. The Malaysian Forestry Department sent their staff to attend tree climbing and tree worker courses to ensure the staff was fed with technical knowledge

S. Pathil (✉)
Tree Care Safety™, Cheras, Malaysia
e-mail: saiffulpathil@yahoo.com

S. Maruthaveeran et al. (eds.), *Urban Forestry and Arboriculture in Malaysia*,
https://doi.org/10.1007/978-981-19-5418-4_11

and skills in 1987. In 2011, Tree Climbers Malaysia (TCM) was initiated by the community of tree climbers in Malaysia to promote tree climbing as a recreational activity for the public and tree lovers. The majority of these climbers do it for the sake of fun and love for trees. In Malaysia, Persatuan Arborist Malaysia (PArM) a.k.a Malaysia Society of Arborist established and founded in 2005. This association has developed and established the arboriculture industry in Malaysia through practising arboriculture work in compliance with international standards and Best Management Practices (BMP) set by the International Society of Arboriculture (ISA). Presently the Malaysia Society of Arborist (PArM) focuses on developing tree climbers for tree worker purposes. The association feeds the tree climber with the knowledge of safety in tree climbing and working in trees. Tree climbers learn more about tree climbing, rigging, pruning and tree biology to improve their knowledge (Fig. 11.1).

Fig. 11.1 Tree climbing activity during Petzl 'Move in Trees' workshop at Dungun, Terengganu, Malaysia

What is Tree Climbing?

Tree climbing is an activity using proper Personal Protective Equipment (PPE) to climb a tree. Most tree climbers around the world climb trees for tree maintenance, tree research, installation work and recreational purposes. In Malaysia, a lot of tree climbers started their pathway in tree climbing from a recreational activity to tree worker activity.

Tree Climbing Championship

The Tree Climbing Championship has been established by the International Society of Arboriculture (ISA). Malaysian tree climbers also have participated a number of times since 2011 as competitors, judges and volunteers. This tree climbing competition is a mock-up of real tree worker activity and expression of tree worker skills in tree climbing. Tree climbing competition provides competitive but educational opportunities for working arborists and tree workers to demonstrate and exchange new climbing techniques and equipment, as well as safe work practice. Generally, this championship has five events (Work Climb, Throwline, Aerial Rescue, Belayed Speed Climb and Speed Ascent/Secured Footlock) in the preliminary round before qualifying for the Master Challenge Event.

Apart from this tree climbing championship organised by the ISA, there are also others in this region such as the Hong Kong Tree Climbing Championship (HKTCC), Singapore Tree Climbing Championship (STCC), New Zealand Tree Climbing Championship (NZTCC) and Australia Tree Climbing Championship (ATCC). Looking at the demand and interest in tree climbing, the Malaysia Society of Arborist (PArM) also took an opportunity to organise Malaysia Tree Climbing Championship (MTCC) recognized by the ISA in 2018 and 2019. Other countries such as Thailand and Taiwan have also organised such an event as well.

In 2019, two tree climbers from Malaysia were qualified to compete in Asia Pacific Tree Climbing (APTCC) in Christchurch, New Zealand and one experienced tree climber/Certified Arborist from Malaysia was involved as Scoring Judge in the Work Climb Event and Master Challenge Event. This is the highest achievement by the Malaysia Society of Arborist (PArM) for tree climbing development in Malaysia (Figs. 11.2 and 11.3).

Tree Climbing for Recreational

Recreational tree climbing is popular worldwide and has been introduced by Peter 'Treeman' Jenkins from Tree Climbers International (TCI), founded in 1983. The main focus of recreational tree climbing is for fun, experience in tree climbing and to

Fig. 11.2 Tree Climbers (Competitor) and Certified Arborist (Judge) from Malaysia at Asia Pacific Tree Climbing Championship (APTCC) 2019, Christchurch, New Zealand

Fig. 11.3 Competitors, judges and volunteers from Malaysia, Singapore, Hong Kong and Sweden at Malaysia Tree Climbing Championship (MTCC) 2019 Putrajaya, Malaysia

appreciate the trees. The Tree Climbers Malaysia (TCM) also involved and played a role in introducing recreational tree climbing in Malaysia from 2011 till the present. The promotion of recreational tree climbing encourages outdoor lovers and youth in Malaysia to start their pathway to become professional tree climbers for tree worker purposes. Recreational tree climbers will learn basic tree climbing techniques such as the 'Body Thrusting' technique to climb high on trees.

Tree Climbing for Tree Worker

Tree worker is not a new profession in Malaysia. Tree workers have been utilised for landscaping work and the arboriculture industry. Tree climbing method has been used to access the trees that are not accessible by the Mobile Elevation Work Platform (MEWP). Tree maintenance such as tree pruning, tree removal and installation of Tree Lightning Protection System (TLPS) is the main task for tree workers. Demand for tree workers in Malaysia is increasing due to maintenance work in urban conditions. Most of these trees in the middle of the city are planted near the buildings and only can be accessed by the tree climbing method. Due to the demand in the industry, tree climbers who started as recreational tree climbers expanded their pathway to tree worker as a profession. The Malaysia Society of Arborist (PArM) takes a lead to encourage and educate the climber to become a tree worker in Malaysia. Due to the demand in the industry in Malaysia, Tree Care Safety (TCS) is a Malaysia training provider and collaboration with the National Institute of Occupational Safety & Health (NIOSH) Malaysia in providing skills training for tree workers such as Tree Climbing, Aerial Rescue, Chainsaw Safety and Tree Worker Safety in-compliance with international standards such as ANSI Z133 for Arboricultural OperationsSafety Requirements. Tree Care Safety (TCS) also has been supported by Petzl Technical Institute (PTI)—Allsports Equipment as established Personal Protective Equipment (PPE) from France in promoting safe work practices in tree climbing using the proper equipment for tree climbing purposes (Figs. 11.4 and 11.5).

Tree Climbing Techniques

Tree climbers have several options in tree climbing techniques from climbing using spurs/spikes to climbing a tree using a rope. Spurs/spikes are acceptable and allowed as a tree access method only for tree removal and aerial rescue purposes. As tree climbers, they need to master tree climbing using a rope because most of the maintenance work involves working around the tree and not only working at the same spot. Tree climbers need to limb walk to perform pruning work and sometimes need to work at the tip of branches.

Moving Ropes System (MRS) previously known as Double Ropes Technique (DdRT) is a common and basic technique in tree climbing using a rope. Climbers

Fig. 11.4 Recreational tree climbing involves ropes, knots and some mechanical devices to increase climbing efficiency at Taman Tasik Shah Alam, Shah Alam, Selangor Darul Ehsan

only use their ropes, prusik and sometimes add a micro-pulley to smooth tending of the climbing system. The Moving Ropes System (MRS) started with the full utilisation of friction knots such as Blake's Hitch to create the system. The earliest technique is called 'Body Thrusting'. Climbers sit on tree climbing saddles /harnesses and are attached to the ropes that tie in at the highest point (suitable branches) on the tree. Tree climbers need to pull themselves up by themselves to climb up the tree using the rope and need a lot of strength to master the technique. Throughout a year and evolution of equipment, the Moving Ropes System (MRS) became more efficient and effortless to master the technique. This includes additional equipment such as foot ascension devices to increase the efficiency of the Moving Ropes System (MRS) (Figs. 11.6 and 11.7).

Fig. 11.5 Tree climbing for tree work (Pruning) at Sireh Park, Johor Baharu, Johor Darul Takzim, Malaysia

Stationary Ropes System (SRS) previously known as Single Rope Technique (SRT) is an advanced technique in tree climbing using a rope. Tree climbers need additional devices to complete the system and need proper training to master the technique. Climbing devices such as Prusik, Hitch Climber Pulley, Auxiliary brake device, hand/foot Ascension device are a must for the system. A lot of advantages of using the Stationary Ropes System (SRS) compare with the Moving Ropes System (MRS) in tree climbing, especially when climbing the tallest tree and moving around the tree. Various options to the anchorage and redirect the system on trees give more advantages when performing the work and aerial rescue process if an emergency happens.

Tree Climbing Step by Step

Firstly, tree climbers need to justify and identify certain things before they start to climb a tree. Tree climbers need to check their Personal Protective Equipment (PPE) to ensure the equipment is fit to use according to the conditions of the equipment and complies with international standards. Tree climbers need to do a visual tree inspection to ensure the tree is safe to climb and tree climbers need to plan work

Fig. 11.6 Moving Ropes System (MRS) suitable for Canopy research during Scientific Expedition at Tawau Hill Park, Sabah, Malaysia

ahead in identifying the best tie in point (suitable branches) as anchorage the ropes on the tree according to the rules of thumb branches size and 45°angle of working around the tree.

Secondly, installing the line using a throwline and throw bag. Tossing techniques need to be mastered by tree climbers to ensure climbers manage to select the best tie-in point to fulfill work requirements such as pruning or installation work. Thirdly, Tree climbers will be setting up climbing lines according to tree climbing techniques either Moving Ropes System (MRS) or Stationary Ropes System (SRS).

Fourthly, tree climbers will now be ready to climb a tree. Tree climbers need to maintain three-point contact (Climbing Line, Positioning Lanyard, Foot/Hand) with the tree whenever to perform the task or work (Pruning, Removal or Installation) to avoid uncontrol pendulum swing and injury by fall/hit the trunk or branches in the tree. Fifthly, tree climbers safely descend and land on the ground. In-control

Fig. 11.7 Stationary Ropes System (SRS) during aerial tree assessment work at Desaru, Johor Darul Takzim, Malaysia

descending using proper and approved climbing systems such as prusik, mechanical prusik and mechanical descending devices. Finally, the tree climber needs to dismantle the climbing line and all climbing equipment from the tree.

Personal Protective Equipment (PPE) for Tree Climbing

Innovation of tree climbing equipment is very dynamic and aggressive due to the evolution of tree climbing techniques. Rock or Wall climbing equipment is different from tree climbing equipment. Tree climbing equipment is designed, rated and approved with certain standards for specific tree climbing purposes. This will ensure the climber safely works, ascend and descend on trees (Table 11.1 and Fig. 11.8).

Table 11.1 List of Personal Protective Equipment (PPE) for Tree Climbing according to Moving Ropes System (MRS) and Stationary Ropes System (SRS)

No	Moving Ropes System (MRS)	Stationary Ropes System (SRS)
1	Climbing helmet	Climbing helmet
2	Tree climbing saddle/harness	Tree climbing saddle/harness
3	Carabiner (double locking mechanism)	Carabiner (double locking mechanism)
4	Climbing rope (arbor rope)	Climbing rope (arbor rope)
5	Prussik/mechanical prussik	Prussik/mechanical prussik
6	Hitch climber pulley	Hitch climber pulley
7	False crotch/friction saver	Foot/hand ascender
8	Foot ascender	Auxiliary brake device
9	Throw bag	Knee ascension devise
10	Throwline	Throw bag
11	–	Throwline

Fig. 11.8 From recreational to tree worker during Pruning work at Terengganu Darul Iman, Malaysia

Conclusion

Tree climbing growth rapidly in Malaysia is in line with worldwide development in the landscaping and arboriculture industry. A lot of opportunities for tree climbers in the career pathway to become professional skilled workers by learning tree climbing. Nowadays, tree climbing skills are expensive and demanded by the industry. Those who are masters in tree climbing techniques have the advantage to create a business opportunity in any related to tree maintenance, academic research and installation work (Canopy walk, Aerial Adventure, Camera Traps, Guying/Bracing and Tree Lightning Protection System). Tree climbing is a combination of skills and knowledge. Learning and understanding safety will create a good attitude and safe working practice. Efforts by association, government agencies and training providers will encourage youth to become involved with tree climbing at an early age and this will create more opportunities for them to expand their skills to work in the industry in the future.

Saifful Pathil born in Muar, Johor Malaysia, holding a credential of ISA Certified Arborist® by the International Society of Arboriculture (ISA) and branding a tree care training company Tree Care Safety™ (TCS) in Malaysia. Tree Care Safety™ (TCS) currently works closely with the National Institute of Occupational Safety & Health (NIOSH) Malaysia for Tree Worker Safety training. As a tree climber, he is the founder of the Tree Climbers Malaysia (TCM) for recreational tree climbing in Malaysia in 2009. Ten years of experience in teaching and conducting training in various fields. Competence Petzl PPE verifier and working closely with Petzl Technical Institute (PTI)—All sports equipment in promoting Tree Care products through workshops and training in Malaysia. Saifful Pathil is currently collaborating with Husqvarna Malaysia in promoting Chainsaw Safety in compliance with international standards and the best practice of arboriculture for the Malaysian Industry.

Part II
Case Studies

Chapter 12
Soil and Water Bioengineering Technique for Urban Forestry and Mitigation of Natural Hazards

Deivaseeno Dorairaj, Nisha Govender, and Normaniza Osman

Abstract Industrialization and rapid urbanization have taken a toll on the environment. The increase in urban population has resulted in land areas being cleared for both residential and infrastructure development. The removal of plant coverage exposes the barren soil to heavy precipitation which subsequently leads to soil erosion and landslides. As part of sustainable development, a growing number of city councils have taken the initiative to integrate green spaces into urban areas. Urban forest parks and gardens are incorporated into the city blueprint to provide urbanites with recreational spaces and a chance to interact with nature. Moreover, the concept of urban forestry provides a balance between environmental degradation and development. Urban trees offer a wide range of services, namely, reduction of air and noise pollution, regulation of atmospheric temperature and mitigation of flood. The success of this concept relies primarily on the right selection of tree species and their rooting properties. Since Malaysia is located on the equator, the hot and humid weather had significantly reduced soil fertility due to chemical weathering and leaching. Hence, the use of phytoremediator or metal tolerant plants is recommended for they could ameliorate contaminated soils and at the same time provide aesthetic landscaping qualities.

Keywords Urban forestry · Urban trees · Rooting · Soil fertility · Phytoremediation

---

D. Dorairaj (✉)
Institute for Environment and Development (LESTARI), Universiti Kebangsaan Malaysia, UKM, 43600 Bangi, Selangor, Malaysia
e-mail: deiva@ukm.edu.my

N. Govender
Institute of Systems Biology (INBIOSIS), Universiti Kebangsaan Malaysia (UKM), 43600 Bangi, Selangor Darul Ehsan, Malaysia

N. Osman
Institute of Biological Sciences, Faculty of Science, University of Malaya (UM), 50603 Kuala Lumpur, Malaysia

S. Maruthaveeran et al. (eds.), *Urban Forestry and Arboriculture in Malaysia*,
https://doi.org/10.1007/978-981-19-5418-4_12

Introduction

Malaysia's population as of 2020 stands at 32.7 million, which is projected to increase to 41.5 million in twenty years' time (DOSM, 2020). As the country moves towards the status of a developed nation, more people will move out of the rural areas to cities for employment opportunities and to improve their socioeconomic status. In 1991, Malaysia became an urbanized nation as the population of urbanites hit 50.4% (DOSM, 2010). Currently, the urban population makes up 77.2% of the total population (The World bank, 2020). The number of urbanites is projected to increase with the rate of urbanization. It is expected that 66% of the world's population will call cities their homes by 2050, compared to 30% in 1950 (United Nations, 2014). High-density cities are expanding to meet the demands of the growing urban population. It is projected that green spaces will continue to shrink with the increasing rate of urbanisation (Aida et al., 2016; Kabisch et al., 2015).

The increasing urban population is the key driver of land clearing for infrastructure and development projects which will eventually cut through forests to accommodate the needs of the people at the expense of human lives (Dorairaj & Osman, 2021). Development of land in urban areas heightens loads on the slopes thus elevating the frequency of slope failures (Szabo, 2003). Tropical regions where Malaysia is situated, experiences torrential rain that elevates the degree of soil erosion and consequently, slope failures (Crozier, 2010). This catastrophe often amounts to huge economic losses and fatalities. According to the literature, the main triggering factors of slope instability are rainfall (57.5%), water level change (35%), loading change (5%), slope geometry and vegetation change (2.5%) (Public Works Department Malaysia, 2009). Rainfall-induced landslides on sloped terrain are results of diminishing soil shear strength (Lu & Godt, 2013). Recent landslides in Bukit Antarabangsa, Selangor, Tanjung Bungah and Paya Terubong in Penang are classic testaments to over-development on slopes that turned into a disaster due to rainfall.

Ultimately there are two main forces that determine slope stability; driving and resisting forces. Driving forces, which refers to gravity in most cases, promote the movement of slope material whilst resisting forces, which are the material's shear strength, oppose the motion (Hughes, 2003). The driving force is influenced by slope steepness, rainfall intensity, load and soil bulk density, presence of impermeable layer and soil thickness, while rock strength influences resisting force, shear strength and vegetation roots strength (Abe & Ziemer, 1991). Down-slope movement takes place at steeper slope angles which increases the shear stress but reduces the shear strength by lowering the cohesion among the particles or the frictional resistance. Shear strength is the capacity of a material to bear the load until failure while the shear stress is the force exerted on a material per unit area. As the soil load increases, it will trigger soil movement since the pushing force on the upper slope is arguably stronger than the resisting force (Abe & Ziemer, 1991). Correspondingly, soil movement usually occurs in the event of rainfall as the water that infiltrates will occupy the soil pore space making it saturated while weakening soil aggregation and reducing soil shear strength (Mulyono et al., 2018).

City councils worldwide have begun integrating and recreating green spaces in the urban areas to provide a balance between development needs and environmental conservation. The effort has intensified and concerted, more so as climate change issues take centre stage (Kanniah, 2017; Kanniah & Ho, 2017). As part of sustainable development, urban parks and urban forests have now become an integral component of town planning. Trees could either be replanted or planted in locations that never existed before. According to FAO (2017), urban forest is defined as networks or systems comprising all woodlands, groups of trees, and individual trees located in urban and peri-urban areas which include trees in gardens and parks, street trees and forests. On the other hand, Escobedo et al. (2011) and Roy et al. (2012) defined an urban forest as having all trees, shrubs, lawns, and pervious soils in urban areas. The concept of urban and peri-urban forestry is multifaceted and involves a multidisciplinary approach.

Urban heating resulting from air-conditioning, traffic and building materials is a predominant atmospheric hazard affecting millions living in the city (Wang et al., 2019). Playing the role of a microclimate moderator, urban trees could mitigate this effect while improving air quality by absorbing pollutants (Lüttge & Buckeridge, 2020). In fact, urban trees have been mentioned in IPCC special report to play a key role in the alleviation of climate change (De Conink et al., 2018). Moreover, in the event of flashflood, these trees could mitigate stormwater (Bartens et al., 2009) and act as efficient windbreaks (Lüttge & Buckeridge, 2020).

Generally, a plant is divided into two major parts, namely, above-ground and belowground. The former mostly intercepts rainfall and is involved in the run-off hydraulics (Morgan & Rickson, 1995) while the latter, though invisible, helps to provide mechanical support by holding the soil particles through its rooting system. Plant roots through mechanical reinforcement, anchoring and compaction increase soil shear strength (Singh, 2010). The use of vegetation to improve soil stability is dependent on the type of plants used, the method of planting and root properties (Huat & Kazemian, 2010). This soil bioengineering approach is apt for soil stabilization against erosion besides lessening the occurrence of landslides (Normaniza & Barakbah, 2011; Liu et al., 2016). Vegetation is self-regenerating, hence highly sustainable and could adapt to its environment, while being cost-effective and needing very little maintenance. In this chapter, we will discuss the mechanical, hydrological and hydraulic effects of vegetation and the use of trees as part of soil bioengineering technique for soil stabilization in urban areas (Fig. 12.1).

Fig. 12.1 Massive soil movement that led to the landslide in Bukit Antarabangsa in 2008 (*Photo credit* Normaniza Osman)

Role of Vegetation

Mechanical

Mechanical root reinforcement is provided by the finger-like root projections which provide strong anchorage that holds the soil particles together to prevent the disintegration of soil structure (Dorairaj & Osman, 2021). This in turn increases the soil shear strength (Gray & Sotir, 1996). Besides, the roots through buttressing and arching, anchor the deep soil layer bolstering the upslope soil mantle (Bruijnzeel, 2004). Further, the direction of root growth matters; roots growing perpendicular to the soil surface provides resistance to shearing forces acting on the soil whereas those extending parallel to the soil reinforces the tensile strength of the soil zone (Jerome, 2010). According to Faisal and Normaniza (2008), the weight of vegetation builds on the load which is the driving force while the roots reinforce the soil thus elevating the resisting force.

Hydrological

Vegetation lessens water runoff by establishing the water cycle of the soil–plant-atmosphere continuum (SPAC) and keeping the soil relatively dry (Normaniza & Barakbah, 2006; Mafian et al., 2009; Normaniza, Saifuddin & Halim, 2014). The tree canopy acts as an umbrella that intercepts rainfall (Llorens & Domingo, 2007) which reduces the water that is infiltrated into the soil (Gonzalez-Ollauri & Mickovski,

2017). Besides that, the roots add to the roughness of the soil and hence the infiltration capacity of the ground (Balzano et al., 2019). In addition, the moisture content of the soil is lessened through evapotranspiration which allows the soil to absorb more water (Dorairaj et al., 2020) and release it into the atmosphere. This in turn reduces pore pressure while increasing the shear strength of soil, thus increasing its resistance (Faisal & Normaniza, 2008).

Hydraulic

The hydraulic mechanism is marked by the reduction in water flow capacity as a result of the contact between the plant via the roughness it creates and the flowing water (Noraini & Roslan, 2008). The stem and roots provide structural support for the soil, hence limiting the loosening and movement of soil sediment (Mulyono et al., 2018) as the inertial force of surface runoff is reduced (Zhao et al., 2019). Furthermore, due to root water uptake, the pore-water pressure is lessened which not only results in increased permeability (Ng et al., 2019), but an increase in the soil shear strength (Liu et al., 2016, Fig. 12.2).

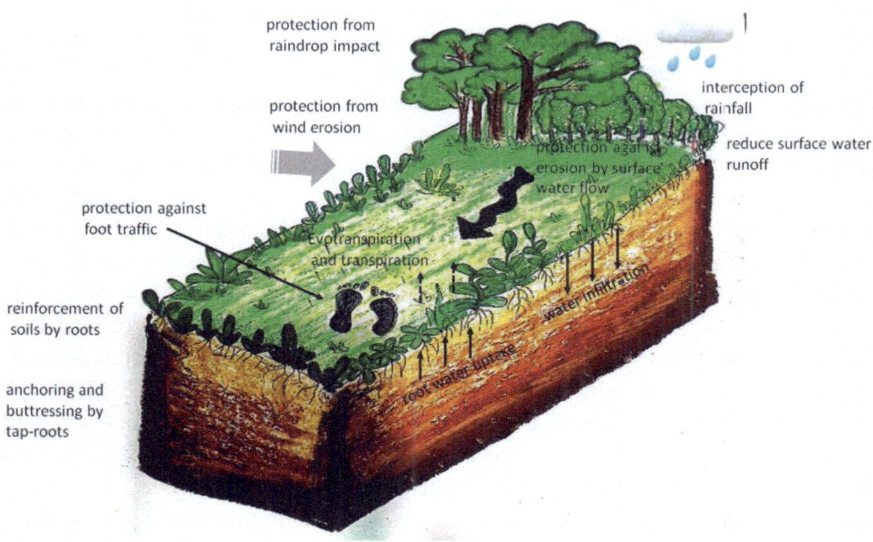

Fig. 12.2 Role of vegetation (*Source* Coppin and Richards (1990), ©CIRIA)

Trees

By definition, trees are woody perennials with single or multiple stems and a more or less definite crown (FAO, 1998, 2004). Meanwhile, according to Stone and Kalisz (1991) trees are mostly evergreen with the roots growing several meters deep and wide, in fact they can reach 53 m deep in rare situations. However, the roots of tropical trees growing in urban areas rarely exceed a depth of about 2 m while the lateral roots may extend to 10 m (Kamal, 2021) or three times the canopy spread (Elmendorf et al., 2005). In fact, most studies reported root biomass to be concentrated in the upper 30 cm of soil (Elmendorf et al., 2005; Gilman, 2003) as the infertile topsoil is usually filled with stones, construction rubble, bricks, and other building materials which hampers root growth (Jin et al., 2011). Extreme rainfall due to climate change is associated with flooding, stormwater runoff and low water quality in urban areas (Beidokhti & Moore, 2021). Being woody, trees play a pivotal role in reducing water erosion by improving water infiltration and stabilizing the soil through root reinforcement (Durán Zuazo & Rodríguez Pleguezuelo, 2008). As for vegetation litter, it acts like a blanket for the soil, hence preventing soil erosion (Li et al., 2014). This litter blanket not only shields the soil against raindrop splashes by intercepting rainfall but also prevents crusting of soil, increases the time of soil infiltration while elevating sediment deposition by increasing the soil surface roughness (Geddes & Dunkerley, 1999; Sayer, 2006).

The selection of urban tree species is vital for it must be adapted to its environment to ensure the success of soil bioengineering practices for soil fixation (Stokes et al., 2014). Among others, stem density, stem bending resistance, root density, root area ratio, the potential to trap sediment and debris, root tensile strength and root morphology are traits of importance (Baets et al., 2008a, b; Bischetti et al., 2014; Giadrossich et al., 2012; Ghestem et al., 2014; Stokes et al., 2009). Generally, the use of native plant species is preferred for they are better acclimatized to the growing conditions, are able to live and thrive among pathogens, and require less maintenance and less fertilizer and water input (Gray & Sotir, 1996). However in the context of the inhospitable urban environment, in presence of abiotic stressors (heat, drought, infertile soil), and also in absence of invasion risks, introduced species are better adapters and generally exhibit better growth and development (Chalker-Scott, 2015; Roloff et al., 2009). In addition, introduced species are capable of fulfilling the ecosystem services and resilience in urban environments when native tree species are limited (Sjöman et al., 2016). Similar to other countries in this region, Malaysia, having been under western colonization, our landscape is adorned with introduced plant species that we assume to be native to the country. The fact is these are introduced plant species that have naturalised to the local environment. For example, trees such as the *Pterocarpus indicus* (Angsana), *Delonix regia* (Flame of the Forest), *Peltophorum pterocarpum* (Yellow Flame), *Jacaranda fillicifolia* (Jacaranda), *Khaya senegalensis* (Khaya), *Tabebuaia rosea* (Tecoma), and *Andira inermis* (Andira) are exotic species that have somewhat became naturalised in Malaysia (Mohd. Shariff & Abu Bakar, 2006). These trees enrich public parks and gardens and are often planted in the cities.

Besides *P. pterocarpum* and *A. inermis*, the Department of Irrigation and Drainage Malaysia (2010) has listed the following as suitable erosion control tree species: *Senna surattensis* (Gelenggang), *Cassia fistula* (Kayu Raja), *Cassia spectabilis*, *Cyrtophyllum fragrans* (Tembusu), *Khaya senegalensis*, *Callerya atropurpurea* (Tualang Daing). Amongst these, *S. surattensis*, *C. fragrans* and *C. atropurpurea* are native to Malaysia. Thus a middle-ground approach should be practiced when it comes to decision-making to assess the advantages of using exotic trees (introduced trees) in urban areas and preventative steps to protect the natural ecosystems from invasion risks (Sjöman et al., 2016). Since most of the tree species suggested by the Department of Irrigation and Drainage Malaysia have no records of further information on their characteristics, the following section will not include a discussion on all the tree species mentioned above.

Rooting Architecture of Selected Urban Trees for Soil Stabilization and Erosion Control

Fast-growing leguminous forest tree species could be planted in urban areas for they not only fix atmospheric nitrogen in the soils but rehabilitate infertile soil (Parrotta, 1993). Besides, they have good rooting properties that enable them to penetrate deep into the soil and serve as anchors (Stokes et al., 2009). The extensive rooting characteristic of both *Leucaena leucocephala* (Petai Belalang) and *P. pterocarpum* provides high root tensile strength and soil shear strength which offer long-term soil reinforcement on slopes (Normaniza et al., 2014). The latter, which is a woody ornamental tree (Saifuddin & Normaniza 2014), has orange-yellow fragrant flowers which are more attractive as compared to the former which possesses cream-coloured globular inflorescence. These trees also provide protection against the wind (Saifuddin & Normaniza 2014). The multi-purpose *L. leucocephala* possesses aggressive taproots which could be anything between 1.8–2.4 m deep and at least 12.7 mm in diameter (Normaniza et al., 2014). Further, it has extensive lateral roots which are ideal for deep-seated stabilization (Ali, 2010) while thriving on steep slopes and in marginal areas with prolonged dry seasons, making it the main choice for forest cover restoration (Normaniza & Barakbah, 2006).

In addition, the majestic Angsana tree (*P. indicus*) is also an urban tree with many uses, namely, as an ornamental, a shade tree, for timber, reforestation, living fences and a windbreak around croplands. Its large canopy could reduce both the impact and velocity of raindrops and thus combat surface runoff and reduce soil erosion. The pea-shaped flowers are bright yellow to orange-yellow while its unique feature is the papery winged fruit (Thomson, 2006). Saifuddin and Normaniza (2016) found that *P. indicus* exhibited a strong tap root whereas its lateral roots grew horizontally and profusely. The plant also showed the highest root biomass in comparison to *L. leucocephala*, *P. pterocarpum*, *A. mangium* and *D. suffruticosa*, thereby providing the greatest anchorage as the roots could hold more soil mass, hence reducing soil erosion and preventing soil displacement (Stokes et al., 2009). Possessing large buttresses and an extensive root spread, this tree provides soil stabilization (Thomson, 2006).

Apart from the previously described species, *Acacia mangium* is often seen growing along roadsides. Similar to *L. leucocephala* and *P. pterocarpum*, *A. mangium* is also an introduced tree species that grows on marginal soil but is prone to heart-rot disease (Asif et al., 2017; Potter et al., 2006). Due to dense foliage, it is known as a shade tree while thriving on roadsides. In addition, *A. mangium* has a great ability to buffer temperature, improve soil organic matters, reduce radiation and improve nutrition (Yang et al., 2009). As the roots of *A. mangium* are predominantly fine roots (90%) which contribute to soil reinforcement, it can prevent surface erosion and shallow landslides (Avani et al., 2014; Lateh et al., 2013). Tensile strength is positively correlated to fine roots, hence having more fine roots means the soil could be fixed more efficiently (). Ishak et al. (2013) reported that the factor of safety (FOS) against slope failure increased by 45.5% as the matric suction increased when *A. mangium* was planted at the toe of a slope. As for *D. regia*, it can grow up to 30 m and being a leguminous tree, it has the ability to restore soil fertility through fixation of atmospheric nitrogen (Karthikeyan et al., 2013; Combalicer et al., 2014). Due to its aggressive and spreading superficial root system, it is planted for erosion control and soil rehabilitation (Rojas-Sandoval et al., 2013). The plant is effective in maintaining soil moisture and reducing soil temperature (Karim, 1987).

Yen (1987) proposed a root system based on the tap, lateral and horizontal roots, classifying them into five types, namely, H, M, R, V and VH. While *L. leucocephala* exhibits H-type, both *P. indicus* and *A. mangium* possess VH-type of root system; both types are deemed suitable for soil reinforcement, slope protection and wind resistance (Reubens et al., 2007). *Pelthoporum pterocarpum* on the other hand shows R-type rooting architecture which was more effective than V-type root in improving soil shear strength (Fan & Chen, 2010; Fig. 12.3).

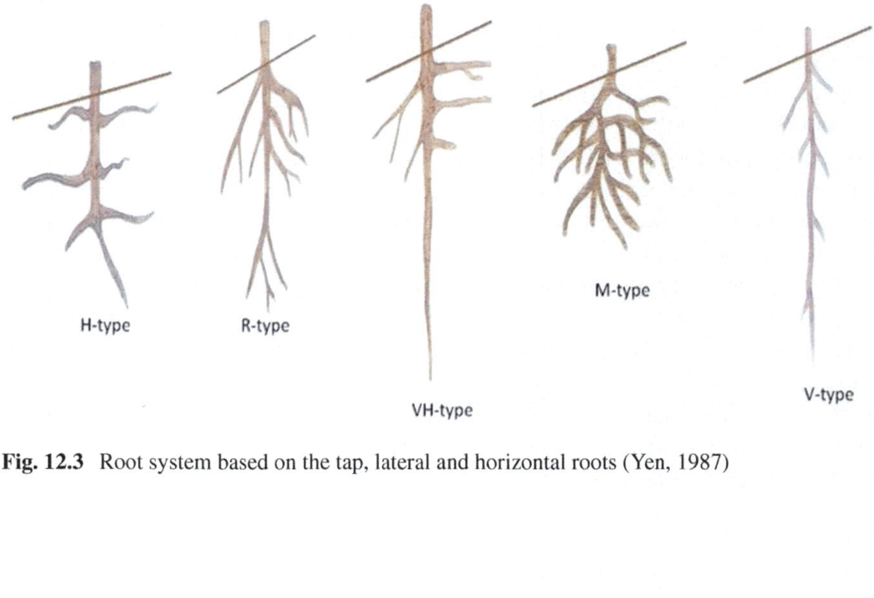

Fig. 12.3 Root system based on the tap, lateral and horizontal roots (Yen, 1987)

Apart from the above, *S. surattensis*, *C. fistula* and *S. spectabilis* too have the potential to increase topsoil infiltration, lessen runoff and reduce soil erosion while *C. atropurpurea* is reported to possess deep penetrating root system that provides soil reinforcement (Duke, 1983; FAO, 2014; Hairiah et al., 2020; Orwa et al., 2009).Besides these legume plants, *Syzygium campanulatum* (Kelat Paya) is also a popular choice for urban forestry due to its hardiness and adaptability (Noor Azlin & Azahari, 2014). The plant, which is also known as wild cinnamon, originates from South East Asia. It has the ability to withstand soil infertility and can increase its biomass over time (Arunbabu et al., 2015). *Syzygium campanulatum* displays R-type rooting which improves soil shear strength and combats slope failures (Dorairaj et al., 2020).

Another candidate for urban forestry is *Dillenia suffruticosa* (Simpoh air) which is a large shrubby tree growing up to 10 m tall. According to Abdullah et al. (2011), it has a heart or M-type root system which means the plant possesses shallow roots that could increase the cohesion at the end of the shear plane. *Dillenia suffruticosa* recorded the highest percentage of root volume and root length density *L. leucocephala* and *A. mangium* which makes it a better surface erosion plant candidate (Abdullah et al., 2011).

Heavy Metal Tolerant Plants

Malaysia's proximity to the equator presents both benefits and drawbacks for urban soils. The annual rainfall of about 2500 mm (Shafie, 2009), transient drought and a warm temperature all year round could only mean acceleration of soil erosion. Both high intensities of rainfall and humidity cause intense chemical weathering and the formation of thick soil profiles (Dorairaj et al., 2020) while rainwater percolation leaches basic elements such as calcium, magnesium, potassium and sodium from the soil profile (Lehmann & Schroth, 2003). Moreover, soils of the tropics are low in base saturation, hence highly acidic, due to weathering and leaching (Foy, 1992). Further, the low buffering capacity of soils in urban areas due to the low proportion of clay does not favour plant establishment.

Moreover, anthropogenic activities namely, traffic emissions, industrial emissions, domestic emissions, weathering of buildings and pavement surfaces, and atmospheric deposition (Wei & Yang, 2010) have not only increased the number of pollutants such as dioxins and persistent organic pollutants (POPs) (Wang et al., 2012), but also contributed to the dispersion of heavy metals into the air. Thereafter, these contaminants and pollutants end up on the urban soil, which then reduces soil fertility (Chen et al., 2005). Soil contamination could be ameliorated via phytoremediation, which in the current context involves the use of plants for reclamation of contaminated soils (Rostami & Azhdarpoor, 2019). For urban forestry, the focus should be on the selection of acid-tolerant trees or phytoremediators. Legumes, for instance, *A. mangium*, *L. leucocephala*, *P. pterocarpum*, *D. regia* and *P. indicus* are prime candidates for they are not only fast-growing but could also form symbiotic relationships with nitrogen-fixing bacteria of the genera *Rhizobium* and *Bradyrhizobium* (Doyle, 1994) that allow them to prosper in contaminated soils. Besides, they play a vital role in soil improvement and are extensively grown for soil rehabilitation and amenity planting (Asif et al., 2017).

Acacia mangium has long been planted in tin tailing sites in Malaysia for it can rehabilitate the soil through the absorption and storage of heavy metals in its leaves, shoots and roots (Muhammad et al., 2017; Veronica et al., 2011). Ahmad and Jeyanny (2018) reported that *A. mangium* ameliorated Al and Fe toxicity in acidic soil that restricts the growth of a plant, especially the belowground biomass. Possessing a high bio-concentration factor, *A. mangium* also remediated heavy metal contamination in sewage sludge as the efficiency of phytoextraction for Zn, Cu and Cd was high (Shibli et al., 2013). Majid et al (2011) too reported that *A. mangium* showed high translocation factor in soils with high heavy metal concentration meaning that it was able to absorb Cd, Cu and Zn from the soil. Since the plant could grow on pebble-laden infertile soil with low water holding capacity (Roszaini et al., 2019), it is a prime candidate for urban forestry. Similar to *A, mangium, L. leucocephala* is also able to grow in the harsh environmental conditions and accumulate heavy metals, thus fixing nutrient-deficient soil (Ssenku et al., 2017). Moreover, it was also grown on metal-contaminated soil for metal stabilization (Shu et al., 2002; Zhang et al, 2001). In addition, Normaniza et al. (2009) observed that the leaves of *L. leucocephala* grown on acidic soil recorded 36% more Al compared to control. Since Al concentration was more than 1000 ppm, the plant was termed an Al-accumulator for it showed high tolerance to Al toxicity. *Leucaena leucocephala* is reported to be a phytostabilizing plant that immobilizes metal in the rhizosphere, hence reducing its bioavailability and making the soil less toxic for the plants to grow (Mendez & Maier, 2008; Saraswat & Rai, 2011; Wong, 2003). As for *P. pterocarpum*, the plant was used to clean up waste oil contamination whereby *P. pterocarpum* treated soil showed significantly lower carbon content (Edwin-Wosu & Nkang, 2016). Likewise, *D. regia* planted soils showed the most hydrocarbon removal ratein which the plant exhibited highest growth in terms of plant height, girth and leaf production (Oyedeji, 2016). Meanwhile, *P. indicus* was found suitable to remediate chromium-contaminated soil (Sunthep et al., 2016; Table 12.1).

Conclusion

Urban trees with suitable rooting architecture are the key to the alleviation of soil erosion and soil mass movement for they provide the much needed mechanical reinforcement. The H-type of root system exhibited by *L. leucocephala* and VH-type possessed by *P. indicus* and *A. mangium* are deemed suitable for soil reinforcement, slope protection and wind resistance. On the contrary, *P. pterocarpum* and *S. campanulatum* which have R-type rooting architecture are more effective in improving soil shear strength. Meanwhile, *Dillenia suffruticosa,* a shrubby tree has M-type rooting which makes it an ideal surface erosion plant candidate.

Table 12.1 List of urban trees with soil-related ecosystem characteristics

Species	Soil stability	Erosion control	Soil remediation	Soil improvement
Acacia mangium	√	√	√	√
Andira inermis		√		
Cassia fistula		√	√	
Cassia spectabilis	√	√		
Senna surattensis		√		
Delonix regia		√	√	√
Dillenia suffruticosa		√		
Cyrtophyllum fragrans		√		
Jacaranda fillicifolia		√		
Khaya senegalensis		√		
Callerya atropurpurea	√	√		
Peltophorum pterocarpum	√	√		
Pterocarpus indicus	√	√	√	√
Syzygium campanulatum	√			
Tabebuaia rosea		√		

Besides root system architecture, stakeholders should also pay attention to the tree's soil amelioration properties. Generally, leguminous trees tick both boxes for they not only possess a suitable root network but could also act as phytoremediators. They can recondition infertile and marginal urban soils due to their nitrogen-fixing ability. Moreover, as discussed above, some of these species have been successfully utilized to rehabilitate tin-tailing sites, hence they could be repurposed as urban trees for alleviation of heavy metal toxicity in urban soils. Do not take a hard stand on exotic or introduced species since Malaysia's biodiversity is enriched by their presence. A softer approach in looking at both the pros and cons of using non-native species or perhaps the use of SWOT analysis could help decision-makers in making the right selection of urban trees since no one size fits all. Although the Department of Irrigation and Drainage Malaysia has released a list of suitable trees for erosion control, no additional information on the characteristics of these plants are published. Thus, researchers, be it in the public or private sectors should shoulder the responsibility of unraveling these data through field trials.

References

Abdullah, M. N., Nomaniza, O., & Ali, F. H. (2011). Soil-root shear strength properties of some slope plants. *Sains Malaysiana, 40*(10), 1065–1073.

Abe, K., & Robert, Z. (1991). Effect of tree roots on a shear zone: Modeling reinforced shear stress. *Canadian Journal of Forest Research., 21*, 1012–1019. https://doi.org/10.1139/x91-139

Ahmad, Z. Y., & Jeyanny, V. (2018). Phytoremediation of heavy metals using acacia mangium in rahman hydraulic tin (RHT) tailings, Klian Intan, Malaysia. *Advances in Plants and Agriculture Research, 8*(3), 247–249. https://doi.org/10.15406/apar.2018.08.00322

Aida, N., Sasidhran, S., Kamarudin, N. & et al. (2016). Woody trees, green space and park size improve avian biodiversity in urban landscapes of Peninsular Malaysia. *Ecological Indicators. 69*, 176–183. https://doi.org/10.1016/j.ecolind.2016.04.025

Ali, F. (2010). Use of vegetation for slope protection: Root mechanical properties of some tropical plants. *International Journal of Physical Sciences, 5*(5), 496–506.

Arunbabu, V., Sruthy, S., Antony, I., et al. (2015). Sustainable greywater management with *Axonopus compressus* (broadleaf carpet grass) planted in subsurface flow constructed wetlands. *Journal of Water Process Engineering, 7*, 153–160.

Asif, M., Dorairaj, D., & Wickneswari, R. (2017). Characterization of natural provenances of Acacia mangium Willd and Acacia auriculiformis A. Cunn. ex Benth. in Malaysia based on phenotypic traits. *Journal of Forest Science, 63*, 562–576.

Asif, M. J., Govender, N. T., Ang, L. H., et al. (2017). Growth performance and lignin content of *Acacia mangium* Willd. and *Acacia auriculiformis* A. Cunn. ex Benth. under normal and stressed conditions. *Journal of Forest Science, 63*, 381–392.

Avani, N., Habibah, L. & Bibalani, H. Ghassem. (2014). Root distribution of Acacia mangium Willd. and Macaranga tanarius L. of rainforest. *Bangladesh Journal of Botany, 43*. https://doi.org/10.3329/bjb.v43i2.21665,

De Baets, S., Poesen, J., Reubens, B., et al. (2008a). Root tensile strength and root distribution of typical Mediterranean plant species and their contribution to soil shear strength. *Plant Soil, 305*, 207–226. https://doi.org/10.1007/s11104-008-9553-0

Baets, S. D., Torri, D., Poesen, J., et al. (2008b). Modelling increased soil cohesion due to roots with EUROSEM. *Earth Surface Processes and Landforms, 33*, 1948–1963.

Balzano, B., Tarantino, A., & Ridley, A. (2019). Preliminary analysis on the impacts of the rhizosphere on occurrence of rainfall-induced shallow landslides. *Landslides, 16*, 1885–1901. https://doi.org/10.1007/s10346-019-01197-5

Bartens, J., Day, S. D., Harris, J. R., et al. (2009). Transpiration and root development of urban trees in structural soil stormwater reservoirs. *Environmental Management, 44*, 646–657. https://doi.org/10.1007/s00267-009-9366-9

Beidokhti, T. A. N., & Moore, T. L. (2021). The effects of precipitation, tree phenology, leaf area index, and bark characteristics on throughfall rates by urban trees: A meta-data analysis. *Urban Forestry and Urban Greening, 60*, 127052.

Bischetti, G. B., Di Fidio, M., & Florineth, F. (2014). On the origin of soil bioengineering. *Landscape Research, 39*, 583–595.

Bruijnzeel, L. A. (2004). Hydrological functions of tropical forests: Not seeing the soil for the trees? (vol 104) (Agriculture, Ecosystems and Environment Elsevier)

Chalker-Scott, L. (2015). Non-native, non-invasive woody species can enhance urban landscape biodiversity. *Arboriculture and Urban Forestry, 41*(4), 173–185.

Chen, T. B., Zheng, Y. M., Lei, M., Huang, Z. C., Wu, H. T., Chen, H., Tian, Q. Z., et al. (2005). Assessment of heavy metal pollution in surface soils of urban parks in Beijing, China. *Chemosphere, 60*(4), 542–551.

Combalicer, M. S., Lee, D. K., Park, Y. D., & Bae, K. K. (2014). The potentials of nitrogen fixing tree species for forest restoration in the Philippines. *International Research Journal of Biological Sciences, 3*(6), 73–82.

Coppin, N. J., & Richards, I. J. (1990). *Use of vegetation in civil engineering*. CIRIA, Butterworths.

Crozier, M. J. (2010). Deciphering the effect of climate change on landslide activity: A review. *Geomorphology, 124*, 260–267. https://doi.org/10.1016/j.geomorph.2010.04.00

De Coninck, H., Revi, A., Babiker, M., Bertoldi, P. & et al. (2018). Strengthening and implementing the global response. In: Masson-Delmotte, V., Zhai, P., Pörtner, H. O., Roberts, D., Skea, J., Shukla, P. R., Pirani, A., Moufouma-Okia, W., Péan, C., Pidcock, R., Connors, S., Matthews, J. B. R., Chen, Y., Zhou, X., Gomis, M. I., Lonnoy, E., Maycock, T., Tignor, M., Waterfield, T. (Eds.), *Global warming of 1.5°C. An IPCC Special Report on the impacts of global warming of 1.5°C above pre-industrial levels and related global greenhouse gas emission pathways, in the context of strengthening the global response to the threat of climate change, sustainable development, and efforts to eradicate poverty*. Intergovernmental Panel on Climate Change (IPCC).

Dorairaj, D., & Osman, N. (2021). Present practices and emerging opportunities in bioengineering for slope stabilization in Malaysia: An overview. *PeerJ, 9*, e10477.

Dorairaj, D., Suradi, M. F., Mansor, N. S. & Osman, N. (2020). Root architecture, rooting profiles and physiological responses of potential slope plants grown on acidic soil. *PeerJ, 8*(2), e9595. https://doi.org/10.7717/peerj.9595

DOSM, (2020). Current population estimates. Department of Statistics Malaysia https://www.dosm.gov.my/v1/index.php?

Doyle, J. J. (1994). Phylogeny of the legume family: An approach to understanding the origins of nodulation. *Annual Review of Ecology and Systematics, 25*, 325–349.

Duke, J. A. (1983). Handbook of energy crops. NewCROPS web site. Purdue University.

Durán Zuazo, V. H., & Rodríguez Pleguezuelo, C. R. (2008). Soil-erosion and runoff prevention by plant covers: A review. *Agronomy for Sustainable Development, 28*, 65–86. https://doi.org/10.1051/agro:2007062

Edwin-Wosu, N. L., & Nkang, A. E. (2016). Evaluation of phytoremediation potential of *Peltophorum pterocarpum* (DC.) Heyne *Leucaena leucocephala* (Lam.) De Wit. and *Crotolaria retusa* Linn for waste oil contaminated soils. *Journal of Applied Sciences and Environmental Management, 20*, 669–678.

Elmendorf, W., Gerhold, H. D. & L. Kuhns. (2005). A guide to preserving trees in development projects (vol. 2009). Penn State College of Agricultural Sciences, Agriculture Research and Cooperative Extension.

Escobedo, F. J., Kroeger, T., & Wagner, J. E. (2011). Urban forests and pollution mitigation: Analyzing ecosystem services and disservices. *Environmental Pollution, 159*(8), 2078–2087.

Faisal, H.A. & Normaniza, O. (2008). Shear strength of a soil containing vegetation roots. *Soils and Foundations 48*, 587–596. https://doi.org/10.3208/sandf.48.587

Fan, C. C., & Chen, Y. W. (2010). The effect of root architecture on the shearing resistance of root-permeated soils. *Ecological Engineering, 36*(6), 813–826. https://doi.org/10.1016/j.ecoleng.2010.03.003

FAO. (1998). The Global Forest Resources Assessment 2000 (FRA 2000): Terms and definitions. FRA Working Paper 1, FAO Forestry Department, Rome.

FAO. (2004). Global Forest Resources Assessment update 2005 (FRA 2005): Terms and definitions. FRA Working Paper 83/E, FAO Forestry Department, Rome. 34 p.

FAO. (2014). EcoCrop FAO online database. Geneva, Switzerland: FAO. http://ecocrop.fao.org/ecocrop/srv/en/home

FAO. (2017). Urban and peri-urban forestry. http://www.fao.org/forestry/urbanforestry/87025

Foy, C. D. (1992). Soil chemical factors limiting plant root growth. In: Hatfield, J. L, Stewart BA, (Eds.), *Advances in Soil Science 19 Limitations to Plant Growth*. Springer-Verlag, (pp. 97–149).

Geddes, N., & Dunkerley, D. (1999). The influence of organic litter on the erosive effects of raindrops and of gravity drops released from desert shrubs. *CATENA, 36*, 303–313. https://doi.org/10.1016/S0341-8162(99)00050-8

Ghestem, M., Cao, K., Ma, W., Rowe, N., et al. (2014). A framework for identifying plant species to be used as 'ecological engineers' for fixing soil on unstable slopes. *PLoS ONE, 9*(8), e95876. https://doi.org/10.1371/journal.pone.0095876

Giadrossich, F., Schwarz, M., Cohen, D., Preti, F., & Or, D. (2012). Mechanical interactions between neighbouring roots during pullout tests. *Plant and Soil, 367*, 391–406.

Gilman, E. F. (2003). *Where are tree roots?* University of Florida.

Gonzalez-Ollauri, A., & Mickovski, S. B. (2017). Hydrological effect of vegetation against rainfall-induced landslides. *Journal of Hydrology, 549*(374), 387. https://doi.org/10.1016/j.jhydrol.2017.04.014

Gray, D. H., & Sotir, R. B. (1996). *Biotechnical and soil bioengineering slope stabilization: A practical guide for erosion control*. John Wiley & Sons Inc.

Hairiah, K., Widianto, W., Suprayogo, D., & Van Noordwijk, M. (2020). Tree roots anchoring and binding soil: Reducing landslide risk in Indonesian agroforestry. *Land*, *9*(8), [256]. https://doi.org/10.3390/LAND9080256

Huat, B. B. K., & Kazemian, S. (2010). Study of root theories in green tropical slope stability. *Electronic Journal of Geotechnical Engineering, 15*, 1825–1834.

Hughes, S. (2003). Slope failure: Landslides, mudflows, earthflows and other mass wasting processes. http://geology.isu.edu/wapi/EnvGeo/EG4_mass_wasting/EG4-MassWasting

Ishak, M. F., Ali, N. & Kassim, A. (2013). The effect of tree induce suction on slope stabilization analysis. *International Journal of Civil Engineering and Geo-Environmental, 4*, 14–22.

Jerome, I. (2010). The role of roots in slope stability. Available at https://open.library.ubc.ca/media/download/pdf/52966/1.0075519/1.

Jin, M. J. W., Ye, H. C., Xu, Y. F., Shen, C. Y., & Huang, Y. F. (2011). Spatial and temporal patterns of soil fertility quality and analysis of related factors in urban-rural transition zone of Beijing. *African Journal of Biotechnology, 10*, 10948–10956.

Kabisch, N., Qureshi, S., & Haase, D. (2015). Human-Environment Interactions In Urban Green Spaces—A systematic review of contemporary issues and prospects for future research. *Environmental Impact Assessment Review, 50*, 25–34.

Kamal, A. (2021). Native trees. Biodiversity Gardens Capacity Building Workshop. Malaysia . https://ubi-my.com/2021/03/17/native-trees-by-adam-kamal/

Kanniah, K. D. (2017). Quantifying green cover change for sustainable urban planning: A case of Kuala Lumpur Malaysia Urban for. *Urban Green, 27*, 287–304.

Kanniah, K., & Ho, C. (2017). Urban forest cover change and sustainability of Malaysian cities. *Chemical Engineering Transactions, 56*, 673–678.

Karim, A. B. (1987). Alley cropping studies in the uplands of Sierra Leone. University of Oxford. xvii 272 pp.

Karthikeyan, A., Chandrasekaran, K., Geetha, M. & Kalaiselvi, R. (2013). Growth response of Casuarina equisetifolia Forst. rooted stem cuttings to Frankia in nursery and field conditions. *Journal of Bioscience*, *38*(4), 741–747. https://doi.org/10.1007/s12038-013-9362-3

Lateh, H., Avani, N., Bibalani, H. G. (2013). Effect of Acacia mangium root properties on shallow landslide and slope stability. *Journal of Life Sciences and Technologies*, 127–131. https://doi.org/10.12720/jolst.1.2.127-131

Lehmann J., & Schroth G. (2003). Nutrient leaching. Trees, crops and soil fertility. Wallingford: CABI Publishing, 151_166.

Li, X., Niu, J., & Xie, B. (2014). The Effect of Leaf Litter Cover on Surface Runoff and Soil Erosion in Northern China. *PLoS ONE, 9*(9), e107789. https://doi.org/10.1371/journal.pone.0107789

Liu, Y., Hu, J., Wang, T., Cai, C., Li, Z., & Zhang, Y. (2016). Effects of vegetation cover and road concentrated flow on fill slope erosion in rainfall and scouring simulation tests in the Three Gorges Reservoir Area, China. *CATENA, 136*, 108–117. https://doi.org/10.1016/j.catena.2015.06.006

Llorens, P., & Domingo, F. (2007). Rainfall partitioning by vegetation under Mediterranean conditions: A review of studies in Europe. *Journal of Hydrology*. https://doi.org/10.1016/j.jhydrol.2006.10.032

Lu, N., & Godt, J. W. (2013). *Hillslope hydrology and stability*. Cambridge University Press.

Lüttge, U., & Buckeridge, M. (2020). Trees: Structure and function and the challenges of urbanization. *Trees*. https://doi.org/10.1007/s00468-020-01964-1

Mafian, S., Huat, B. B., Rahman, N. A., & Singh, H. (2009). Potential plant species for live pole application in tropical environment. *American Journal of Environmental Sciences*, *5*(6), 759–764 https://doi.org/10.3844/ajessp.2009.759.764
Majid, N. M., Islam, M. M., Justin, V., Abdu, A., & Ahmadpour, P. (2011). Evaluation of heavy metal uptake and translocation by *Acacia mangium* as a phytoremediator of copper contaminated soil. *African Journal of Biotechnology, 10*, 8373–8379.
Mendez, M. O., & Maier, R. M. (2008). Phytoremediation of mine tailings in temperate and arid environments. *Reviews in Environmental Science and Biotechnology, 7*, 47–59. https://doi.org/10.1007/s11157-007-9125-4
Mohd. Shariff, M.K. & Abu Bakar, S. (2006). Invasive plants in the Malaysian landscape. *ALAM CIPTA, International Journal on Sustainable Tropical Design Research and Practice*, *1*(1), 41–48.
Morgan, R. P. C., & Rickson, R. J. (1995). *Slope stabilization and erosion control: A bioengineering approach.*
Muhammad, A. J., Ong, S. S., & Ratnam, W. (2017). Characterization of mean stem density, fibre length and lignin from two Acacia species and their hybrid. *Journal of Forestry Research, 29*, 549–555.
Mulyono, A., Subardja, A., Ekasari, I., Lailati, M., Sudirja, R. & Widya. N. (2018). The hydromechanics of vegetation for slope stabilization. *IOP Conference Series: Earth and Environmental Science*, 118. 012038. https://doi.org/10.1088/1755-1315/118/1/012038
Ng, C. W. W., Leung, A., & Ni, J. (2019). *Plant–soil slope interaction*. CRC Press.
Noor Azlin, Y. & Azahari, M.Y. (2014). *Syzygium campanulatum*: The pretty screen. https://www.frim.gov.my/colour-of-frim/syzygium-campanulatum-the-pretty-screen
Noraini, M. T., & Roslan, H. (2008). An annotated field guide to tropical eco-engineering, Monograph Series 10. Kuala Lumpur: Institute of Ocean and Earth Sciences (IOES), University of Malaya.
Normaniza, O., & Barakbah, S. S. (2006). Parameters to predict slope stability-soil water and root profiles. *Ecological Engineering*, 28,90–95. https://doi.org/10.1016/j.ecoleng.2006.04.004
Normaniza, O., & Barakbah, S. S. (2011). The effect of natural succession on slope stability. *Ecological Engineering*, *37*(2), 139–147. https://doi.org/10.1016/j.ecoleng.2010.08.002. Chapman & Hall.
Normaniza, O., Faizal, H. A., & Barakbah, S. S. (2009). The role of pioneer vegetations in accelerating the process of natural succession. *American Journal of Environmental Sciences*, *5*(1), 7–15. https://doi.org/10.3844/ajessp.2009.7.15
Normaniza, O., Saifuddin, M., & Halim, A. (2014). Contribution of vegetation to alleviate slope's erosion and acidity, environmental risk assessment of soil contamination, Maria C. Hernandez-Soriano. IntechOpen.
Orwa, C., Mutua, A., Kindt, R., Jamnadass, R., & Anthony, S. (2009). Agroforestree Database: A tree reference and selection guide version 4.0 (http://www.worldagroforestry.org/sites/treedbs/treedatabases.asp)
Oyedeji, A.A. (2016). Impacts of selected leguminous tree species and kaolinite pre-amendment on oil-contaminated soil for bioremediation in the oil-bearing region of Nigeria. Thesis PhD University of Wolverhampton.
Parrotta, J. A. (1993). Secondary forest regeneration on degraded tropical lands. In *Restoration of tropical forest ecosystems* (pp. 63–73). Springer.
Potter, K., Rimbawanto A., Beadle C. (Eds.). (2006). Heart Rot and Root Rot in Tropical *Acacia* Plantations. Canberra, Australian Centre for International Agricultural Research, 92.
Public Works Department Malaysia, (PWD). (2009). Public Works Department Malaysia homepage. Available at http://www.kkr.gov.my/en/node/14668
Reubens, B., Poesen, J., Danjon, F., Geudens, G., & Muys, B. (2007). The role of fine and coarse roots in shallow slope stability and soil erosion control with a focus on root system architecture: a review. *Trees: Structure and Function, 21*(4), 385–402. https://doi.org/10.1007/s00468-007-0132-4

Rojas-Sandoval, J., & Acevedo-Rodriguez, P. (2013). *Delonix regia* (Flamboyant). Invasive Species Compendium, CABI Publishing.

Roloff, A., Korn, S., & Gillner, S. (2009). The Climate-Species-Matrix to select tree species for urban habitats considering climate change. *Urban Forestry and Urban Greening, 8*(4), 295–308.

Rostami, S., & Azhdarpoor, A. (2019). The application of plant growth regulators to improve phytoremediation of contaminated soils: A review. *Chemosphere, 220*, 818–827. https://doi.org/10.1016/j.chemosphere.2018.12.203

Roszaini, K., Wan-Rashidah, A. K., Mohamad-Fakri, I., & Ang, L. H. (2019). *Acacia mangium, Hopea odorata* and *Khaya ivorensis* grown in ex-mining and bris plantation plots: Resistance against subterranean termites. *Journal of Tropical Forest Science, 31*(3), 362–370. https://www.jstor.org/stable/26740153

Roy, S., Byrne, J., & Pickering, C. (2012). A systematic quantitative review of urban tree benefits, costs, and assessment methods across cities in different climatic zones. *Urban Forestry and Urban Greening, 11*(4), 351–363.

Saifuddin, M., & Normaniza, O. (2014). Evaluation of hydro-mechanical properties and root architecture of plants for soil reinforcement. *Current Science, 107*, 845–852.

Saifuddin, M., & Normaniza, O. (2016). Rooting characteristics of some tropical plants for slope protection. *Journal of Tropical Forest Science, 28*(4), 469–478.

Saraswat, S., & Rai, J. (2011). Prospective application of *Leucaena Leucocephala* for phytoextraction of Cd and Zn and nitrogen fixation in metal polluted soils. *International Journal of Phytoremediation, 13*, 271–288. https://doi.org/10.1080/15226514.2010.483259

Sayer, E. J. (2006). Using experimental manipulation to assess the roles of leaf litter in the functioning of forest ecosystems. *Biological Reviews, 81*, 1–31. https://doi.org/10.1017/S1464793105006846

Shafie, A. (2009). A case study on floods of 2006 and 2007 in Johor, Malaysia. Master's Thesis, Colorado State University, Fort Collins, CO, USA.

Shibli, N., Majid, N., Shazili, M., Azhar, N., & Abdu, A. (2013). Growth performance, biomass and phytoextraction efficiency of acacia mangium and Melaleuca cajuputi in remediating heavy metal contaminated soil. *American Journal of Environmental Sciences, 9*, 310–316. https://doi.org/10.3844/ajessp.2013.310.316

Shu, W., Xia, H. P., Zhang, Z. Q., Lan, C. Y., & Wong, M. (2002). Use of vetiver and three other grasses for revegetation of Pb/Zn mine tailings: Field experiment. *International Journal of Phytoremediation, 4*, 47–57. https://doi.org/10.1080/15226510208500072

Singh, A. (2010). Bioengineering techniques of slope stabilization and landslide mitigation. *Disaster Prevention and Management, 19*(3): 384–397. https://doi.org/10.1108/09653561011052547

Sjöman, H., Morgenroth, J., Sjöman, J. D., Sæbø, A., & Kowarik, I. (2016). Diversification of the urban forest—Can we afford to exclude exotic tree species? *Urban Forestry and Urban Greening, 18*, 237–241.

Ssenku, J. E., Ntale, M., Backeus, I., & Oryem-Origa, H. (2017). Phytoremediation potential of *Leucaena leucocephala* (Lam.) de Wit. for heavy metal-polluted and heavy metal-degraded environments. In *Phytoremediation potential of bioenergy plants* (pp. 189–209). https://doi.org/10.1007/978-981-10-3084-0_7

Stokes, A., Atger, C., Bengough, A. G., Fourcaud, T., & Sidle, R. C. (2009). Desirable plant root traits for protecting natural and engineered slopes against landslides. *Plant and Soil, 324*(1–2), 1–30. https://doi.org/10.1007/s11104-009-0159-y

Stokes, A., Douglas, G. B., Fourcaud, T., Giadrossich, F., Gillies, C., et al. (2014). Ecological mitigation of hillslope instability: Ten key issues facing researchers and practitioners. *Plant and Soil, 377*(1–2), 1–23. https://doi.org/10.1007/s11104-014-2044-6

Stone, E. L., & Kalisz, P. J. (1991). On the maximum extent of tree roots. *Forest Ecology and Management, 46*, 59–102. https://doi.org/10.1016/0378-112790245-Q

Suntep, S., Duangrat, S., Prayad, P., & Kraichat, T. (2016). Phytoremediation of cadmium by selected leguminous plants in hydroponics culture. *Reseach and Reviews: Journal of Ecology and Environmental Sciences, 4*(1), 1–7.

Szabó, J. (2003). The relationship between landslide activity and weather: Examples from Hungary. *Natural Hazards and Earth System Sciences, 3*, 43–52.

The World Bank. (2020). World development indicators. http://data.worldbank.org/data-catalog/world-developmentindicators

Thomson, L. A. J. (2006). *Pterocarpus indicus* (narra). In Elevitch, C. R (Ed.), *Species profiles for Pacific Island agroforestry*. Permanent Agriculture Resources (PAR) Hawai'i.

United Nations. (2014). World Urbanization prospects: 2014 revision, department of economic and social affairs, population division,

Veronica, J., Majid, N. M., Islam, M. M., et al. (2011). Assessment of heavy metal uptake and translocation in *Acacia mangium* for phytoremediation of cadmium contaminated soil. *Journal of Food Agriculture and Environment., 2011*(9), 588–592.

Wang, X. P., Sheng, J. J., Gong, P., Xue, Y. G., Yao, T. D., & Jones, K. C. (2012). Persistent organic pollutants in the Tibetan surface soil: Spatial distribution, air–soil exchange and implications for global cycling. *Environmental Pollution, 170*, 145–151.

Wang, K., Aktas, Y. D., Stocker, J., et al. (2019). Urban heat island modelling of a tropical city: Case of Kuala Lumpur. *Geoscience Letter, 6*, 4. https://doi.org/10.1186/s40562-019-0134-2

Wei, B., & Yang, L. (2010). A review of heavy metal contaminations in urban soils, urban road dusts and agricultural soils from China. *Microchemical Journal, 94*(2), 99–107.

Wong, M. H. (2003). Ecological restoration of mine degraded soils, with emphasis on metal contaminated soils. *Chemosphere, 50*(6), 775–780. https://doi.org/10.1016/s0045-6535(02)00232-1

Yang, L., Liu, N., Ren, H., & Wang, J. (2009). Facilitation by two exotic Acacia: *Acacia auriculiformis* and *Acacia mangium* as nurse plants in South China. *Forest Ecology and Management, 257*(8), 1786–1793.

Yen, C. P. (1987). *Tree root patterns and erosion control* (pp. 92–111). In S. Jantawat (Ed.), *Proceedings of the International Workshop on Soil Erosion and its Counter Measures*, 11–19 November 1984.

Zhao, B., Zhang, L., Xia, Z., Xu, W., Xia, L., Liang, Y., & Xia, D. (2019). Effects of rainfall intensity and vegetation cover on erosion characteristics of a soil containing rock fragments slope. *Advances in Civil Engineering, 2019*, 1–14. https://doi.org/10.1155/2019/7043428

Zhang, Z. & Shu, Wen & Lan, C., Wong, M. (2001). Soil seed bank as an input of seed source in revegetation of lead / zinc mine tailings. *Restoration Ecology*. 9. 378 - 385. https://doi.org/10.1046/j.1526-100X.2001.94007.x

Deivaseeno Dorairaj is a Post-Doctoral Research Fellow attached to Institute for Environment and Development (LESTARI) at the National University of Malaysia (UKM). Her research expertise is in the field of abiotic stress of rice plant especially plant nutrition and plant physiology besides ecophysiology and soil bioengineering. At present, her focus is on assessing ecosystem services and socio-ecological system impacts and outcomes of mangrove habitat protection and restoration in Malaysia. She is also interested on green synthesis of nanomaterial for seed priming and plant establishment.

Nisha Govender is a research fellow at the Institute of Systems Biology (INBIOSIS), UKM. Her research expertise includes plant sciences, management of agricultural crop diseases and yield gain using multi-modal omics approaches along with high-throughput computational tools.

Normaniza Osman is a professor affiliated to the Institute of Biological Sciences, Faculty of Science, University of Malaya. She is currently the Deputy Dean of Innovative Industry and Sustainability (IISS) Research Cluster. Her expertise is in the field of plant eco-physiology, slope bioengineering and soil science. Her research focus is on the use of suitable slope plants to improve slope stability and root reinforcement.

Chapter 13
Effects of Tree Shading in Modifying Tropical Microclimate and Urban Heat Island Effect

Mohd Fairuz Shahidan

Abstract Trees are a natural feature with a unique potential to reduce the consequences of climate change on the environment, such as urban heat islands (UHI). Tree canopies can help to alleviate these issues by providing shade that can change the microclimate, which is especially important in tropical climates. Shades that form tree canopies respond significantly to microclimate indicators such as solar radiation and air temperature because of their relationship with their canopy form and density. The shade intensity of the tree canopy, on the other hand, varies between tree species and even within the same species. Thus, this vital knowledge is enlightened on how the variation could modify urban microclimate and consequently mitigate the UHI effect. A guideline in promoting optimum cooling from the tree shading effect is also highlighted. The theoretical, process and application evidence are explained in general with the details to implement in real-world practice. Comprehensive interpretation of design recommendations is established for readers and designers who are concerned with the microclimate control, UHI mitigation and sustainable environment for guidance in designing landscapes such as parks, open spaces, residential areas, and parking spaces.

Keywords Tree shade · Tree form · Microclimate modification · UHI · Tree planting arrangement · Tropical climate

Introduction

Villages, towns, and cities have prioritised the requirements of their citizens in terms of society, economy, culture, and comfort over time as a consequence of a desire to explore new living environments and establish new settlements that better fit human needs. Cities have expanded on increasingly larger scales to meet human requirements. As a result, enormous areas of urbanisation occupied by new building structures and materials have changed the climatic features of urban spaces (Roth,

M. F. Shahidan (✉)
Department of Landscape Architecture, Faculty of Design and Architecture, Universiti Putra Malaysia (UPM), 43400 Serdang, Selangor Darul Ehsan, Malaysia
e-mail: mohdfairuz@upm.edu.my

S. Maruthaveeran et al. (eds.), *Urban Forestry and Arboriculture in Malaysia*,
https://doi.org/10.1007/978-981-19-5418-4_13

2002). In addition, man-made urbanisation has had a significant detrimental influence on the natural ecology and landscape (Akbari, 2002). These changes have a direct influence on urban climate, causing air and surface temperatures to increase above those in rural regions; this phenomenon is known as the 'urban heat island' or UHI (Emmanuel, 2005; Gartland, 2008; Wong & Yu, 2009; Landsberg, 1981). According to these research, inadequate design planning, rising urbanisation, and fast changes in the external environment are the major drivers of the UHI phenomenon (Roth, 2002). Changes in roadway surface materials and a decrease in green space are mostly to blame for the rise in urban temperature (Papadakis et al., 2001; Shashua-Bar et al., 2004; Takahashi et al., 2004; Wong et al., 2007).

A single tree may greatly moderate the environment in hotter regions, and the combined influence of multiple urban trees can increase urban thermal comfort (Akbari, 2002; Jauregui, 1990, 1991). In fact, in tropical climes, the cooling impact of urban trees is most obvious (Shashua-Bar & Hoffman, 2000). Trees have a lot of promise when it comes to reducing environmental issues like heat stress. Many studies have been undertaken to monitor and assess how trees might improve the microclimatic performance of built environments, adjust patterns to climate change, and reduce energy consumption and temperature control. The physical aspect of trees (e.g., density, height, shape, and species variations) is one of the most important aspects to consider when regulating microclimates, according to most of these research (Abreu-Harbich et al., 2015; Dimoudi & Nikolopoulou, 2003; Picot, 2004).

Direct shading and evapotranspiration are the two processes through which trees in a city may provide cooling advantages (Klemm et al., 2015). The impact, however, is not due to the air being cooled, but rather to the air being warmed less (Dimoudi & Nikolopoulou, 2003). The density of trees has a significant influence on their shading effect (Coutts et al., 2016). Because the most influenced climatic variable in a tropical environment is radiation, the shade provided by a tree canopy's foliage geometry might minimise glare and filter diffuse light from the sky and surrounding surfaces, modifying the heat exchange between building and ground surfaces (Berry et al., 2013; Coutts et al., 2016; Klemm et al., 2015). The evapotranspiration process of plants and heat blocking by the canopy will both be moderated by this circumstance (Konarska et al., 2014). Aside from the reduced heat from the tree canopy obstruction and the development of solid shade, the area's air and surface temperatures might be reduced (Abreu-Harbich et al., 2015). Consequently, the ambient temperature, as well as the shaded hard surfaces, may be kept low.

Shade trees' potential to improve tropical outdoor conditions is based on their ability to reduce heat by blocking radiation and creating shade (Abreu-Harbich et al., 2015). By absorbing, reflecting, and transmitting solar radiation, this aids in radiation filtering (Kotzen, 2003). Large and small branches or limbs, as well as the amount of leaf cover, combine to form this shadow (Abreu-Harbich et al., 2015). The number of limbs and leaf cover, on the other hand, varies greatly amongst tree species and even within the same species. It should be noted that each species' shadow performance varies, and their ability to filter radiation will have an impact on microclimate alteration (Brown & Gillespie, 1995; Shahidan et al., 2010).

Tree Canopy and Shade

Trees have many canopy forms, and it can be claimed that there are differences according to species. In general, canopy form can be classified as columnar, pyramidal, oval, rounded, spreading, vase (upright spreading) and weeping (Fig. 13.1).

The tree canopies create shade according to their form. The form of the canopy is influenced by the branching system, quantity and size of leaves, tree height and size. These criteria affect the area and quality of shade. The quantity of radiation that is absorbed, reflected and transmitted by the tree canopy influences the quality of the shade. The quality of shade is determined by how well the tree blocks sunlight and how much heat is absorbed and transmitted beneath the canopy. As a result, the tree's shade affects the heat and air temperature beneath it (Kotzen, 2003; Shahidan et al., 2010).

In the context of tree height and form in tropical climate, it was found that the effects of shade and evapotranspiration were greatest in the afternoon (12 p.m.–1 p.m.). Due to the near proximity of the solar height angle to 90 degrees, the majority of shadow and evaporation occurs just around the tree canopy (Shahidan et al., 2010). When tree canopy forms and shapes are chosen based on width (form) rather than height (height) in this situation, the impacts will be significantly stronger (Fig. 13.2). This is because canopies with a wider breadth (shape) (i.e., oval/round/spreading) create more shadow and a greater shadowed area most of the time.

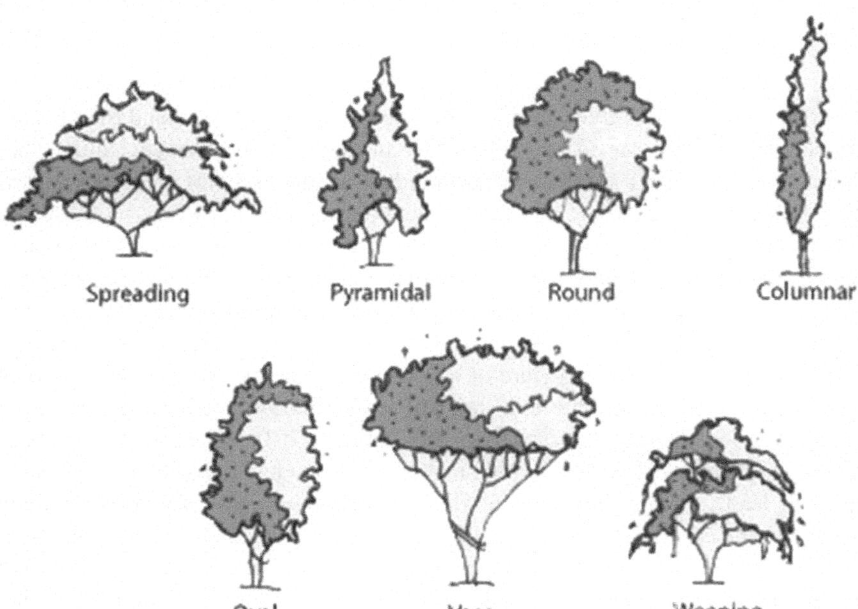

Fig. 13.1 Types of tree canopy form

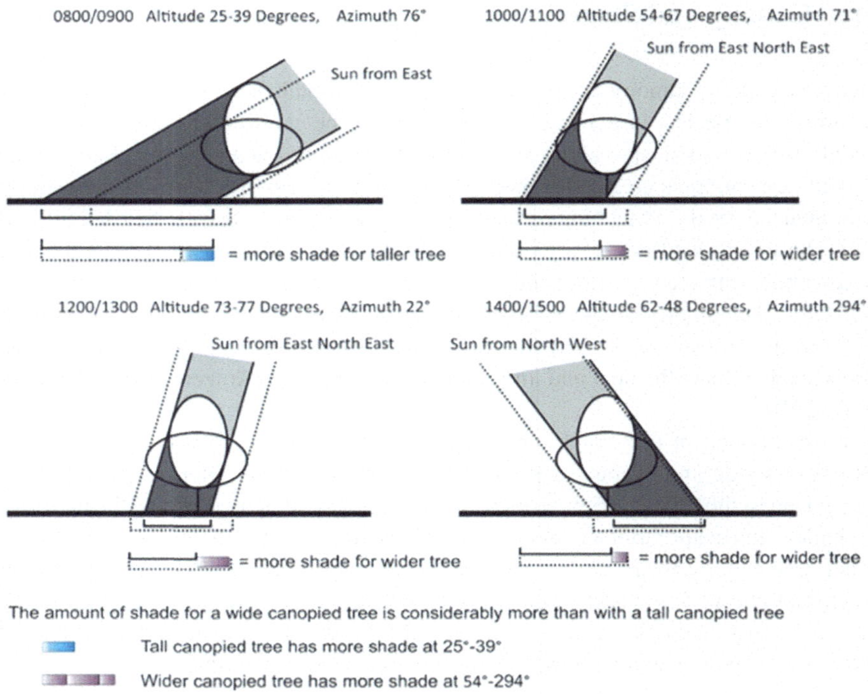

Fig. 13.2 Comparison of the quantity of shade generated by broad/wide canopied trees versus tall canopied trees. *Source* Adapted from Shahidan et al. (2016)

In this context, wider tree canopy forms such as spreading, oval and rounded are suitable tree canopy forms in providing wider and larger areas of shade. Unlike other taller canopy forms such as columnar, pyramidal, vase and weeping creating a narrow and smaller area of shades at all times, except during the early in the morning due to the canopy form and angle of the sun. Thus, the size of the shaded area created by the canopy form will influence the size of the modified area of the microclimate. The greater the shaded area, the better the microclimate improvement.

The density of tree canopy also plays an important role in increasing the intensity of tree canopy shade. The structure of tree canopy form such as rounded and oval form provide better density due to the branching and limb system that arranges leaves cumulatively and densely (Kotzen, 2003; Tukiran et al., 2016). The canopy's arrangement of leaves allows for more radiation absorption and less radiation transmission. This will lead to a high intensity of shade with low amount of heat transmitted underneath the canopy (Shahidan et al., 2010; Takács et al., 2016). Thus, the intensity of tree shade can be evaluated by knowing the amount of heat we experience. On the other hand, high intensity of shade from dense tree canopy will improve air and surface temperature by decreasing the heat transmission to the immediate air, ground, building wall and roof area. The magnitude of shading intensity will be different for each species depending on the density of tree canopy (Shahidan et al., 2010, 2016; Takács et al., 2016).

Direct shadow intensity may reduce glare and prevent diffuse light from the sky and surrounding surfaces in some instances. This could be effective for screening indoor spaces with trees in the external environment as natural shade devices. Furthermore, densely planted trees in parking spots limit heat transfer and block glare onto vehicles parked beneath the tree.

By having suitable tree species, the reduction of heat transmitted into a car can be controlled due to the understanding of tree shading intensity. For instance, a study done by Shahidan et al. (2010) has shown that by having *Mesua ferrea* trees covering parking spaces, almost 97% of radiation was absorbed by high-density canopy. Hence, only 3% heat was transferred underneath the canopy which had a cooling effect to the car parked underneath the tree. Meanwhile, the less dense tree, such as *Hura crepitans*, was able to provide limited intensity of shade with only 78% radiation absorbed and 28% transferred underneath the canopy (Fig. 13.3). Thus, this type of tree species provides less cooling to the car parked beneath the tree. Obviously, choosing tree species with appropriate canopy density is critical when deciding which tree species to plant on our allotted area and how much microclimate adjustment is required. Thus, it is important to understand how much radiation filtration that each individual species can offer in minimizing the heat and directly cooling the surrounding environment.

In maximising the effect of shading, the quantity of planted trees should be increased. The shade effect would be significantly larger in circumstances when the sun is directly above if trees were placed in clusters of linear, continuous, or zigzag configurations, depending on the size and position of the trees where they are planted (Fig. 13.4) (Zhao et al., 2018). The impact of shading and the intensity will be greater if the trees were planted in continuous or zigzag arrangements

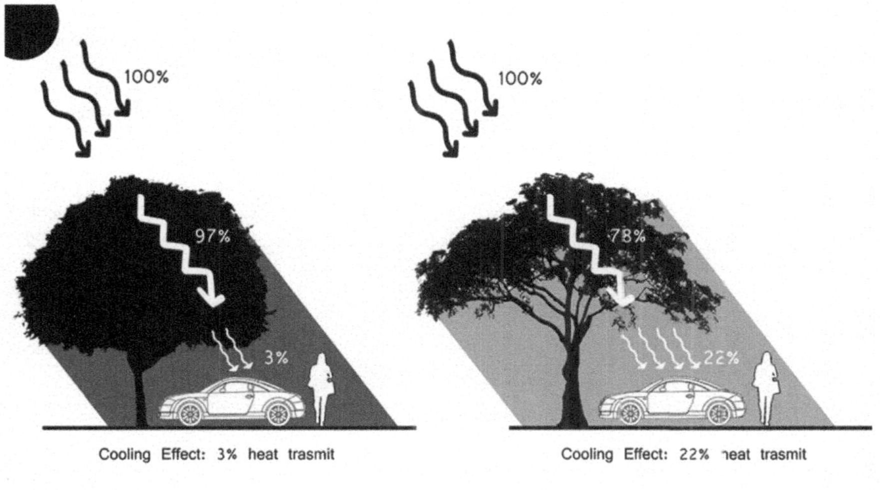

Fig. 13.3 The effect of different types of trees shading on humans, vehicles and ground surfaces

where the double planting creates higher intensity when planted in clusters. Thus, the shaded area will be optimised by overlapping shadows and variations of tree intensity (Fig. 13.5). Thus, the quality of shading will improve the immediate environment through effective reduction of air and surface temperature.

The linear arrangement can be designed ideally for pedestrian areas and street tree plantings. Meanwhile, the continuous and zigzag planting arrangements can be designed for car parking areas, wider pedestrian walkways and streets, parks, plazas, squares, residential areas or any spaces that require optimum shading to improve the microclimate. For a larger shaded area, the wide and spreading canopy form with medium tree height is recommended. Since the canopy of the medium tree is closer to the ground and the canopy angle is larger, it will effectively shade the spaces (Kotzen, 2003; Shahidan et al., 2010). This shape of the canopy correlates well with the orientation of the sun and can give a wider area of shading (Kotzen, 2003; Shahidan et al., 2010). In this situation, it is possible to enhance the microclimate of the landscaped area. Consequently, the modified environment will improve human thermal comfort due to less heat gain by reducing the heat transmission and finally contributing in mitigating the urban heat island effect.

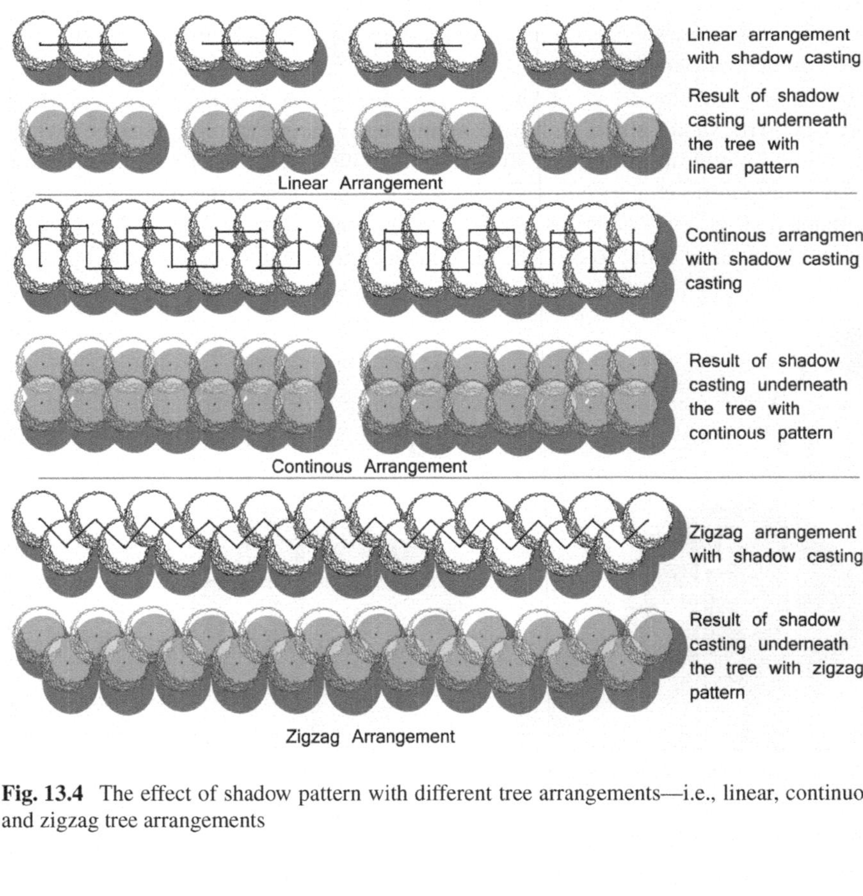

Fig. 13.4 The effect of shadow pattern with different tree arrangements—i.e., linear, continuous and zigzag tree arrangements

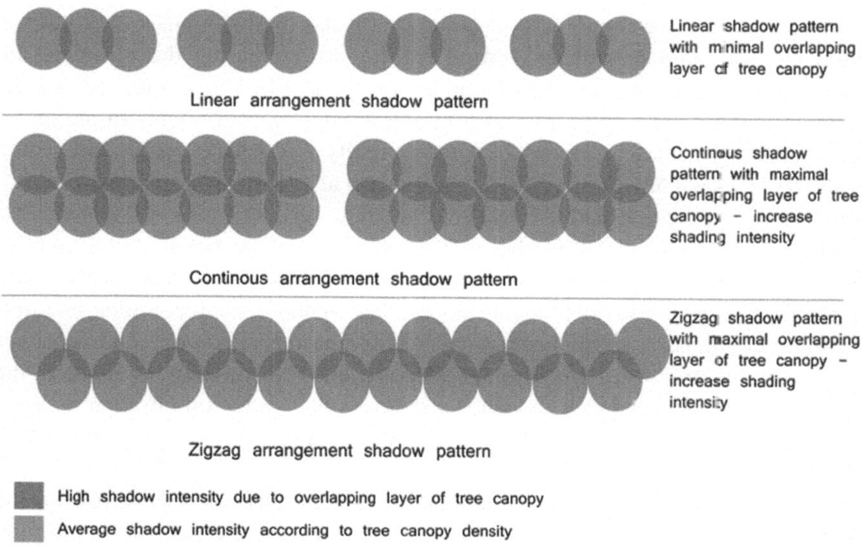

Fig. 13.5 Variation of tree shading intensity due to the arrangement and overlapping tree canopy

Evaluation of Trees and Heat Filtration

Several factors must be considered in order to use trees that can respond to climate change and maximise their shading benefits to humans and the environment. Direct shading, evapotranspiration, and radiation are three mechanisms that must be considered when designing landscapes for the environment. During the planting design phase, there are four primary evaluation elements to consider These are based on the functions of tree design as well as their effects on the environment. Height, canopy form, leaf area index, and radiation filtration are all factors to consider.

The density of a tree canopy affects shade intensity, evapotranspiration rate, and radiation filtration. The thermal efficiency and shadow efficacy of a tree are determined by its foliage properties as well as its mature shape, i.e. total height and canopy form (Fahmy et al., 2010; Shahidan et al., 2010). While evapotranspiration depends on the amount of vegetation that can absorb the energy and convert it to water vapour. Both are closely related to the leaf area index (LAI)—'leaf area per unit ground area;' which is considered a key metric used to recognize and compare tree canopies and one of the measurement criteria for calculating tree canopy density (Meir et al., 2000; Steven et al., 1986; Shahidan et al., 2010). The value is given as an integer scale ranging from 0 to 8 or more which represents sparse to dense canopy. We may estimate the quantity and density of the leaves and the branching mechanism in a single tree by knowing the value. We will therefore precisely understand the value of the intensity of shade, evapotranspiration and filtration rate that each individual tree species provides.

Furthermore, the change of solar radiation caused by high-density tree canopy might increase human outdoor thermal comfort by increasing the mean radiant temperature beneath the canopy (Shahidan et al., 2010). Thus, alteration of the radiation is the most significant factor in determining improvements to the area's microclimate. The air temperature, relative humidity and surface temperature would be affected (Brown & Gillespie, 1995; Shahidan et al., 2010). When trees are photosynthesising, radiation filtration may occur. A single layer of leaves will absorb 50% of solar radiation, while 30% will be reflected and 20% will be transmitted (Fig. 13.6) (Brown & Gillespie, 1995; Shahidan et al., 2010). Hence, more leaf layers would result in greater efficiency in reducing solar radiation under the tree. Radiation filtration can be defined as a blockage of solar radiation through leaves and branches of a tree by the process of absorption, reflection, and transmission that reduces the energy reaching the ground. The value used to calculate radiation filtration is W/m^2 (watt per square metre), and the total value below the canopy is known as the overall filtration value.

The list of example species of tree with the radiation filtration value reflects the intensity and efficiency of the tree shading (Table 13.1). The table shows the top five types of tropical tree species from a previous study which provide the greatest radiation filtration due to greater density tree canopy (i.e., LAI) are *Mesua ferrea, Syzygium campanulatum, Diospyros blancoi, Cinnamomum iners* and *Dillenia indica* with average solar radiation filtration from 93 to 97% with density from 3.5 to 6.1 LAI value (Shahidan et al., 2010, 2016). These trees have an ability to filter radiation up to 941 W/m^2 from the actual daily solar radiation of 969 W/m^2. Due to this reason, the shading intensity is very high and the quality of shadow is greater. If a person stands or sits under the canopies for longer periods of time, they receive only very limited heat varying from 28 to 63 W/m^2. The selection is very good for providing the optimal cooling to humans and therefore to the immediate surroundings. The effect of high-density tree cooling would be greater when the tree is planted in a cluster generating higher shading intensities. It is also important to recognise that

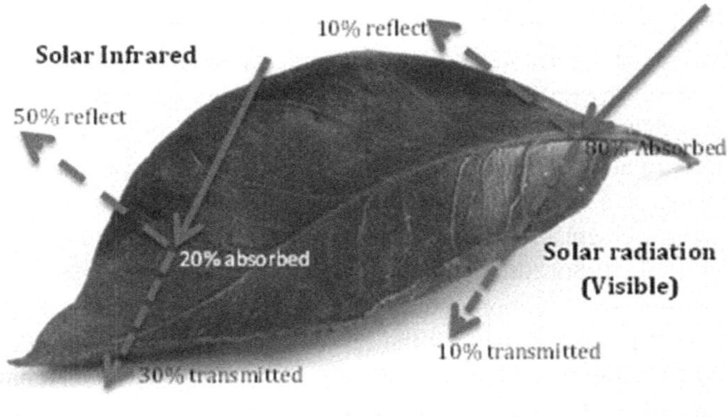

Fig. 13.6 Percentage of solar radiation absorbed transmitted and reflected by leaf. Adapted from Brown and Gillespie (1995)

other species with higher filtration abilities may have a comparable or even bigger effect than the ones listed.

Irrespective of their type of canopy, unlike *Pometia pinnata, Dalbergia oliveri, Tamarindus indica, Erythrina fusca* and *Cassia fistula*, radiation filtration is relatively poor due to less leaves and loose canopy. From 0.8 to 1.5 canopy density, these tree canopies received more heat under the canopy from 115 to 243 W/m^2. The loose canopy feature allows more sunlight to reach the ground, while having less shading intensity. Therefore, shading efficiency is lower and microclimate alteration is gaining less.

Conclusion and Recommendations

Understanding the effect of tree shade, particularly in a tropical environment, is essential for providing a greater cooling effect and thus contributing to alleviating the urban heat island effect. New knowledge on Malaysia's native tree species ability in filtering solar radiation with shading intensity will act as a reference to the designers in selecting suitable native plant species for appropriate function. It is recommended that five species are significant in providing higher filtration, shading intensity and improving microclimate due to high-density canopy, and effective radiation filtration. These include *Mesua ferrea, Syzygium campanulatum, Diospyros blancoi, Cinnamomum iners* and *Dillenia indica.* However, it is not limited to this selected number of tree species. The species that have similar characteristics of high-density canopy and canopy form can be considered in improving the microclimate.

These are recommended for outdoor space designers when designing landscapes such as parks, open spaces, residential area, parking space, green roof, vertical landscaping that concerns more on microclimate moderation, human thermal comfort, building energy performances. The study also recommends that the performance of these trees would be greater if the plantings are arranged in clusters with a continuous and zigzag arrangement. Besides, the information generated will aid designers or public to apply different compositions using different types of trees' filtration performances that could provide for better shading quality, filtration performances, and the ability to moderate urban microclimate and contribute towards mitigating urban heat island effect.

Table 13.1 Tropical trees with different heat filtration abilities

Name of tropical tree species	Average leaf area index (LAI)	Average solar radiation Filtration in Percentage (%)	Average total solar radiation filtration (W/m^2)	Average total solar irradiance received underneath canopy (W/m^2)	Type of canopy form
Mesua ferrea (Penaga Lilin)	6.1	97	941	28	Round
Syzygium campanulatum (Kelat Paya)	5.1	97	940	29	Oval
Diospyros blancoi (Mentega)	4.4	96	929	40	Oval
Cinnamomum iners (Kayu Manis)	3.5	94	906	63	Oval
Dillenia indica (Simpoh)	3.5	93	903	66	Spreading
Peltophorum pterocarpum (Batai Laut)	3.1	88	853	88	Wide Spreading
Juniperus chinensis (Juniper Cina)	2.8	91	882	87	Conical
Callerya atropurpurea (Tulang Daing)	2.2	89	858	111	Round
Mimusops elengi (Bunga Tanjung)	2.1	87.5	848	121	Conical
Filicium decipiens (Kiara Payung)	2.0	83	804	165	Oval
Cyrtophyllym fragans (Tembusu)	2.0	91	886	83	Vase
Lagerstroemia floribunda (Kedah Bungor)	1.9	88	850	119	Oval to conical-rounded

(continued)

Table 13.1 (continued)

Name of tropical tree species	Average leaf area index (LAI)	Average solar radiation Filtration in Percentage (%)	Average total solar radiation filtration (W/m^2)	Average total solar irradiance received underneath canopy (W/m^2)	Type of canopy form
Lagerstroemia speciosa (Bungor Raya)	1.9	87	840	129	Round
Syzygium grande (Jambu Laut)	1.5	87	843	126	Irregular Oval
Hura crepitans (Payung Indonesia)	1.5	85	824	145	Wide Spreading
Pometia pinnata (Kasai)	1.5	88	854	115	Irregular Round
Dalbergia oliveri (Tamalan)	1.3	80	777	192	Vase
Tamarindus indica (Asam Jawa)	1.2	78	756	213	Round
Erythrina fusca (Dedap Merah)	1.0	73	707	262	Irregular Round
Cassia fistula (Kayu Raja, Pancuran Emas)	0.8	75	726	243	Irregular

Source Shahidan et al. (2016)

References

Abreu-Harbich, L. V. D., Labaki, L. C., & Matzarakis, A. (2015). Effect of tree planting design and tree species on human thermal comfort in the tropics. *Landscape and Urban Planning, 138*, 99–109.

Akbari, H. (2002). Shade trees reduce building energy use and CO2 emissions from power plants. *Environmental Pollution, 116*(Supplement 1), 119–126.

Berry, R., Livesley, S. J., & Aye, L. (2013). Tree canopy shade impacts on solar irra-diance received by building walls and their surface temperature. *Building and Environment, 69*, 91–100.

Brown, R. D., & Gillespie, T. J. (1995). *Microclimate landscape design: Creating thermal comfort and energy efficiency*. Wiley.

Coutts, A., White, E., Tapper, N., Beringer, J., & Livesley, S. (2016). Temperature and human thermal comfort effects of street trees across three contrasting street canyon environments. *Theoretical and Applied Climatology*. https://doi.org/10.1007/s00704-015-1409-y

Dimoudi, A., & Nikolopoulou, M. (2003). Vegetation in the urban environment: Microclimatic analysis and benefits. *Energy and Buildings, 35*(1), 69–76.

Emmanuel, M. R. (2005). *An urban approach to climate-sensitive design strategies for the tropics*. Taylor and Francis.

Fahmy, M., Sharples, S., & Yahiya, M. (2010). LAI based trees selection for mid latitude urban developments: A microclimate study in Cairo, Egypt. *Building and Environment, 45*(2), 345–357.

Gartland, L. (2008). *Heat islands understanding and mitigating heat in urban areas*. Cromwell Press.

Jauregui, E. (1990/91). Influence of a large urban park on temperature and convective precipitation in a tropical city. *Energy and Buildings, 15*(3–4), 457–463.

Klemm, W., Heusinkveld, B. G., Lenzholzer, S., Jacobs, M. H., & van Hove, B. (2015). Psychological and physical impact of urban green spaces on outdoor thermal comfort during summertime in The Netherlands [Special Issue: Climate adaptation in cities]. *Building and Environment, 83*, 120–128. https://doi.org/10.1016/j.buildenv.2014.05.013

Konarska, J., Lindberg, F., Larsson, A., Thorsson, S., & Holmer, B. (2014). Transmissivity of solar radiation through crowns of single urban trees—Application for outdoor thermal comfort modelling. *Theoretical and Applied Climatology, 117*, 363–376.

Kotzen, B. (2003). An investigation of shade under six different tree species of the Negev Desert towards their potential use for enhancing micro-climatic conditions in landscape architectural development. *Journal of Arid Environments, 55*, 231–274.

Landsberg, H. E. (1981). *The urban climate*. Academic Press.

Meir, P., Grace, J., & Miranda, A. C. (2000). Photographic method to measure the vertical distribution of leaf area density in forests. *Agricultural and Forest Meteorology, 102*(2–3), 105–111.

Papadakis, G., Tsamis, P., & Kyritsis, S. (2001). An experimental investigation of the effect of shading with plants for solar control of buildings. *Energy and Buildings, 33*(8), 831–836.

Picot, X. (2004). Thermal comfort in urban spaces: Impact of vegetation growth case study: Piazza Della Scienza, Milan, Italy. *Journal of Energy, 36*, 329–334.

Roth, M. (2002, January 23–25). *Effects of cities on local climates*. Paper presented at the Workshop of IGES/APN Mega-City Project. Kitakyushu, Japan.

Shahidan, M. F., Shariff, M. K. S., Jones, P., Salleh, E., & Abdullah, A. M. (2010). A comparison of *Mesua ferrea* L. and *Hura crepitans* L. for shade creation and radiation modification in improving thermal comfort. *Landscape and Urban Planning, 97*, 168–181.

Shahidan, M. F., Shariff, M. K. S., Jones, P., Salleh, E., & Qi, B. H. J. (2016). *Tropical tree for urban environment: Their microclimatic properties*. Penerbit Universiti Putra Malaysia (UPM).

Shashua-Bar, L., & Hoffman, M. E. (2000). Vegetation as a climatic component in the design of an urban street: An empirical model for predicting the cooling effect of urban green areas with trees. *Energy and Buildings, 31*(3), 221–235.

Shashua-Bar, L., Swaid, H., & Hoffman, M. E. (2004). On the correct specification of the analytical CTTC model for predicting the urban canopy layer temperature. *Energy and Buildings, 36*, 975–978.

Steven, M. D., Biscoe, P. V., Jaggard, K. W., & Paruntu, J. (1986). Foliage cover and radiation interception. *Field Crops Research, 13*, 75–87.

Takács, A., Kiss, M. A., Hof, A., Tanács, E., Gulyás, Á., & Kántor, N. (2016). Microclimate modification by urban shade trees—An integrated approach to aid ecosystem service based decision-making. *Procedia Environmental Sciences, 32*, 97–109.

Takahashi, K., Yoshida, H., Tanaka, Y., Aotake, N., & Wang, F. (2004). Measurement of thermal environment in Kyoto city and its prediction by CFD simulation. *Energy and Buildings, 36*(8), 771–779.

Tukiran, J. M., Jamel Ariffin, J., & Ghani, A. N. A. (2016). Cooling effects of two types of tree canopy shape in Penang, Malaysia. *International Journal of GEOMATE, 11*, 2275–2283.
Wong, N. H., Kardinal Jusuf, S., Aung La Win, A., Kyaw Thu, H., Syatia Negara, T., & Xuchao, W. (2007). Environmental study of the impact of greenery in an institutional campus in the tropics. *Building and Environment, 42*(8), 2949–2970.
Wong, N. H., & Yu, C. (2009). *Tropical urban heat islands: Climate, buildings and greenery*. Taylor and Francis.
Zhao, Q., Sailor, D. J., & Wentz, E. A. (2018). Impact of tree locations and arrangements on outdoor microclimates and human thermal comfort in an urban residential environment. *Urban Forestry & Urban Greening, 32*, 81–91.

Mohd Fairuz Shahidan has been an Associate Professor in the Department of Landscape Architecture in the Faculty of Design and Architecture, Universiti Putra Malaysia (UPM) since 2004. His research interests are in architecture and landscape energy-efficient design and planning. His study focuses on the use of landscape and green materials to address environmental issues including the Urban Heat Island (UHI) phenomenon in tropical climates. He's also interested in the study of climate-responsive design, which has a big impact on outdoor thermal comfort and building energy efficiency in terms of delivering sustainable development. He has authored numerous papers and served as a reviewer for a number of prestigious national and international journals. He has been invited as a speaker at several local and international workshops and conferences. He published several books, specialising in his focused research and design.

Chapter 14
Influence of Roadside Trees and Road Orientation on Outdoor Thermal Environment: Case Study in Kuala Lumpur, Malaysia

Sheikh Ahmad Zaki

Abstract Malaysia is experiencing rapid population growth and development, which can greatly affect outdoor thermal environments. Therefore, the influence of roadside trees and road orientation on outdoor thermal environment were studied on four different roads in Kuala Lumpur, Malaysia. Field measurements were carried out to evaluate outdoor thermal environments, where the selection of sites was based on different roadside tree morphological features and road orientations. Outdoor air temperature, relative humidity, globe temperature, wind speed, and wind direction were measured. Absolute humidity was estimated based on relative humidity and air temperature. Planting dense canopy trees with an average sky view factor of 0.07 reduced the mean radiant temperature by 35% and the physiological equivalent temperature (PET) by 25%. East–West (E–W) and Northwest–Southeast (NW–SE) oriented roads had high PET values of 41 °C and 43 °C, respectively. North–South and Northeast–Southwest oriented roads had lower PET values (37 °C), providing improved outdoor microclimate. Roadside trees provided greater cooling potential in E–W and NW–SE oriented roads. The findings are beneficial for urban road design in Malaysia to improve the outdoor thermal environment.

Keywords Field measurements · Roadside trees · Road orientation · Thermal environment · Physiological equivalent temperature

Introduction

An improvement in the outdoor thermal environment can reduce heat-related illness and mortality while elevating the comfort level of pedestrians. These changes can also benefit the physical, environmental, economic, and social aspects of a city (Hass-Klau, 1993). Studies of outdoor thermal environments have been conducted in both temperate (Kikuchi et al., 2007; Narita et al., 2008; Tsiros, 2010) and tropical regions

---

S. A. Zaki (✉)
Malaysia-Japan International Institute of Technology, Universiti Teknologi Malaysia (UTM), Jalan Sultan Yahya Petra, 54100 Kuala Lumpur, Malaysia
e-mail: sheikh.kl@utm.my

S. Maruthaveeran et al. (eds.), *Urban Forestry and Arboriculture in Malaysia*,
https://doi.org/10.1007/978-981-19-5418-4_14

(Shahidan et al., 2012; Wong & Jusuf, 2010; Wong et al., 2016) where trees and urban green areas are popular solutions for mitigating the urban heat island (UHI) effect (Fan et al., 2019; Yang et al., 2020; Yu et al., 2018). Urban vegetation provides shelter and shade while decreasing the temperature of urban areas (Givoni, 1991; Gonçalves et al., 2019; Roth, 2012; Takács et al., 2016). However, the value and potential role of roadside trees in mitigating UHI effects are often underestimated (Pauleit, 2003). Moreover, planting roadside trees in limited and constrained spaces is expected to result in insufficient soil, and lack of growth space (Thaiutsa et al., 2008).

In the capital city of Malaysia, Kuala Lumpur, it has been proposed that the thermal environment of the city's urban areas could be improved by planting roadside trees. While research regarding the mitigating effects of urban greenery has taken place, and guidelines for improving the outdoor thermal environment have been published, most of these studies were conducted in temperate climates (Ali & Patnaik, 2018; Cheung et al., 2021; Claris Fisher et al., 2021; Kántor et al., 2018). These studies hypothesized that the findings could not be easily adopted in tropical countries like Malaysia due to distinct climate differences, including higher outdoor temperatures and humidity. Changes in the solar angle in tropical regions are also small compared to regions of higher latitudes. Therefore, the main objective of this study was to investigate the effects of different roadside tree configurations and road orientations on the outdoor thermal environment based on field measurements.

Materials and Method

Measurement Sites and Periods

The selection of urban roads for this study was based on roadside tree canopy coverage and tree height. In addition, different road orientations were selected, including the main orientations of East–West (E–W) and North–South (N–S) and intermediate orientations of Northwest–Southeast (NW–SE) and Northeast–Southwest (NE–SW), to incorporate the most common orientations. Four roads were chosen as measurement locations—Jalan Raja Muda Abdul Aziz (R1) (E–W), Jalan Produktiviti (R2) (N–S), Jalan Perdana Utama (R3) (NW–SE), and Jalan Sultan Yahya Petra (R4) (NE–SW). All roads within this study are situated within a 7 km radius of Kuala Lumpur city centre (Fig. 14.1).

All streets contained two tree species, Angsana (*Pterocarpus indicus*; a broad-leaved deciduous tree) and Rain tree (*Samanea saman*; a wide-canopied tree with a large symmetrical umbrella-shaped crown) (Fig. 14.2). The location of roadside trees was randomly selected within close proximity of the urban city center of Kuala Lumpur. This was due to the difficulty of finding an ideal location for the real measurement in an emerging tropical country.

As the focus of this study was on urban roads with roadside trees, all selected road segments were located approximately 24.9–33.1 m from surrounding buildings to

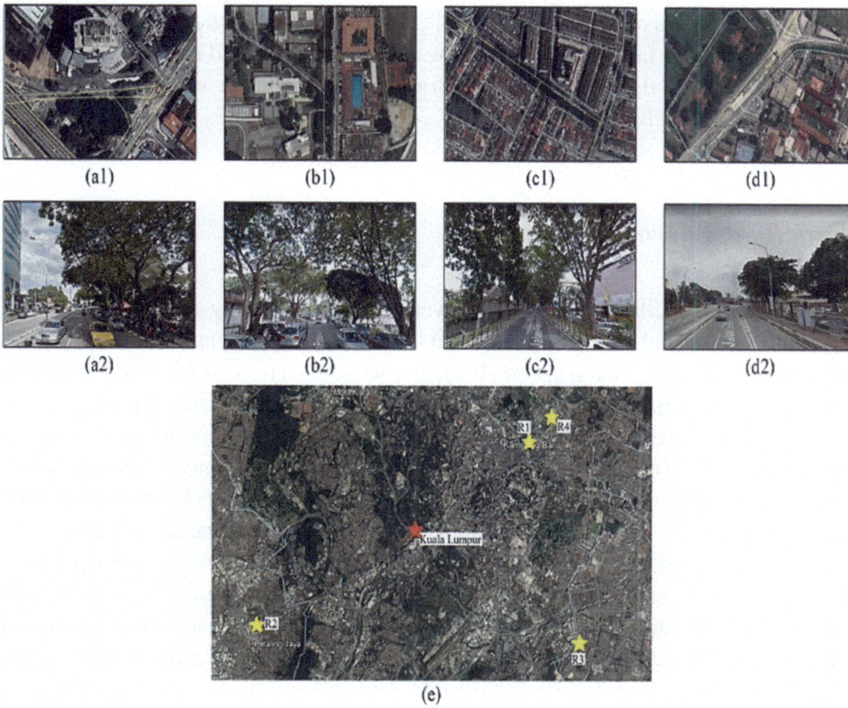

Fig. 14.1 Photographs of survey locations (a) R1, (b) R2, (c) R3, and (d) R4. The top row (a1, b1, c1, d1) shows plan views (Google Earth screenshot taken 19 December 2017), and the bottom row (a2, b2, c2, d2) shows the corresponding Google Street view images. (e) Satellite map showing the four survey locations (Zaki et al., 2020)

Fig. 14.2 Photographs of the (a) *Pterocarpus indicus* and (b) *Samanea saman* trees located at the studied sites. The images were supplied by the Forest Research Institute Malaysia (Zaki et al., 2020)

ensure that they were not shading the roads. This ensured that the data collected was not affected by other factors, apart from the roadside trees. Field measurements were conducted to collect data from each site for two selected clear sunny days between April 2015 and March 2016.

Microclimate Measurements

Five outdoor microclimate parameters were measured during each field study to assess outdoor thermal comfort: outdoor air temperature (T_a), relative humidity (*RH*), globe temperature (T_g), wind speed (*WS*), and wind direction (*WD*). Three measurement stations were placed along each selected road to capture the various values of tree canopy coverage. The first station (ST1) had all of the aforementioned measurement sensors, where one setup measured *WS* and *WD*, and a second setup consisted of T_a, *RH*, and T_g sensors. Due to limited instruments, the second (ST2) and third (ST3) stations only had T_a, *RH*, and T_g sensors. As there was only one station measuring *WS* and *WD*, these data could not be compared for different configurations of roadside trees along the same road. However, these factors were used as one of the main inputs to estimate the mean radiant temperature (T_m) and physiological equivalent temperature (PET). All instruments were placed 1.5 m above ground level (Costa et al., 2007) and were left for 30 min prior to data collection to equilibrate with local conditions. The measurements were taken for R1, R2, and R3 between 09:30 and 13:30. The measurements at R4 were divided into early morning (07:30–09:30) and evening (16:00–18:00) sessions to investigate the effect of roadside trees on outdoor microclimate conditions for both early morning and evening hours.

Measurement of Tree Canopy Coverage and Tree Height

The tree canopy coverage at the measurement locations was represented by the sky view factor (SVF), which was obtained using a digital single-lens reflex camera equipped with a fisheye lens. This enabled hemispherical or fisheye photos to be taken at all monitoring stations (ST1, ST2 and ST3) for each selected location of roadside trees. All SVF measurements were performed on the same day as the accompanying field survey. The photos were imported into RayMan software (version 1.2, Freiburg, Germany) to calculate the SVF (Costa et al., 2007). Here, the roads were divided into three categories depending on the SVF: low SVF (R-LS) (SVF < 0.10) represents dense tree canopy; medium SVF (R-MS) ($0.10 \leq$ SVF < 0.79) represents sparse tree canopy; and high SVF (R-HS) (SVF ≥ 0.80). The SVF represents tree canopy coverage, except for a reference location with minor sky view obstruction by trees and other anthropogenic features, such as buildings and lampposts. The sky view obstruction was based on tree-canopy elements only (e.g., branches, leaves, and twigs). Therefore, to investigate the effects of different tree canopy densities on the thermal environment, R-HS with low sky obstruction (SVF $\geq$ 0.80) was chosen as a reference location for comparison with R-LS and R-MS (with tree canopies). Unless measurements were taken in the middle of a road, or in a location at a great distance from the site, it was very difficult to find a reference location where SVF = 1. The SVF measurement locations are denoted throughout the study as R1-LS (low SVF on R1), and every SVF level was available for all roads, except for R4 where only R4-LS and R4-HS were selected for measurements due to a lack of suitable R-MS tree canopy. The average SVF values for R-LS, R-MS, and R-HS are 0.07, 0.31, 0.86, respectively (Fig. 14.3).

Tree height (H_T) was also investigated as a factor to fully consider the shading effects on the road thermal environment. The H_T values were measured using a laser range finder on the same day as the field survey. The total H_T was obtained from the sum of the crown depth, H_c and height of the tree trunk, H_{tt} (Fig. 14.4), and the results varied from 6 to 21 m. The trees were categorized into the following four groups: short (6–9 m), medium (10–13 m), tall (14 to 17 m), and very tall (18–21 m) (Table 14.1).

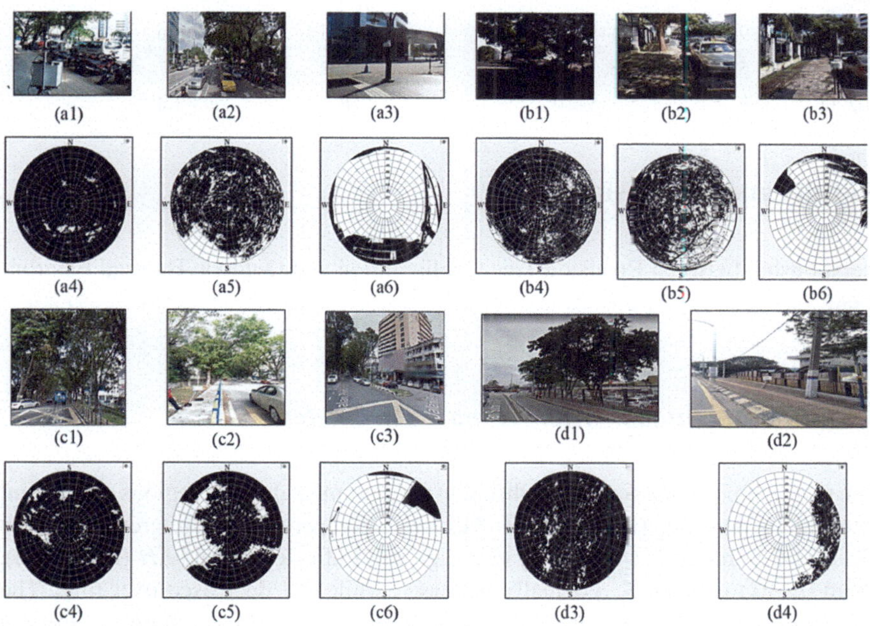

Fig. 14.3 Street view and fisheye photos of measurement locations (a) R1, (b) R2, (c) R3, and (d) R4. At R1, (a1) to (a3) are the street view photos, SVF value for (a4) (R1 -LS) is 0.04, (a5) (R1 – MS) is 0.28, and (a6) (R1 – HS) is 0.08. At R2, (b1) to (b3) are the street view photos, SVF value for (b4) (R2 -LS) is 0.08, (b5) (R2 -MS) is 0.35, and (b6) (R2 -HS) is 0.85. At R3, (c1) to (c3) are the street view photos, SVF value for (c4) (R3 -LS) is 0.08, (c5) (R3 -MS) is 0.29, and (c6) (R3-HS) is 0.92. At R4, (d1) and (d2) are the street view photos, SVF value for (d3) (R4 -LS) is 0.09, and (d4) (R4 -HS) is 0.88 (Zaki et al., 2020)

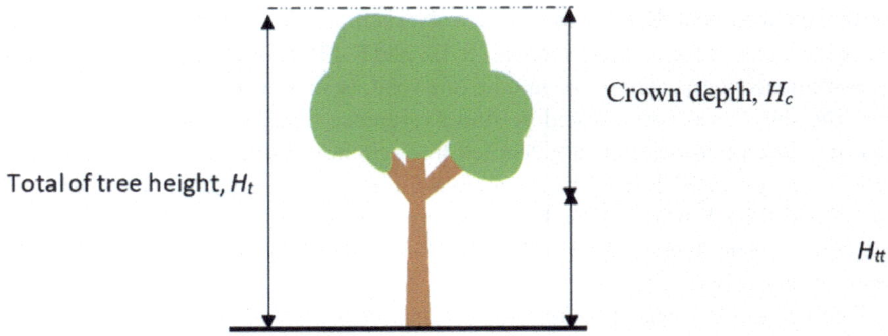

Fig. 14.4 Illustration of the summation of the total of tree height

Table 14.1 Categories of roadside trees based on tree height (Zaki et al., 2020)

Crown Depth (m)	Number of Trees	Trunk Height (m)	Number of Trees	Total Tree Height (m)	Category	Number of Trees
4–5	3	2–3	1	6–9	Short	2
6–7	9	4–5	8	10–13	Medium	9
8–9	5	6–7	8	14–17	Tall	7
10–11	3	8–10	3	18–21	Very Tall	2

Results and Discussion

Relationship between Roadside Tree Configuration and Outdoor Thermal Environment.

Variation in Outdoor Microclimate Parameters

The microclimate data for the two different measurement days for regions of different tree canopy densities (R -LS and R -MS) were compared to the reference location (R -HS). The roadside tree canopy coverage at R1 affected T_a and *AH* (Fig. 14.5a). From 09:30 to 13:30, T_a gradually increased while *AH* decreased over time. The T_a values did not vary significantly for different roadside tree conditions, where a maximum T_a difference between R1 -LS and R1 -HS of 2.1 °C was observed (8 April at 13:30) (Fig. 14.5a). Similarly, the difference in *AH* values between different tree canopy densities was small (< 9%) (Fig. 1.5a). Conversely, T_g showed a large dependence on tree canopy coverage. For R1 -LS and R1 -MS, T_g values were much lower than measured at R1 -HS, with a maximum difference of 11.3 °C (Fig. 14.5a). It was also observed that T_g was higher than T_a for all tree canopy densities.

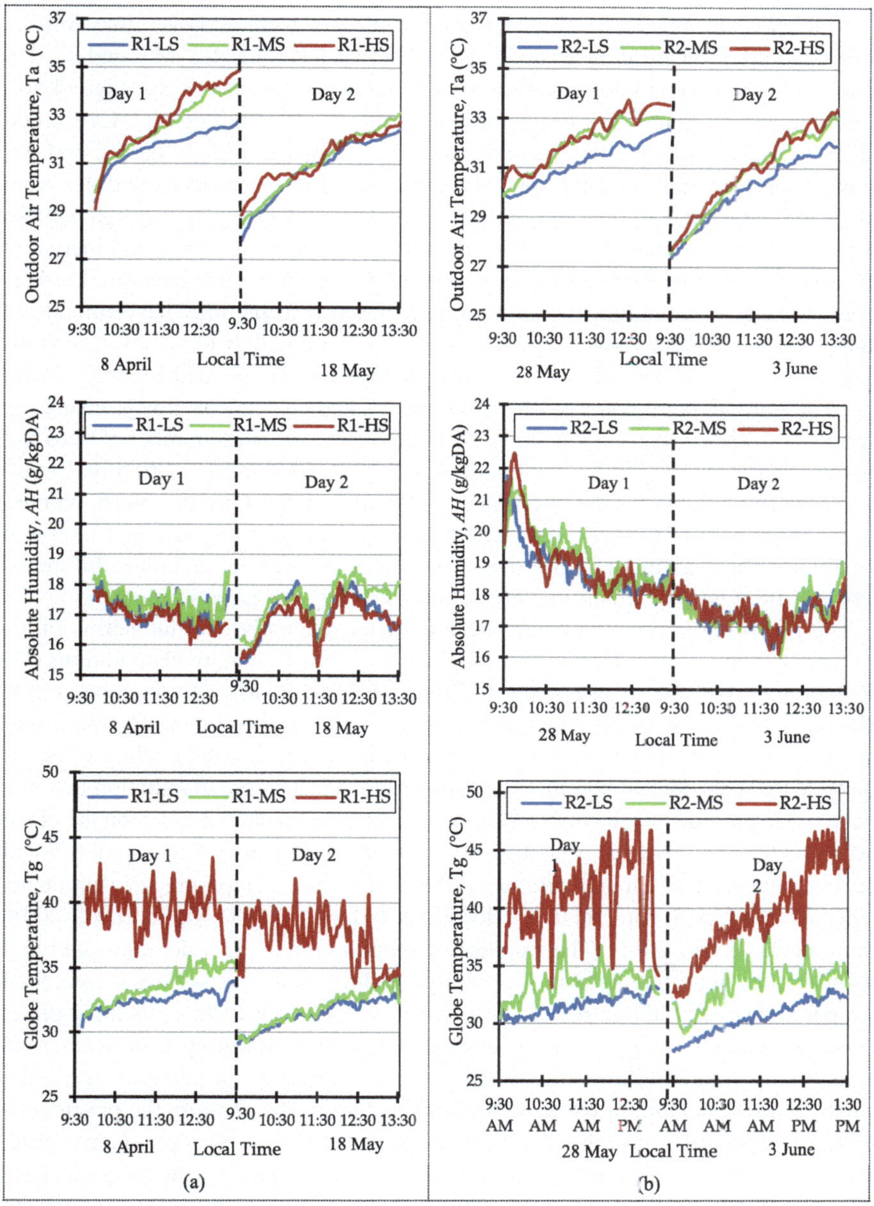

Fig. 14.5 Variation of outdoor microclimate parameters for locations (a) R1 and (b) R2. The vertical dotted lines separate the two measurement days (Zaki et al., 2020)

At R2 (Fig. 14.5b), T_a steadily increased, while AH decreased over time. Similar to results for R1, the differences in T_a and AH between R2 -LS, R2 -MS, and R2 -HS were small, with maximum variations of 1.8 °C and 6%, respectively (Fig. 14.5b). However, variations in T_g were large, with a maximum difference of 15.3 °C recorded on 28 May. The microclimate variation at R3 is shown in Fig. 14.6a. T_a and AH showed similar trends as other roads. However, the differences in T_a and AH were noticeably larger, 4.3 °C and 16%, respectively. This may be due to the high number of roadside trees on both sides of R3, which reduced the wind speed, and increased differences in AH between locations with trees and the reference location. Furthermore, because surrounding buildings were blocked off, this may have influenced changes in wind effect, as well as changes due to the traffic flow of vehicles, and thus may not be due to the absolute effect from the high number of roadside trees. Wind speed is a very dynamic microclimatic parameter, and the value is always changing. Among all selected roads in this study, R3 contained the greatest number of trees and a large difference in T_g, with a maximum of 14.1 °C on 19 March (11:06) (Fig. 14.6a).

Figure 14.6b shows the variations in T_a, AH and T_g at R4. On day 1, the maximum T_a recorded at R4-HS was 32.4 °C, while under the same spatial and temporal conditions, the maximum T_a on day 2 was 34.8 °C. The temperature difference between different SVF conditions was minimal during the early morning.

The following implications have been concluded. Firstly, the fluctuation of T_a was small, with a similar trend and difference of < 7% across all road conditions (R1 -R4). Higher AH values were measured for R -LS and R -MS sites compared to R -HS ones, likely due to the effect of tree transpiration (release of water vapor into the atmosphere). Comparison of the T_g values indicated that the cooling effect of roadside trees during the daytime was mainly due to a decrease in radiation flux due to the shade provided by the trees. Although the effect of wind reduction on the globe thermometer was observed, radiation effects were dominant. Several studies noted the insignificance of air temperature differences when comparing measurements under tree canopies with reference locations (Ahmed, 2003; Narita et al., 2008; Shashua-Bar et al., 2011). The findings of (Narita et al., 2008) demonstrated that the difference in air temperature was negligible throughout the day when compared to a parallel street without tree crowns. However, there was a clear shielding effect for solar radiation and downward long wave radiation, resulting in a lower road surface temperature. Additionally, the night-time temperatures would also receive the benefit of trees shading in reducing the solar radiation absorbed by the pavements, influencing positively to relieve the effects of the UHI. For example, Gupta et al. (2015) had done the study in Singapore. It indicates that the zone with better greenery has a lower night-time temperature due to the roles played by trees shading the pavements, decreasing the average and minimum air temperature. The research found that the released heat will be influenced by the amount of trees shading, declining the radiation absorption and reradiation to the air. This implies that the main reason pedestrians feel thermally comfortable under tree crowns was not due to lower T_a values but were instead due to radiation effects. Our results indicate that the density of roadside trees affected T_g, which in turn affected the outdoor thermal environment.

Relationship between Road Orientation and Outdoor Thermal Environment.

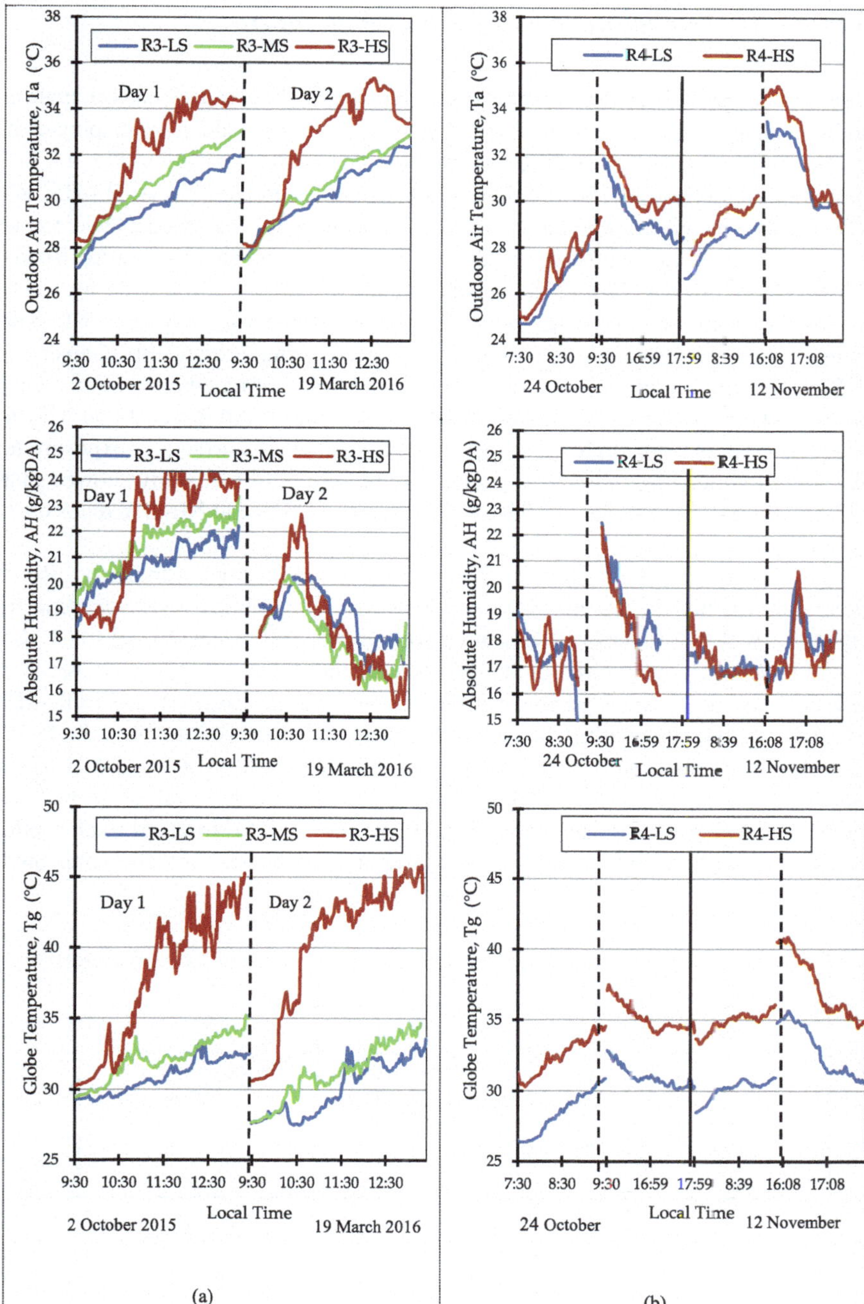

Fig. 14.6 Variation in outdoor microclimate parameters at locations (a) R3 and (b) R4, where the error bars represent standard deviation. The vertical dotted lines separate the two measurement days or measurement periods on the same day (Zaki et al., 2020)

Outdoor Air Temperature and Globe Temperature

T_a data was collected over eight sunny days to investigate the effect of road orientation and roadside trees on the outdoor thermal environment. The values are displayed as the overall average of T_a for different roadside tree canopy densities (Fig. 14.7a). The measurement results showed that road orientation affected T_a. At R3-HS, R3 (NW–SE) had the highest average T_a of 31.8 °C, while R3 -LS showed a lower value of 30.4 °C, indicating the effect of roadside trees on lowering air temperature. Roads with E–W orientation showed the second-highest average T_a, followed by roads with N–S orientation, while the lowest T_a was measured for roads with NE–SW orientation. These results are consistent with previous simulation results (Cao et al., 2015), where the average temperature was lowest for 45°-oriented (NE–SW) streets and highest for 90°-oriented (E–W) streets. According to Ali-Toudert and Mayer (2006) E–W streets are slightly warmer than N–S streets due to longer exposure to solar radiation, leading to higher T_a. Overall, the results of this study showed that NW–SE and E–W oriented roads have higher T_a, while N–S and NE–SW roads have lower T_a.

Figure 14.7b shows the average T_g for various tree canopy conditions. Road orientation affected T_g. At the reference location, T_g exhibited a similar trend to that of T_a, with the highest T_g measured for the NW–SE oriented road (38.1 °C) and the lowest value measured on the NE–SW oriented road (35.2 °C). However, a steep decrease in T_g was observed with increasing canopy cover (R -HS to R -MS to R -LS). At the sites with R-LS, a similar trend was observed for both the E–W and NW–SE oriented roads, where the highest T_g values were 31.6 °C and 31.5 °C, respectively, while the N–S and NE–SW roads showed lower T_g values of 30.4 °C and 30.8 °C, respectively. The higher T_g at E–W and NW–SE oriented roads could be correlated with the higher exposure to solar radiation. Shishegar (2013) indicated

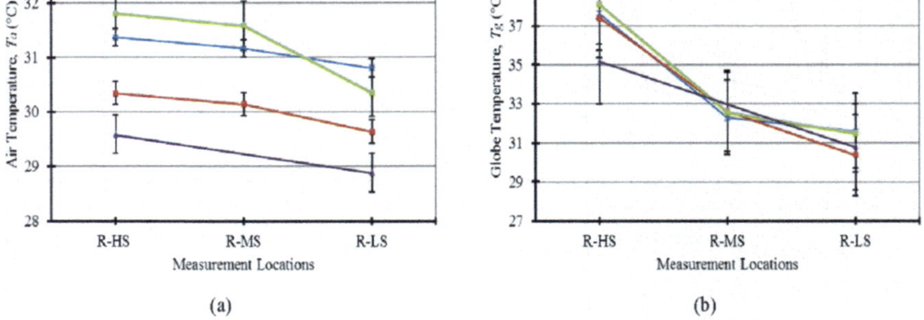

Fig. 14.7 Overall average (a) outdoor air temperature and (b) globe temperature for four selected urban roads with different orientations and tree canopy densities. The standard deviations are represented by the error bars. The measurement periods were 8 April, 18 May, 28 May, 2 October, 24 October, 12 November 2015, and 19 March 2016 (Zaki et al., 2020)

that NW–SE oriented roads are exposed to direct solar radiation for a longer period of time, especially in the afternoon hours during intense sunlight. In summary, the N–S and NE–SW oriented roads displayed better thermal performance, while E–W and NW–SE oriented roads had higher outdoor air and globe temperatures.

Effect of Roadside Tree Configuration

The average T_{mrt} and PET values at R -LS, R -MS, and R -HS areas were calculated from microclimatic measurement data (Fig. 14.8). The data show that T_{mrt} calculated with the actual wind speed had a significantly greater range (29.3–67.0 °C) than those calculated with the average wind speed (31.2–61.1 °C), although both showed similar average values (Fig. 14.8a and b). This implies that the wind speed influenced the T_{mrt} value and, consequently, the thermal comfort of pedestrians. The trends in the T_{mrt} curve were notably similar to those for T_g. Theoretically, faster wind speeds over the globe thermometer result in T_g approaching T_a. In addition, we confirmed that roadside trees provide significant microclimate benefits (by reducing T_{mrt}), as reported in other similar studies (Johansson & Emmanuel, 2006; Sanusi et al., 2017).

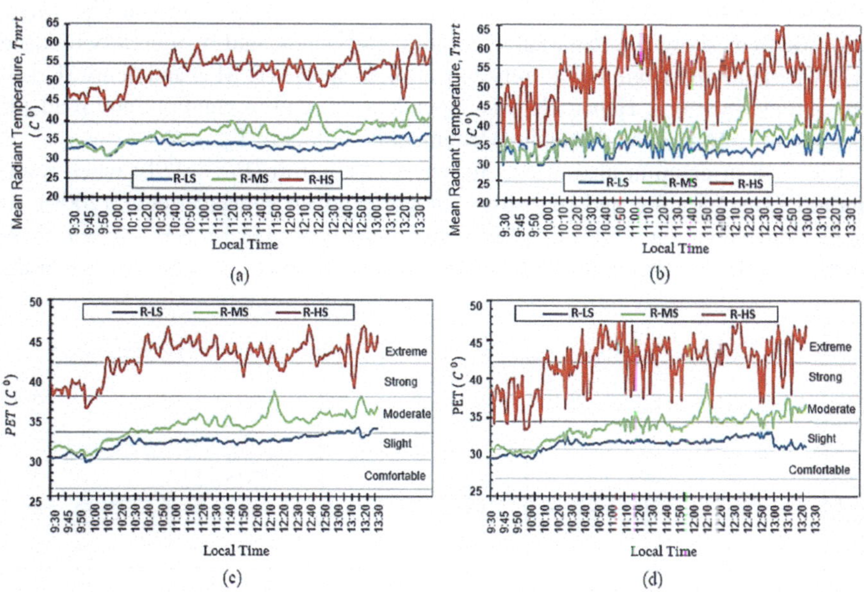

Fig. 14.8 T_{mrt} calculated using (a) averaged wind speed and (b) actual wind speed. Physiological equivalent temperature (PET) calculated using (c) averaged wind speed and (d) actual wind speed. The level of heat stress is indicated on the right-hand side of the figure, where the horizontal lines indicate the boundaries for each heat stress level (Zaki et al., 2020)

During the daytime, PET values at the reference location with R -HS conditions exceeded the upper thermal comfort range limit of 30 °C, and hence, exhibited poor thermal comfort (Fig. 14.8c and d). The PET values were rarely classified as moderate heat stress, except for a short period around 09:50. For the remainder of the measurement period, they fluctuated between strong and extremely strong heat stress. In contrast, PET values under sparse and dense tree canopies with R -MS and R -LS conditions, respectively, provided much better thermal comfort as the heat stress was reduced to a moderate level. The R -LS location showed the lowest PET values, which corresponded to slight heat stress throughout the warmest period of the day. These findings indicate that the presence of roadside trees was able to improve outdoor thermal comfort tremendously by effectively shielding the incoming solar radiation.

Effect of Road Orientation

The estimated T_{mrt} and PET values for the four different road orientations are shown in Fig. 14.9. T_{mrt} showed similar trends to T_g, with the highest T_{mrt} at the reference location (R-HS) occurring for the NW–SE oriented road (51.9 °C), while the lowest value was found for the NE–SW oriented road (46.6 °C), as shown in Fig. 14.9a. Roadside trees showed a remarkable cooling effect, indicated by a greatly reduced T_{mrt}, when moving from the reference location to sparse (R-MS) and dense (R-LS) tree canopy locations. Hence, T_{mrt} is dependent on road orientation, and roadside trees greatly improve the thermal environment by reducing T_{mrt}.

The road orientation also affects the thermal comfort (Fig. 14.9b) when referring to thermal comfort and heat stress categories given by Mayer et al. (2009). At the reference location, PET values were significantly greater for NW–SE and E–W oriented roads. For example, PET values reached extreme heat stress levels in the

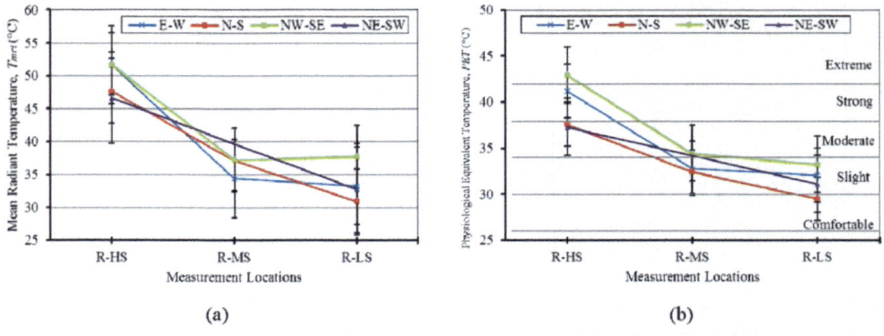

Fig. 14.9 Average (a) T_{mrt} and (b) PET for the four selected urban roads with different orientations. The standard deviations are represented by the error bars. The level of heat stress is shown on the right-hand side of (b), where the horizontal lines indicate the boundaries for each heat stress level (Zaki et al., 2020)

NW–SE oriented road and strong heat stress in the E–W oriented road. However, for R -MS conditions, the PET values dropped to moderate heat stress levels for the NW–SE oriented road and slight heat stress levels for the other roads. Under R -LS conditions, the N–S oriented road provided the best thermal comfort with PET values corresponding to comfortable or no heat stress levels, while the other roads maintained a slight heat stress level. In addition, roadside trees had a greater cooling potential in NW–SE and E–W oriented roads. By comparing PET at R -LS and R -HS, average PET reductions of 9.7 °C and 9.2 °C were observed for NW–SE and E–W oriented roads, respectively, while the lowest PET reductions of 8.1 °C and 6.2 °C were for the N–S and NE–SW oriented roads, respectively. This demonstrates the dominant effect of the sun's zenith, which has the same orientation as E–W roads, allowing dense tree canopy to reduce solar radiation exposure during the daytime. Roads with E–W and NW–SE orientation have greater potential for high human heat stress, while the cooling potential and microclimate benefits provided by trees are greater. Therefore, it is important to prioritize roadside tree planting for such roads as the PET can be greatly reduced.

Effect of Tree Canopy Coverage

Roadside trees provided microclimate and outdoor thermal comfort benefits. On average, planting trees with a dense tree canopy (average SVF of 0.07) can reduce T_{mrt} significantly by 18.7 °C (35%) and PET by 10.6 °C (25%) compared to the reference location (average SVF of 0.86). Although not as effective as a dense tree canopy, the sparse tree canopy (average SVF of 0.31) also reduced T_{mrt} and PET considerably, by 15.7 °C and 8.5 °C, respectively, compared to the reference condition. Thermal benefits of roadside trees were revealed in this study, particularly for the hot and humid tropical climate of Kuala Lumpur. Our results are consistent with those of several similar studies that investigated the influences of urban vegetation and roadside trees on microclimates (Costa et al., 2007; Sanusi et al., 2017). Other studies also reported the reduction of PET by trees. For example, in Freiburg, Germany, PET was reduced by 4.6 °C (Mayer et al., 2009) whereas in Campinas, Brazil, tree canopies reduced midday summer PET by 16 °C (de Abreu-Harbich et al., 2015).

Conclusion

There are three main conclusions based on the research objectives and findings of this study. First, mitigation effects of roadside trees on outdoor meteorological parameters were shown, with T_a, and T_g values decreasing on average by 4, and 16%, respectively, for dense roadside tree canopies compared to the reference location. Under sparse tree canopy, the reduction in T_a, and T_g values were slightly lower, with values of 3, and 14%, respectively. In contrast, higher AH values were observed for areas of higher tree density. Despite the tendency of previous similar studies to focus on the reduction of T_a and T_g were observed, suggesting that radiation was the dominant factor affecting the thermal environment, where the tree canopy effectively blocks the incoming direct solar radiation and cools the environment. In addition, taller and more mature trees provided a larger shade area than smaller trees, resulting in a larger cooling effect.

Second, different road orientations were found to influence the road thermal environment. T_a, T_g, and T_{mrt} were all lower for N–S and NE–SW oriented roads, whereas E–W and NW–SE oriented roads had the highest temperatures at the studied latitude and solar path. Therefore, urban planners could consider building fewer E–W and NW–SE oriented roads in new urban development areas. However, the microclimate benefits of roadside trees were still significant, irrespective of the road orientation. For example, the presence of trees with high canopy density along N–S roads reduced midday T_g by a considerable 7.0 °C, compared to the reference location. T_{mrt} showed similar trends to other parameters (i.e., lower values for N–S and NE–SW roads).

Finally, we confirmed that roadside trees contributed to lower temperatures and better thermal environments by investigating the thermal comfort level expressed by the PET, which showed a decrease from strong heat stress to slight heat stress with the presence of roadside trees. Specifically, PET decreased by 25 and 20% under dense and sparse canopies, respectively, compared to the reference location. Although N–S and NE–SW roads had lower PET values, the cooling effects of roadside trees were observed for NW–SE and E–W roads, with an average PET reduction of 23%, compared to 18 and 17% for N–S and NE–SW oriented roads, respectively.

The findings of this study are expected to be beneficial and applicable for planning similar future studies, as they provide evidence for urban planners and policymakers on the extent and magnitude of the cooling effects of roadside trees. The proposed measure of planting roadside trees is useful for both new urban developments and existing urban areas in Malaysia. However, this study has limitations in terms of daily measurement periods and could be expanded and improved to incorporate more elements. This could include incorporating and comparing different tree species with distinct features, such as leaf size, branching arrangement, and canopy shape. In addition, various road widths and road pavement surface materials could be considered to investigate their effects on the road thermal environment.

Acknowledgements This work was supported by the Ministry of Education (MOE) through Fundamental Research Grant Scheme (FRGS/1/2019/TK07/UTM/02/5), and Universiti Teknologi Malaysia under Matching Grant (Vot 01M46).

References

Ahmed, K. S. (2003). Comfort in urban spaces: Defining the boundaries of outdoor thermal comfort for the tropical urban environments. *Energy and Buildings, 35*(1), 103–110. https://doi.org/10.1016/S0378-7788(02)00085-3

Ali, S. B., & Patnaik, S. (2018). Thermal comfort in urban open spaces: Objective assessment and subjective perception study in tropical city of Bhopal, India. *Urban Climate, 24*(October 2017), 954–967. https://doi.org/10.1016/j.uclim.2017.11.006

Ali-Toudert, F., & Mayer, H. (2006). Numerical study on the effects of aspect ratio and orientation of an urban street canyon on outdoor thermal comfort in hot and dry climate. *Building and Environment, 41*(2), 94–108. https://doi.org/10.1016/J.BUILDENV.2005.01.013

Cao, A., Li, Q., & Meng, Q. (2015). Effects of orientation of urban roads on the local thermal environment in Guangzhou City. *Procedia Engineering, 121*, 2075–2082. https://doi.org/10.1016/J.PROENG.2015.09.209

Cheung, P. K., Livesley, S. J., & Nice, K. A. (2021). Estimating the cooling potential of irrigating green spaces in 100 global cities with arid, temperate or continental climates. *Sustainable Cities and Society, 71*(April), 102974. https://doi.org/10.1016/j.scs.2021.102974

Claris Fisher, J., Emmerson Bicknell, J., Nesbitt Irvine, K., Fernandes, D., Mistry, J., & Georgina Davies, Z. (2021). Exploring how urban nature is associated with human wellbeing in a neotropical city. *Landscape and Urban Planning, 212*, 104119. https://doi.org/10.1016/j.landurbplan.2021.104119

Costa, A., Labaki, L., & Araújo, V. (2007). A methodology to study the urban distribution of air temperature in fixed points. *2nd PALENC Conference and 28th AIVC Conference on Building Low Energy Cooling and Advanced Ventilation Technologies, 1*(September), 227–230.

de Abreu-Harbich, L. V., Labaki, L. C., & Matzarakis, A. (2015). Effect of tree planting design and tree species on human thermal comfort in the tropics. *Landscape and Urban Planning, 138*, 99–109. https://doi.org/10.1016/j.landurbplan.2015.02.008

Fan, H., Yu, Z., Yang, G., Liu, T. Y., Liu, T. Y., Hung, C. H., & Vejre, H. (2019). How to cool hot-humid (Asian) cities with urban trees? An optimal landscape size perspective. *Agricultural and Forest Meteorology, 265*, 338–348. https://doi.org/10.1016/J.AGRFORMET.2018.11.027

Givoni, B. (1991). Impact of planted areas on urban environmental quality: A review. *Atmospheric Environment. Part B. Urban Atmosphere, 25*(3), 289–299. https://doi.org/10.1016/0957-1272(91)90001-U

Gonçalves, A., Ribeiro, A. C., Maia, F., Nunes, L., & Feliciano, M. (2019). Influence of green spaces on outdoors thermal comfort-structured experiment in a Mediterranean climate. *Climate, 7*(2). https://doi.org/10.3390/cli7020020

Gupta, S., Anand, P., & Shashwat, S. (2015). Improvement of outdoor thermal comfort for a residential development in Singapore. *International Journal of Energy & Environment, 6*(6), 567–586.

Hass-Klau, C. (1993). A review of the evidence from Germany and the UK. *Transport Policy, 1*(1), 21–31. https://doi.org/10.1016/0967-070X(93)90004-7

Johansson, E., & Emmanuel, R. (2006). The influence of urban design on outdoor thermal comfort in the hot, humid city of Colombo, Sri Lanka. *International Journal of Biometeorology, 51*(2), 119–133. https://doi.org/10.1007/S00484-006-0047-6

Kántor, N., Chen, L., & Gál, C. V. (2018). Human-biometeorological significance of shading in urban public spaces—Summertime measurements in Pécs, Hungary. *Landscape and Urban Planning, 170*(November 2016), 241–255. https://doi.org/10.1016/j.landurbplan.2017.09.030

Kikuchi, A., Hataya, N., Mochida, A., Yoshino, H., Tabata, Y., Watanabe, H., & Jyunimura, Y. (2007). Field study of the influences of roadside trees and moving automobiles on turbulent diffusion of air pollutants and thermal environment in urban street canyons. *IAQVEC 2007 Proceedings - 6th International Conference on Indoor Air Quality, Ventilation and Energy Conservation in Buildings: Sustainable Built Environment, 3*, 137–144.

Mayer, H., Kuppe, S., Holst, J., Imbery, F., & Matzarakis, A. (2009). Human thermal comfort below the canopy of street trees on a typical Central European summer day. *Berichte des Meteorologischen Instituts der Albert-Ludwigs-Universität Freiburg, 18*, 211–219. http://www.researchgate.net/publication/228503675_Human_thermal_comfort_below_the_canopy_of_street_trees_on_a_typical_Central_European_summer_day/file/d912f5072b20905674.pdf

Narita, K., Sugawara, H., & Honjo, T. (2008). *Effects of roadside trees on the thermal environment within a street canyon.* Geographical Reports of Tokyo Metropolitan University, 43, 41–48—References—Scientific Research Publishing. Department of Geography, Tokyo Metropolitan University, Geographical Reports of Tokyo Metropolitan University. http://hdl.handle.net/10748/3807

Pauleit, S. (2003). Urban street tree plantings: Identifying the key requirements. *Proceedings-Institution of Civil Engineers Municipal Engineer, 156*, 43–50.

Roth, M. (2012). Urban heat islands. In *Handbook of environmental fluid dynamics* (Vol. 2, pp. 160–177). CRC Press. https://doi.org/10.1201/B13691-13

Sanusi, R., Johnstone, D., May, P., & Livesley, S. J. (2017). Microclimate benefits that different street tree species provide to sidewalk pedestrians relate to differences in Plant Area Index. *Landscape and Urban Planning, 157*, 502–511. https://doi.org/10.1016/J.LANDURBPLAN.2016.08.010

Shahidan, M. F., Jones, P. J., Gwilliam, J., & Salleh, E. (2012). An evaluation of outdoor and building environment cooling achieved through combination modification of trees with ground materials. *Building and Environment, 58*, 245–257. https://doi.org/10.1016/J.BUILDENV.2012.07.012

Shashua-Bar, L., Pearlmutter, D., & Erell, E. (2011). The influence of trees and grass on outdoor thermal comfort in a hot-arid environment. *International Journal of Climatology, 31*(10), 1498–1506. https://doi.org/10.1002/JOC.2177

Shishegar, N. (2013). Street design and urban microclimate: Analyzing the effects of street geometry and orientation on airflowand solar access in urban canyons. *Journal of Clean Energy Technologies,* 52–56. https://doi.org/10.7763/JOCET.2013.V1.13

Takács, Á., Kiss, M., Hof, A., Tanács, E., Gulyás, Á., & Kántor, N. (2016). Microclimate modification by urban shade trees—An integrated approach to aid ecosystem service based decision-making. *Procedia Environmental Sciences, 32*, 97–109. https://doi.org/10.1016/J.PROENV.2016.03.015

Thaiutsa, B., Puangchit, L., Kjelgren, R., & Arunpraparut, W. (2008). Urban green space, street tree and heritage large tree assessment in Bangkok, Thailand. *Urban Forestry and Urban Greening, 7*(3), 219–229. https://doi.org/10.1016/J.UFUG.2008.03.002

Tsiros, I. X. (2010). Assessment and energy implications of street air temperature cooling by shade tress in Athens (Greece) under extremely hot weather conditions. *Renewable Energy, 35*(8), 1866–1869. https://doi.org/10.1016/J.RENENE.2009.12.021

Wong, N. H., & Jusuf, S. K. (2010). Study on the microclimate condition along a green pedestrian canyon in Singapore. *Architectural Science Review, 53*(2), 196–212. https://doi.org/10.3763/ASRE.2009.0029

Wong, P. P. Y., Lai, P. C., Low, C. T., Chen, S., & Hart, M. (2016). The impact of environmental and human factors on urban heat and microclimate variability. *Building and Environment, 95*, 199–208. https://doi.org/10.1016/J.BUILDENV.2015.09.024

Yang, G., Yu, Z., Jørgensen, G., & Vejre, H. (2020). How can urban blue-green space be planned for climate adaption in high-latitude cities? A seasonal perspective. *Sustainable Cities and Society, 53*, 101932. https://doi.org/10.1016/J.SCS.2019.101932

Yu, Z., Xu, S., Zhang, Y., Jørgensen, G., & Vejre, H. (2018). Strong contributions of local background climate to the cooling effect of urban green vegetation. *Scientific Reports, 8*(1), 1–9. https://doi.org/10.1038/s41598-018-25296-w

Zaki, S. A., Toh, H. J., Yakub, F., Saudi, A. S. M., Ardila-Rey, J. A., & Muhammad-Sukki, F. (2020). Effects of roadside trees and road orientation on thermal environment in a tropical City. *Sustainability (Switzerland), 12*(3). https://doi.org/10.3390/su12031053

Sheikh Ahmad Zaki is an Associate Professor at the Malaysia-Japan International Institute of Technology, Universiti Teknologi Malaysia (UTM) Kuala Lumpur.

Chapter 15
Effect of *Ficus pumila*, *Ficus pumila* 'Variegata' and *Vernonia elliptica* on Sound Absorption Performance of Green Wall Fences

Zaiton Haron, Khairulzan Yahya, Zanariah Jahya, Nadirah Darus, Yap Zhen Shyong, and Herni Halim

Abstract Currently there is a very high trend among Malaysians to use creeping plants for decorating the walls of houses. Popular plants include *Ficus pumila*, *Ficus pumila* 'Variegata' and *Vernonia elliptica* which creep on fences around the house effectively making a 'green wall fence'. This study examines whether *Ficus pumila*, *Ficus pumila* 'Variegata' and *Vernonia elliptica* on wall fences can be used to reduce noise from traffic and overcome canyon effects. The effectiveness on noise reduction were carried out through the sound absorption coefficient measured using two microphone impedance tubes on a series of green wall fence specimens. Porous concrete and brick wall with *Vernonia elliptica* 10–50 mm thick specimens were the best green wall fencing since the value of all sound absorption coefficient value ranged from greater than 0.3 up to 0.9 for 500–1600 Hz frequency range, hence can reduce all categories of noise emanating from road traffic and the associated canyon effect. *Vernonia elliptica* specimen with 10–50 mm thick coverings on a porous wall has sound absorption coefficient better than rockwool for 200–630 Hz frequency range.

Keywords Noise pollution · Green façade · *Ficus pumila* · *Vernonia elliptica* · Urban forestry

Introduction

Today there is a trend among Malaysians to use creeping plants to decorate the walls of houses. Popular plants include *Ficus pumila* and *Vernonia elliptica* which creep on the fence around houses effectively making a 'green wall fence'. The aim is to create an aesthetic and calm atmosphere while reducing the cost of concrete wall maintenance that are susceptible to fungi growth. The concept is similar to the

Z. Haron (✉) · K. Yahya · Z. Jahya · N. Darus · Y. Z. Shyong · H. Halim
Department of Environmental, School of Civil Engineering, Universiti Sains Malaysia, 11800 George Town, Penang, Malaysia
e-mail: zaitonharon@utm.my

S. Maruthaveeran et al. (eds.), *Urban Forestry and Arboriculture in Malaysia*,
https://doi.org/10.1007/978-981-19-5418-4_15

production of a green façade (Galagoda et al., 2018; May Tzuc et al., 2021), whose function is to achieve the concept of sustainable infrastructure.

Ficus pumila is a beautiful plant and is widely grown. *Ficus pumila* is native to East Asia, and thrives in warm or tropical regions, such as Malaysia. The specific epithet '*pumila*' is derived from the Latin word for 'dwarf'. This refers to its small, ovate or heart-shaped growth of no more than 2.5 cm (Suzuki et al., 2020). Some *Ficus pumila* has variegated leaves and these are known as *Ficus pumila* 'Variegata'. *Ficus pumila* and *Ficus pumila* 'Variegata' are usually planted on the ground and they climb up and spread across the concrete / brick wall. The stems and leaves of this plant are reported to contain antioxidants and are even used in traditional Chinese medicine as a tonic and as a treatment for fever (Groot et al., 2003; Qi et al., 2021; Salehi et al., 2020; Trinh et al., 2018). *Ficus pumila* leaves are also used by the Japanese in beverages to treat diabetes and high blood pressure (Suzuki et al., 2020).

In contrast, *Vernonia elliptica* or Lee Kuan Yew does not grow on walls like *Ficus* but it can cover walls as a curtain plant. *Vernonia elliptica* plants usually grow in pots planted atop the wall and grow downward, hanging so that they can cover the walls. This plant has been widely used for vertical gardens in offices and tall buildings to cover glass walls from sunlight and also on balconies acting as a 'green wall fence' (Fig. 15.1). The plants are creeping and have silvery leaves (Lau & Frohlich, 2012; Thamapan et al., 2020). It originated in southeast Asia, in countries including India, Myanmar, and Thailand (Keeley et al., 2007). In Thailand *Vernonia* are used as medicine by the Thais (Siriyong et al., 2020). Vernon*ia elliptica* is the popular choice for green façade plants especially in Singapore, Malaysia and Indonesia (Kumar et al., 2019).

Physically, these two types of plants are very interesting and can fulfil key roles in urban ecology, but the effect on traffic noise control is still unknown. In densely populated urban areas and congested roads, traffic noise poses a problem for the community. Traffic noise is very noticeable at frequencies between 500 to 1600 Hz, and this is within the human hearing range of between 20 Hz to 20 kHz. At a distance

Fig. 15.1 Green wall fences: Concrete wall covered by *Ficus pumila* and balcony covered by *Vernonia elliptica*

of 15 m from a busy road, usually the traffic noise exceeds 90 dB and is disturbing to the human ear. Using hard surface walls introduces sound reflection and results in a canyon effect as well as adding to the overall sound to the urban environment. Excessive noise can have a negative impact on human health and well-being as it contributes to sleep disorders (Fyhri & Aasvang, 2010; Popp et al., 2015; Seidler et al., 2021; Skrzypek et al., 2017; Sygna et al., 2014), mental illness (Hegewald et al., 2020; Ma et al., 2018; Stansfeld & Shipley, 2015; Sygna et al., 2014), cardiovascular disease (Fyhri & Aasvang, 2010; Seidler et al., 2021) and can even lead to Alzheimer disease (Cui & Li, 2013).

Green wall fences may be used to reduce noise from traffic and overcome canyon effects. Thus, the purpose of this study is to examine the effects of using *Ficus pumila, Ficus pumila* 'Variegata' and *Vernonia elliptica* plants that are the current trend-green wall fences, in absorbing noises from the road. Impedance tubes with two microphones were used to characterise the sound absorption coefficient, reflection, surface impedance modulus and admittance of sound waves onto specimens. The tests were carried out separately using leaf specimens, bare wall specimens, and leaf together with wall specimens to investigate the effect of leaves and wall fences.

Literature Review

Reducing urban noise naturally and sustainably, especially noise caused by traffic has received widespread attention now. Among the methods in focus are the use of plants, including planting barrier trees, shrubs and vines. Recently, researchers have investigated the use of green walls and their effects on indoor heating (Galagoda et al., 2018; Jim, 2015a; Moya et al., 2019), air pollution and even outside noise blocking (Azkorra et al., 2015; Benkreira et al., 2011; Thomazelli et al., 2017). Most studies target the walls of houses or buildings; they use light or porous substrate layers and various attractive foliage plants as green walls. The design for the comprehensive green wall was discussed by (Jim, 2015b). Among the plants used for green walls are creeping plants where these plants can increase the absorption of noise compared to arboreal plants. However, the absorption of sound by green walls is still not fully understood, although recently (Paull et al., 2020) found that plants and substrates in 12 green walls can reduce noise by more than 10 dB.

In general, research shows that plants can absorb sound more effectively than hard, solid walls. This is because the leaves absorb high frequency sounds, exceeding 1000 Hz while the twigs and stems absorb sound between 500 and 1000 Hz (Paull et al., 2020). Plant density is a major contributing factor in reducing noise. Yet the leaves of the plants themselves have a very low mass to significantly function as an insulating material. However, there are several related theories that cause the leaves to have a significant effect: (1) leaves and twigs reduce noise by reflecting and scattering the sounds (Li & Kang, 2020; Paull et al., 2020); (2) the layer of thermoviscous air that surrounds the leaf converts sound energy into mechanical vibration, thus sound is converted to heat (Azkorra et al., 2015).

The sound absorption of *Euonymus japonicus* plants covering a wall has been previously studied (Attal et al., 2019). *Euonymus japonicus* 16 mm thick with 95% porosity reaches a sound absorption coefficient of 0.2 between 300 and 1000 Hz for green facades, i.e., plants in front of the wall, 0.2 between 200 and 1000 Hz for continuous living wall system, i.e., plants and walls separated by air space, 0.9 between 300 and 1000 Hz for modular living wall systems, i.e., plants, substrates, air space and walls. Attal et al. described that when the sound strikes the surface of leaf specimens, the sound absorption coefficient spectrum consists of a resonance and anti-resonance which can be related with surface impedance modulus. Maximum resonance (quarter-wavelength) is associated with the lowest surface impedance modulus.

The plant species used for green wall construction that have undergone investigation for their ability to absorb sound include *Euonymus japonicus, Callisia repens, Geranium zonale, Hedera helix, Pieris japonica, Primula vulgaris, Cyperus rotundus, and Bellis perennis* (D'Alessandro et al., 2015; Horoshenkov et al., 2013; Yu et al., 2015). All of these are shrubs, except *Hedera helix* a creeping plant, therefore in this study the potential of plants such as *Ficus pumila*, *Ficus pumila* 'Variegata' and *Vernonia elliptica* were studied for their ability in absorbing sound and their potential for use as green wall fences.

Methodology

Sound Absorption Test

Sound absorption coefficient (α) of specimens was obtained by using an impedance tube Type 4206-A, from Bruel and Kjaer, which is in accordance with ASTM E1050-98 (American Society for Testing & Materials, 2019). The sound absorption coefficient is the capacity of material to absorb sound at normal incidence. Sound absorption coefficient has value from 0 to 1 in which 0 is fully reflecting and 1 is fully absorbent. Sound absorption coefficient is relatively low if the value is than 0.35 (Attal et al., 2019).

The sound absorption coefficient was determined using a transfer-function in a two-microphone method by placing the specimen at one end of the tube. The method involves the decomposition of a broadband stationary random signal into incident sound, P_i and reflected sound, P_r. The transfer function compensates for the possible gain and phase mismatch of the two microphones, then the measurement is repeated by interchanging the two channels. The complex reflection coefficient R is calculated by Eq. 15.1:

$$R = \left[\frac{H_1 - H_{IN}}{H_R - H_1}\right] e^{j2k(l+s)} \tag{15.1}$$

where H_I is the frequency response function; H_{IN} is the frequency response function associated with the incident component; H_R is the frequency response function associated with the reflected component; j is defined as $\sqrt{-1}$, k is wave number, l is the distance to the first microphone location from the specimen and s is spacing between the microphones.

The normalised surface impedance ratio of specimen, $(\frac{z}{\rho c})$ and α can be calculated by Eq. 15.2 and Eq. 15.3, respectively;

$$\frac{z}{\rho c} = \frac{1+R}{1-R} \tag{15.2}$$

$$\alpha = 1 - |R|^2 \tag{15.3}$$

z is the surface impedance modulus of specimen, which is obtained by calculating the characteristic air impedance, ρc. Surface impedance implies the resistance of the specimen surface to the sound energy. In this study ρc for a temperature of 25 °C is 409 Rayls. Admittance (Ad) measures how easily the sound energy flows through specimens and can also be obtained by calculating the inverse of surface impedance (Eq. 15.4).

$$Ad = inverse\left(\frac{z}{\rho c}\right) \tag{15.4}$$

Using this technique, the bare wall, leaves and green wall fence specimens' acoustic absorptions were separately measured, and this was carried out by inserting the specimens in the impedance tubes and measuring the acoustic absorption of the whole system.

Acoustic Absorption of Plant Types and Thickness Specimens

Specimens of plants consist of *Ficus pumila*, *Ficus pumila* 'Variegata' and *Vernonia elliptica*. All were cultivated in polybags in Universiti Teknologi Malaysia. *Ficus pumila* and *Ficus pumila* 'Variegata' were assumed to cover the whole porous wall and brick wall with 10 mm thick, while *Vernonia elliptica* was assumed draping with 25 mm air spacing from the concrete porous wall and brick wall. *Ficus pumila* and *Ficus pumila* 'Variegata' specimens had a density of 101.91 kg/m^3 and the 10 mm thickness represented the thickness of such a plant. The thickness of *Vernonia elliptica* draping was assumed as 10 mm, 25 mm and 50 mm with a density of 91.63 kg/m^3.

In order to replicate the actual situation, the plant is cut along with the stalk and inserted into a specimen holder 100 mm in diameter and the desired thickness (Fig. 15.2). The *Ficus pumila* stalk size is small at 0.6 mm compared to *Vernonia*'s

Fig. 15.2 Preparation of plants' leaves for sound absorption performance measurement

2 mm. The size of *Ficus pumila* leaves is also comparatively small. *Vernonia* specimens contain 40% of the stalk compared to *Ficus pumila* and *Ficus pumila* 'Variegata' at 30%. Thus, the porosity of *Ficus pumila* and *Ficus pumila* 'Variegata' are lower at 42% compared to *V. elliptica*'s 65%. Porosity was measured by water displacement. In other words, *Ficus pumila and Ficus pumila* 'Variegata' contain greater number of leaves.

Bare Wall Fences Specimens

A 100 mm thick bare concrete brick wall and porous wall were used to represent the wall. Cylindrical specimens of 100 mm diameter × 100 mm thickness of these walls were fabricated to test the sound absorption coefficient of specimens with plants and bare walls. The concrete bricks with porosity 1.4% were fabricated using 1:4 mix proportion (one part cement to four parts fine aggregate by weight). Porous walls with porosity of 12.0% were made using 50 mm thick concrete brick combined with a 50 mm thick layer of porous concrete. During construction, the porous layer will be facing the road. Porous layer concrete was fabricated using crushed 5–10 mm granite aggregate, combined with cement and water in a proportion of 1:4 for cement to coarse.

Green Wall Fence Specimens

Bare wall fence specimens were covered by *Ficus pumila*, *Ficus pumila* 'Variegata' and *Vernonia elliptica* as green wall fence specimens (Fig. 15.3). Green wall fences

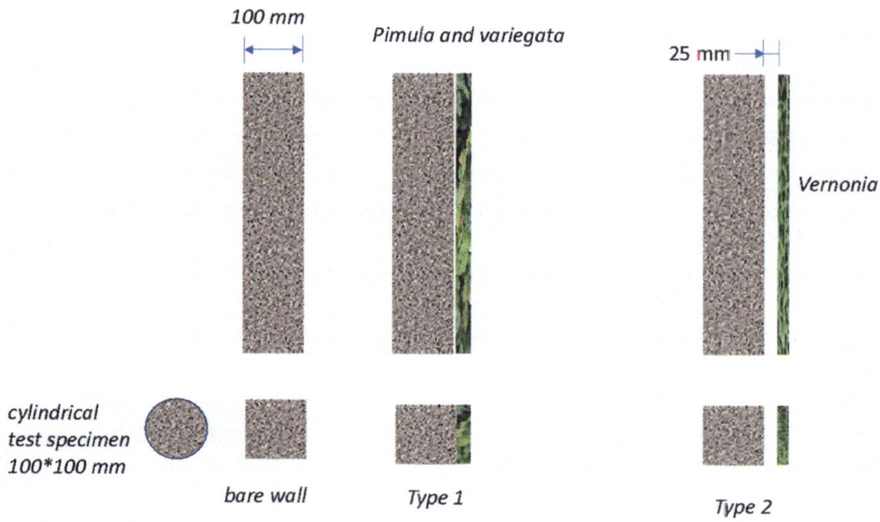

Fig. 15.3 Green wall specimen's series Type 1 and Type 2

are divided into two types; Type 1 is a wall with plants climbing on its surface which is similar to green façade geometry—a single layer of plants fixed on a building wall. Type 2 is wall and plant separated with a 25 mm air gap. It is similar to a continuous living wall system where the plant layer is fixed on a support separated from the building wall. Table 15.1 shows the characterisation of wall fence specimens in this study.

Results of Measurement

Types of Plants

Sound absorption coefficient spectrum and the surface impedance modulus of *Ficus pumila*, *Ficus pumila* 'Variegata' and *Vernonia elliptica* plant specimens with the same 10 mm thick rigid backing condition are given in Fig. 15.4. Variation of the surface impedance modulus versus frequency displays no resonant or anti-resonant effects, therefore sound absorptions were also neither resonant nor anti-resonant (Fig. 15.4b). For both *Ficus* leaves, surface impedance was higher than air resistance, causing the sound absorption coefficient by both *Ficus* leaves to have low sound absorption (<0.35). For *Vernonia* leaves, surface impedance at 1250 Hz becomes reduced compared to the characteristic air impedance causing sound absorption at 1250 Hz and above increases to 0.42 at 1600 Hz. Acoustic absorption is relatively low up to 800 Hz (<0.35) due to weak admittance (Fig. 15.4c) and high reflection in these frequency ranges. Attal et al. also obtained an average sound absorption

Table 15.1 Characterisation of wall fence specimens

Characterisation	Specimen name	Type
Bare concrete brick	Brick	–
Concrete brick covered by *Ficus pumila*	Brick + *pumila*	1
Concrete brick covered by *Ficus pumila* 'Variegata'	Brick + Variegata	1
Concrete brick covered by 10 mm thick *Vernonia elliptica* separated by air gap 25 mm	Brick + *Vernonia*-10 mm	2
Concrete brick covered by 25 mm thick *Vernonia elliptica* separated by air gap 25 mm	Brick + *Vernonia*-25 mm	2
Concrete brick covered by 50 mm thick *Vernonia elliptica* separated by air gap 25 mm	Brick + *Vernonia*-50 mm	2
Bare porous concrete wall	Porous	–
Porous wall covered by *Ficus pumila*	Porous + *pumila*	1
Porous wall covered by *Ficus pumila* 'Variegata'	Porous + Variegata	1
Porous wall covered by 10 mm thick *Vernonia elliptica* separated by air gap 25 mm	Porous + *Vernonia*-10 mm	2
Porous wall covered by 25 mm thick *Vernonia elliptica* separated by air gap 25 mm	Porous + *Vernonia*-25 mm	2
Porous wall covered by 50 mm thick *Vernonia elliptica* separated by air gap 25 mm	Porous + *Vernonia*-50 mm	2

coefficient value of 0.2 with an 8 mm thick *Euonymus japonicus* specimens (Attal et al., 2019).

Effect of Thickness of Plants

The effect of *Vernonia elliptica* specimen 10 mm–50 mm thick on the sound absorption coefficient, surface impedance modulus, admittance and reflections in rigid support conditions is given in Fig. 15.5. *Vernonia elliptica* leaves 50 mm thick have a maximum absorption at 800 Hz (sound absorption coefficient = 0.75) and 25 mm thick at 1250 Hz (sound absorption coefficient = 0.55). This is due to the variation of the surface impedance modulus relative to the frequency display minima is associated with the quarter-wavelength resonance at 630 Hz for $h = 50$ mm, and also to admittance and low reflection (Fig. 15.5c and d). For 10 mm thick specimens, the acoustic absorption is relatively low for 800 Hz (<0.35) due to poor admittance (Fig. 15.5c) and higher reflection in this frequency range. The 50 mm specimen reached peak sound absorption coefficient at 800 Hz due to absorption dominated by *Vernonia elliptica* twigs/stems because 40% of the specimen contained stems that highly scatters the sound at 650–850 Hz (Fig. 15.5c). This is in line with previous research showing that twigs absorb more sound at a frequency range of 400–1000 Hz (Paull et al., 2020). Generally, the effect of specimen thickness on sound absorption is

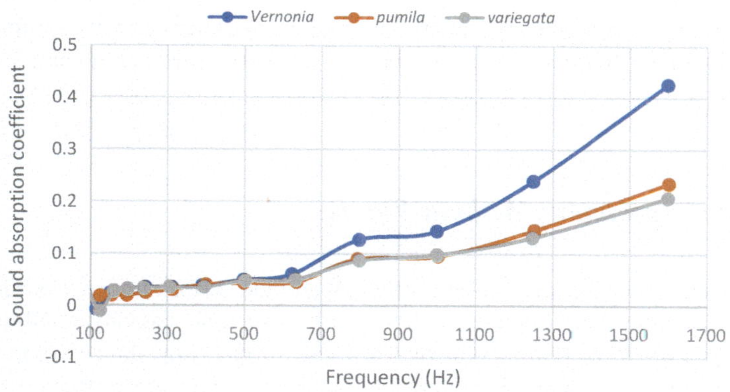

(a)

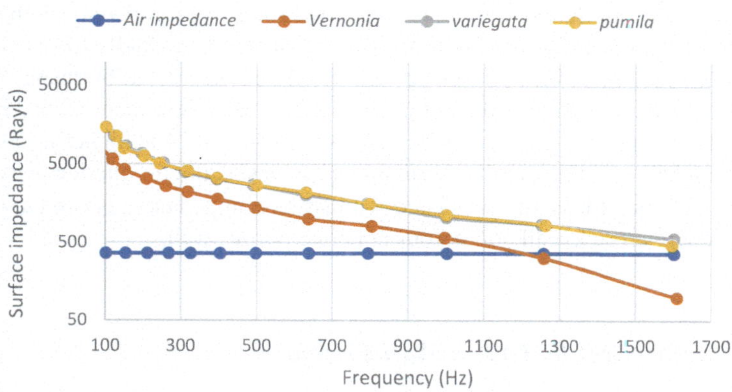

(b)

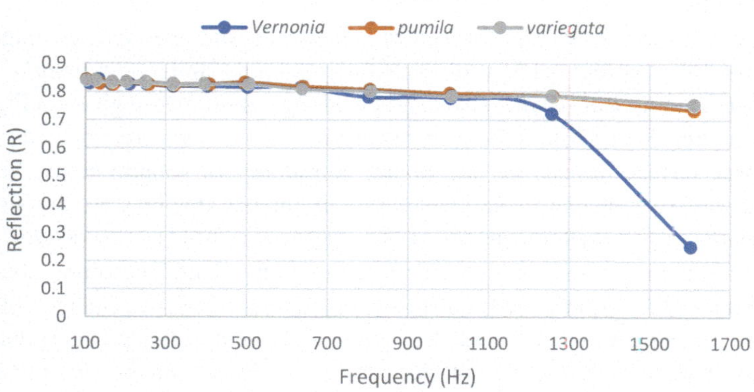

(c)

Fig. 15.4 Acoustic absorption properties 10 mm thickness of *Ficus pumila*, *Ficus pumila* 'Variegata' and *Vernonia elliptica*. (a) Sound absorption coefficient vs frequency experienced by the specimens, (b) Surface impedance of the specimens, (c) Reflection of surface specimens similar with that of concrete specimens with various thickness; the thicker the specimen the dominant frequency porous concrete shifts to a lower frequency (Kim & Lee, 2010).

Wall Fence Specimens Without Plants

Frequency spectra of the absorption coefficient and the surface impedance modulus of porous concrete and concrete brick specimens in rigid backing condition are displayed in Fig. 15.6. Porous concrete has less surface impedance compared to brick walls and higher in air characteristic impedance. Thus, absorption coefficient for porous concrete exhibits a more resonance-like behaviour than brick walls in frequency similar to the one previously reported for porous concrete (Wang et al., 2016). Absorption coefficient of both specimens increases gradually from 0.1 at low 400 Hz frequency up to 0.5 at frequency 630 Hz (brick) and 0.7 at frequency 630 Hz (porous wall) at absorption peak and decreases to a value less than 0.2 at 1250 Hz. Afterwards, that porous wall specimen increases close to 0.76 at higher frequencies. As for 25 and 50-mm thickness of *Vernonia elliptica* specimens, quarter-wavelength resonances of specimens are associated with surface impedance minima at 630 Hz for both concrete brick wall and porous concrete porous wall.

Sound Absorption by Green Wall Fences

- **Brickwall**
  Frequency spectra of the absorption coefficient and the surface impedance modulus of concrete brick wall with *Ficus pumula* and *Ficus pimula* 'Variegata' specimens with 10 mm thick rigid backing condition are given in Fig. 15.7, showing that *Ficus* shifted the peak absorption to the lower frequency as a result of lower surface impedance and higher admittance of sound energy. Variation of the surface impedance modulus versus frequency displays are associated to the quarter-wavelength resonance of the specimen at 500 Hz, respectively; while for sound admittance, it is higher than without any plant. The sound absorption coefficient then experienced anti-resonance and gradual increment with *Ficus pumila* reaching a higher sound absorption coefficient of 0.82. This is due to the surface impedance reduction until it decreases to lower than the air's impedance. Maximum of the surface impedance modulus (anti resonance) at 630 Hz is nearly at a quarter-wavelength of 10 mm thick specimen. Due to weak attenuation (Fig. 15.7b), acoustic absorption remains relatively low (< 0.35) in the frequency range less than 400 Hz. Highest absorption coefficients occur near specimen quarter wave resonance where surface impedance of the specimen and the air characteristic impedance are better matched.

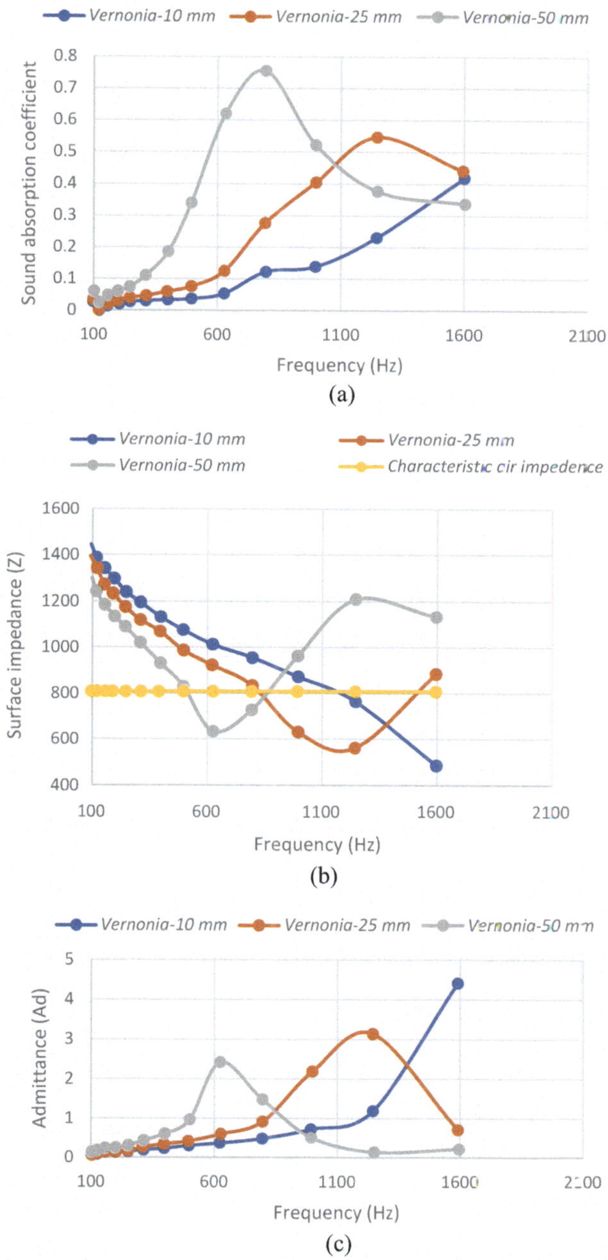

Fig. 15.5 The effect of thickness on acoustic absorption properties. (a) Sound absorption coefficient vs frequency experienced by the specimens, (b) Surface impedance of the specimens, (c) Surface admittance of the specimens, (d) Surface reflection

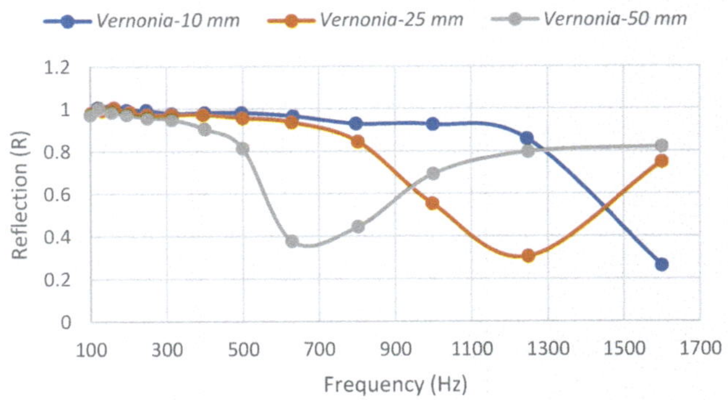

(d)

Fig. 15.5 (continued)

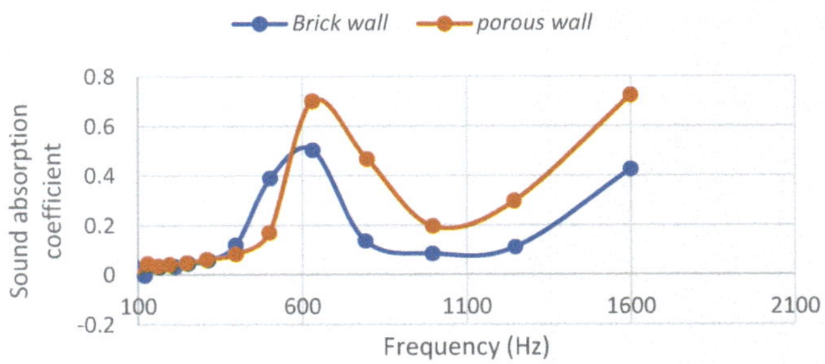

(a)

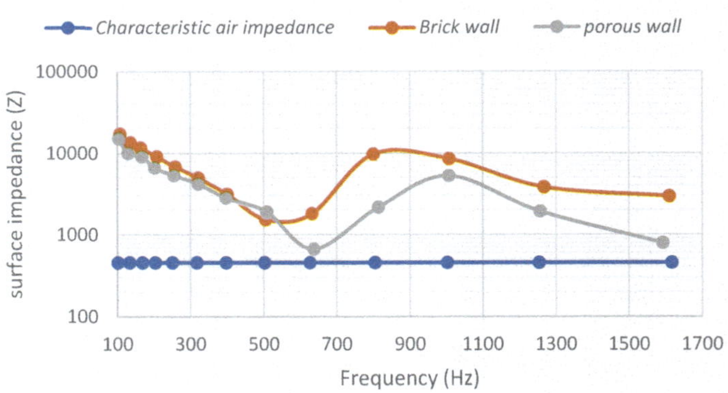

(b)

Fig. 15.6 Acoustic properties of bare wall fences specimens. (a) Sound absorption coefficient, (b) Surface impedance of specimens

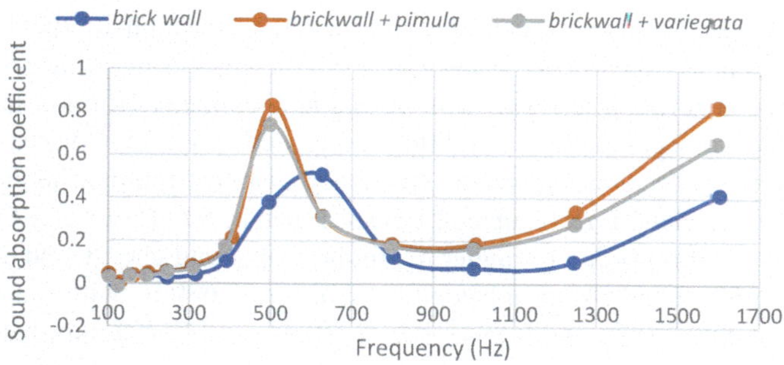

(a)

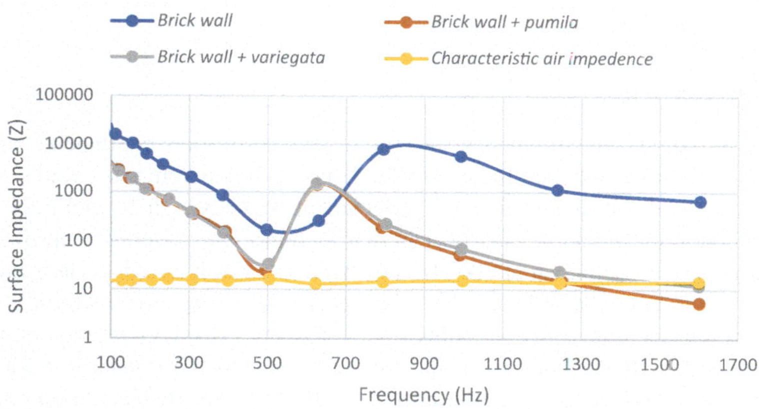

(b)

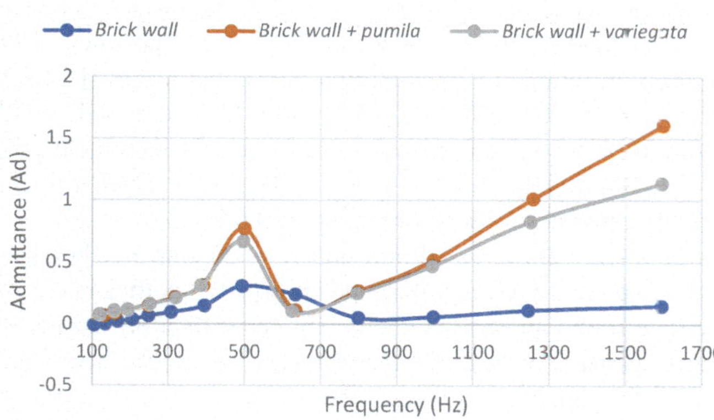

(c)

Fig. 15.7 Acoustic properties of wall fences specimens (a) Sound absorption coefficient, (b) Surface impedance of specimens, (c) Admittance of sound wave

Overall, *Vernonia elliptica* increases the sound absorption of the brick wall system. Frequency spectra of the absorption coefficient and the surface impedance modulus of concrete brick wall with *Vernonia* specimens 10 mm, 25 mm and 50 mm thick in rigid backing condition are given in Fig. 15.8. Variation of the surface impedance modulus versus frequency displays minimal association to the quarter-wavelength resonance of the specimen at 500 Hz (h = 10 mm), and 630 Hz (h = 50 mm) due to higher admittance of sound energy and 25 mm air gap. Maximum sound absorption coefficient occurs at 630 Hz with sound absorption coefficient of 0.71 and 0.89, for h = 10 mm and 25/50 mm, respectively. Maximum surface impedance modulus (anti resonance) at 1250 Hz for specimen brick wall with Vernonia 50 mm is nearly at its quarter-wavelength of thick *Vernonia* specimen.

- **Porous concrete**

Figure 15.9 shows the frequency spectra of the absorption coefficient, surface impedance modulus and admittance of sound energy of concrete brick wall covered with *Ficus pumila* and *Ficus pumila* 'Variegata' specimens with 10 mm thick rigid backing condition. It can be clearly seen that 10 mm *Ficus* coverage improved the sound absorption coefficient at 600 Hz and above as a result of reduction in surface impedance modulus that enables the sound energy admittance to increase. Variation of the surface impedance modulus versus frequency displays association to the quarter-wavelength resonance of the specimen at 630 Hz for *Ficus pumila* 'Variegata' and *Ficus pumila*. Due to low admittance (Fig. 15.9c), acoustic absorption remains relatively low (< 0.35) in the frequency range less than 400 Hz.

For the absorption coefficient frequency spectra, the surface impedance modulus of porous concrete and admittance of *Vernonia elliptica* specimens 10 mm, 25 mm and 50 mm thick in rigid backing condition are shown in Fig. 15.10. Generally, it can be seen that plants covering the porous wall had increased sound absorption coefficient due to reduction of surface impedance and increment of sound wave admittance. The lowest impedance is given by *Vernonia elliptica* thickness 10 mm at 1000 Hz with sound absorption coefficient of 0.64.

Vernonia elliptica with h = 10 mm and h = 25 mm reaches its peak sound absorption at 630 Hz and 500 Hz with sound absorption coefficients of 0.80 and 0.94, respectively. The sound absorption coefficient curve then has no resonance due to lower surface impedance at 1000 Hz, that is lower than air characteristic impedance. Except for h = 50 mm, highest sound absorption coefficients were obtained near specimen quarter wave resonance where a better matching is obtained between the surface impedance of the specimen and air characteristic impedance. For h = 50 mm, sound absorption coefficient peaks at 630 Hz and reaches anti-resonant afterwards due to specimen's high surface impedance compared to that of air impedance that causes very low sound energy penetration ability (Fig. 15.10c). It can be noted that h = 25 mm and 50 mm improved sound absorption coefficient of wall system at 500 Hz greater than 0.35 and all specimens have sound absorption coefficient greater than 0.35 for h = 10 –50 mm in the range of 1000–1250 Hz.

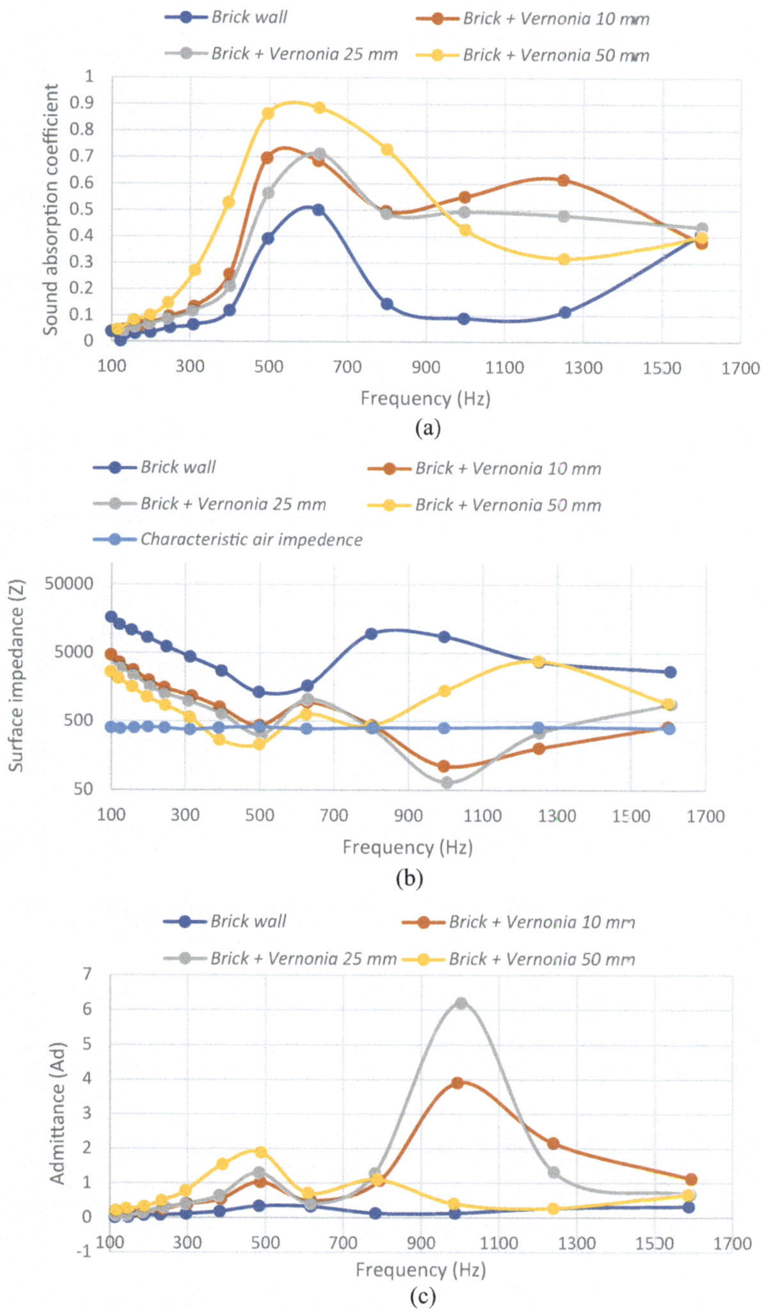

Fig. 15.8 Acoustic properties of wall fences specimens -Type 2 (a) Sound absorption coefficient, (b) Surface impedance of specimens, (c) Admittance of sound wave

Fig. 15.9 Acoustic properties of brick wall fences specimens Type 1 (a) Sound absorption coefficient, (b) Surface impedance of specimens, (c) Admittance of sound

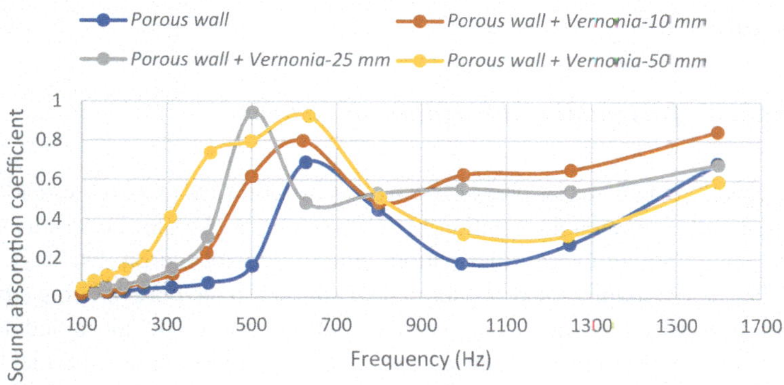

(a)

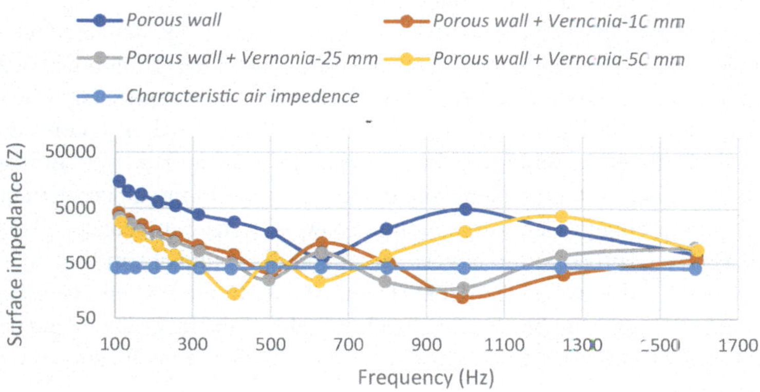

(b)

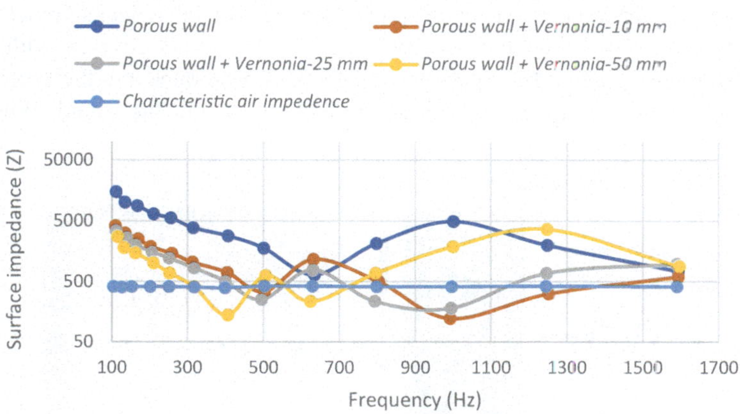

(c)

Fig. 15.10 Acoustic properties of porous wall fences specimens Type 2. (a) Sound absorption coefficient, (b) Surface impedance of specimens, (c) Admittance of sound

Discussion

Mechanism of Acoustic Absorption of Leaves

In porous walls, when sound pressure waves at different frequencies spread to the pores filled with saturated air, friction between the air molecules occurs inside the pores, causing a sound energy dissipation called viscosity-inertia discharge effect (Fig. 15.11). The remaining sound energy then spreads to the pore opening structure and develops high sound wave energy as the friction between the air molecules and the pore walls produces heat exchange. The dissipation of viscous and thermal layers, as well as the remaining sound energy occurring through the porous concrete will cause different sound absorption effects at different frequencies.

When the structural system consists of a porous concrete layer covered with plant material, for example *Vernonia elliptica* leaf specimens of $h = 25$ mm with a 25 mm air gap, the sound penetrates the leaf; leaves and twigs vibrate and scatter the sound. Also as explained, the Thermovisc air layer around the leaf will also convert sound energy into mechanical vibrations, thus dissipating as heat. This system resonates when the frequency of the incident sound wave is equal to the inherent frequency, and the damping effect reaches maximum at lowest surface impedance. This yield sound energy is absorbed to the largest extent and the sound absorption coefficient reaches the peak value at 500 Hz–1000 Hz and above. In the system, the sound will also penetrate the pores in the specimen and spread into the inner layer, and transmit to the 25 mm air space, turning into heat, and partially penetrate the porous concrete and the same process takes place.

On the other hand, a concrete brick wall has less pores, hence when sound pressure waves at different frequencies spread, sound will be reflected much more than a porous wall. When the concrete brick wall layer is covered with *Ficus pumila/Vernonia elliptica* leaf specimens, the sound penetrates into the specimen's structural system, but leaves will vibrate, reflect and scatter the sound. When the

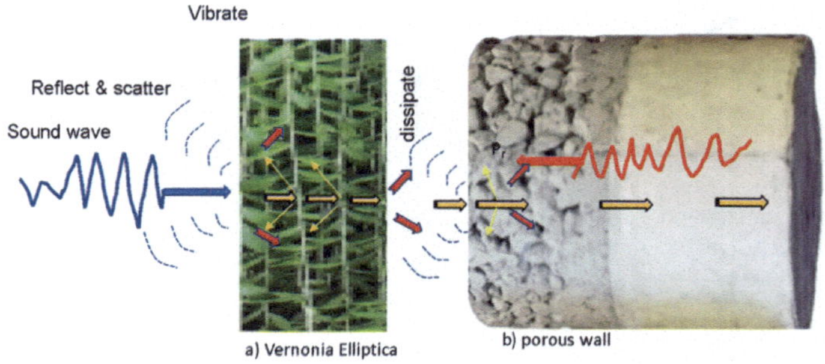

Fig. 15.11 Dissipation of sound energy in green porous wall system in the impedance tube

frequency of the incident sound wave is equal to the inherent frequency of the system, and the damping effect reaches maximum at lowest surface impedance and yield significant sound energy reduction peak value at 1600 Hz.

Optimisation of Green Wall Fences for Traffic Noise Attenuation

In reality the disturbing traffic noise sources are 500 Hz by heavy vehicle, 1600 Hz by mid-sized vehicle and at 1250 Hz by light vehicle (Wang et al., 2016, 2018). The noise generated from tire-pavement contacts at less than 500 Hz is by structure borne noise, 500–1,000 Hz by sidewall vibration, 1000–1500 Hz by air pumping and resonance from 1500 to 2,000 Hz. Hence, sound absorption coefficient in 500–1600 Hz range would be considered for the noise barrier construction.

Considering sound absorption coefficient for traffic noise, it can be noted that porous concrete and brick wall with 10–50 mm thick *Vernonia elliptica* is the best green wall fencing since the value of all sound absorption coefficient values are greater than 0.35 to reduce all categories of noise emanating from road traffic, which consequently reduces the canyon effect. Porous walls covered with *Ficus pumila* and *Ficus pumila* 'Variegata' may reduce noise in the frequency range greater than 500 Hz such as noises from side wall vibration, air pumping and resonance from tire to pavement contact. While brick walls covered by *Ficus pumila* have good sound absorption coefficients at 500 Hz and 1600 Hz that may reduce noise from engine and air pumping.

Comparison with Glass Wool and Rock Wool

Comparisons were carried out between the acoustic performance of green wall fence specimens with rockwool and fibreglass absorber in previous research (Haron et al., 2020) (Fig. 15.12). It can be seen that *Vernonia elliptica* specimens 10–50 mm thick on a porous wall have sound absorption coefficient better than rockwool for frequencies ranging between 200 to 630 Hz. The same is observed for a concrete brick wall with 50 mm *Vernonia elliptica*. While it was found only porous concrete with 50 mm *Vernonia elliptica* has a sound absorption coefficient exceeding glass wool for the same frequency range. Rockwool and fibreglass are synthetic fibre and normally used to infill noise barriers on highways, thus *Vernonia elliptica* can also be used to reduce noise from traffic especially from the heavy vehicle, where the noise energy peak frequency is 500 Hz.

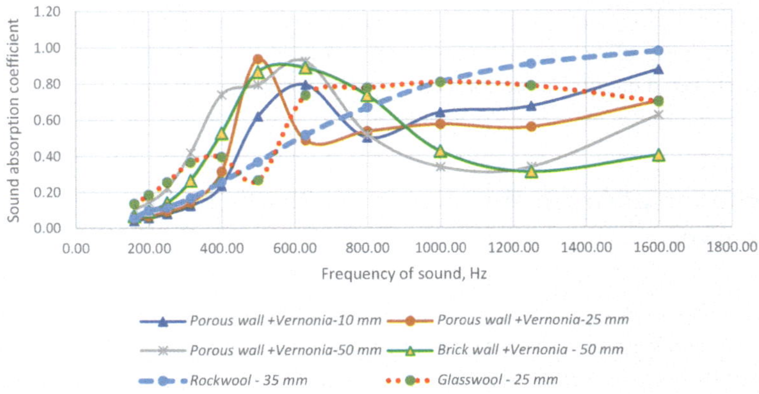

Fig. 15.12 Comparison of sound absorption with different types of green wall fences & synthetic materials

Conclusion

This study examines whether *Ficus pumila*, *Ficus pumila* 'Variegata' and *Vernonia elliptica* plant species on wall fences can be used to reduce noise from traffic and overcome canyon effects. The effect on sound absorption was measured using two microphone impedance tubes. Using this technique, the specimens' wall, leaves and green wall fence systems' acoustic absorptions were measured separately. The following are the results obtained from this study and the conclusions are:

i. Specimens of *Ficus pumila* and *Ficus pumila* 'Variegata' leaves 10 mm thick have no resonance and anti-resonance sound absorption spectrum and have low sound absorption (<0.35) due to high surface impedance modulus than air resistance. Applying *Ficus pumila* and *Ficus pumila* 'Variegata' on brick wall shifted the peak absorption to a lower frequency caused by lower surface impedance and higher sound energy admittance. The maximum absorption reaches 0.82–0.84 at 500 Hz compared to that of plain concrete block with maximum sound absorption coefficient of 0.5 at 630 Hz. On the other hand, *Ficus* on porous concrete walls has demonstrated an even higher sound absorption coefficient of 0.84–0.96 occurring at a higher frequency, 630 Hz, for both *Ficus pumila* and *Ficus pumila* Variegata.
ii. *Vernonia elliptica* on porous wall specimens separated with 25 mm air void had increased sound absorption coefficient from 0.7 to 0.9 when *Vernonia elliptica* thickness is 50 mm, due to reduction of surface impedance and increased admittance of sound wave. The lowest impedance is given by 10 mm *Vernonia elliptica* at frequency 1000 Hz with sound absorption coefficient 0.64, compared to that of porous wall with only sound absorption value of 0.19. Similarly, *Vernonia elliptica* also increases the sound absorption of a brick wall fence system. Variation of the surface impedance modulus versus frequency displays minimal association to the quarter-wavelength resonance of specimen at 500 Hz (h = 10 mm),

and 630 Hz ($h = 50$ mm), due to higher admittance of sound energy. Maximum sound absorption coefficient occurs at 630 Hz with sound absorption coefficient of 0.71 and 0.89, for $h = 10$ and 25/50 mm, respectively.

iii. *Vernonia elliptica* leaf and porous concrete separated by a 25 mm air gap demonstrates significant increase in sound absorption due to: (a) when sound penetrates, leaves and twigs will bounce and spread the sound, hence the noise reduction is quite significant at 500 Hz–1000 Hz, and above. (b) The sound will also enter the air space like pores due to the porosity of the specimen and spread into the inner layer, the 25 mm air space, which then then turn into heat, and partially penetrate the porous concrete and the same process takes place. (c) the thermo-viscous air layer around the leaf will also convert sound energy into mechanical vibrations, consequently disappearing as heat.

iv. Porous concrete and brick wall with draped *Vernonia elliptica* 10–50 mm thick are the best green wall fencing since the value of all sound absorption coefficient value ranges are greater than 0.35 up to 0.9 for 500–1600 Hz frequency range. Hence, they reduce all categories of noise emanating from road traffic and the associated canyon effect.

v. *Vernonia elliptica* specimens with 10–50 mm thick coverings on a porous wall with 25 mm air gap have sound absorption coefficient better than rockwool for 200 to 630 Hz frequency range.

References

American Society for Testing and Materials. (2019). *ASTM E1050: Standard test method for impedance and absorption of acoustical materials using a tube, two microphones and a digital frequency analysis system*. Astm International.

Attal, E., Côté, N., Shimizu, T., & Dubus, B. (2019). Sound absorption by green walls at normal incidence: Physical analysis and optimization. *Acta Acustica United with Acustica*. https://doi.org/10.3813/AAA.919313

Azkorra, Z., Pérez, G., Coma, J., Cabeza, L. F., Bures, S., Álvaro, J. E., Erkoreka, A., & Urrestarazu, M. (2015). Evaluation of green walls as a passive acoustic insulation system for buildings. *Applied Acoustics*. https://doi.org/10.1016/j.apacoust.2014.09.010

Benkreira, H., Horoshenkov, K. V., Khan, A., Mandon, A., & Rohr, R. (2011). The effect of drying on the acoustic absorption of novel green noise insulation. *European Drying Conference-EuroDrying 2011*.

Cui, B., & Li, K. (2013). Chronic noise exposure and Alzheimer disease: Is there an etiological association? *Medical Hypotheses*, *81*(4). https://doi.org/10.1016/j.mehy.2013.07.017

D'Alessandro, F., Asdrubali, F., & Mencarelli, N. (2015). Experimental evaluation and modelling of the sound absorption properties of plants for indoor acoustic applications. *Building and Environment*. https://doi.org/10.1016/j.buildenv.2015.06.004

Fyhri, A., & Aasvang, G. M. (2010). Noise, sleep and poor health: Modeling the relationship between road traffic noise and cardiovascular problems. *Science of the Total Environment*, *408*(21). https://doi.org/10.1016/j.scitotenv.2010.06.057

Galagoda, R. U., Jayasinghe, G. Y., Halwatura, R. U., & Rupasinghe, H. T. (2018). The impact of urban green infrastructure as a sustainable approach towards tropical micro-climatic changes and human thermal comfort. *Urban Forestry and Urban Greening*. https://doi.org/10.1016/j.ufug.2018.05.008

Groot, E. P., Sweeney, E. J., & Rost, T. L. (2003). Development of the adhesive pad on climbing fig (Ficus pumila) stems from clusters of adventitious roots. *Plant and Soil*. https://doi.org/10.1023/A:1022342025447

Haron, Z., Yahya, K., Fasli, N. M., Darus, N., Galib, S., Halim, H., & Mat, T. N. F. T. (2020). The potential of napier grass leaf fibres as an acoustic absorber. *Journal of Environmental Treatment Techniques, 8*(2), 797–803.

Hegewald, J., Schubert, M., Freiberg, A., Starke, K. R., Augustin, F., Riedel-Heller, S. G., Zeeb, H., & Seidler, A. (2020). Traffic noise and mental health: A systematic review and meta-analysis. *International Journal of Environmental Research and Public Health, 17*(17), 6175. https://doi.org/10.3390/ijerph17176175

Horoshenkov, K. V., Khan, A., Benkreira, H., Smyrnova, Y., Rehioui, K., & Kang, J. (2013). Acoustic properties of low growing plants. *42nd International Congress and Exposition on Noise Control Engineering 2013, INTER-NOISE 2013: Noise Control for Quality of Life*. https://doi.org/10.1121/1.4798671

Jim, C. Y. (2015a). Cold-season solar input and ambivalent thermal behavior brought by climber greenwalls. *Energy*. https://doi.org/10.1016/j.energy.2015.07.127

Jim, C. Y. (2015b). Greenwall classification and critical design-management assessments. *Ecological Engineering*. https://doi.org/10.1016/j.ecoleng.2015.01.021

Keeley, S. C., Forsman, Z. H., & Chan, R. (2007). A phylogeny of the "evil tribe" (Vernonieae: Compositae) reveals old/new world long distance dispersal: Support from separate and combined congruent datasets (trnL-F, ndhF, ITS). *Molecular Phylogenetics and Evolution*. https://doi.org/10.1016/j.ympev.2006.12.024

Kim, H. K., & Lee, H. K. (2010). Acoustic absorption modeling of porous concrete considering the gradation and shape of aggregates and void ratio. *Journal of Sound and Vibration*. https://doi.org/10.1016/j.jsv.2009.10.013

Kumar, C. R. S., Ghosh, S., Gumber, S., Mishra, R., Thangaraja, J., & Yang, H. (2019). Carbonaceous particle scavenging and thermal comfort augmentation with an extended green facade draped with Vernonia elaeagnifolia. *EnvironmentAsia*. https://doi.org/10.14456/ea.2019.46

Lau, A., & Frohlich, D. (2012). New plant records from O 'ahu for 2009. *Bishop Museum Occasional Papers, 113*, 7–26.

Li, M., & Kang, J. (2020). Influence of leaf physical properties on single-leaf vibrational response to sound. *Forests*. https://doi.org/10.3390/f11010115

Ma, J., Li, C., Kwan, M. P., & Chai, Y. (2018). A multilevel analysis of perceived noise pollution, geographic contexts and mental health in Beijing. *International Journal of Environmental Research and Public Health, 15*(7). https://doi.org/10.3390/ijerph15071479

May Tzuc, O., Rodríguez Gamboa, O., Aguilar Rosel, R., Che Poot, M., Edelman, H., Jiménez Torres, M., & Bassam, A. (2021). Modeling of hygrothermal behavior for green facade's concrete wall exposed to nordic climate using artificial intelligence and global sensitivity analysis. *Journal of Building Engineering*. https://doi.org/10.1016/j.jobe.2020.101625

Moya, T. A., van den Dobbelsteen, A., Ottelé, M., & Bluyssen, P. M. (2019). A review of green systems within the indoor environment. *Indoor and Built Environment*. https://doi.org/10.1177/1420326X18783042

Paull, N. J., Krix, D., Torpy, F. R., & Irga, P. J. (2020). Can green walls reduce outdoor ambient particulate matter, noise pollution and temperature? *International Journal of Environmental Research and Public Health*. https://doi.org/10.3390/ijerph17145084

Popp, R. F. J., Maier, S., Rothe, S., Zulley, J., Crönlein, T., Wetter, T., Rupprecht, R., & Hajak, G. (2015). Impact of overnight traffic noise on sleep quality, sleepiness, and vigilant attention in long-haul truck drivers: Results of a pilot study. *Noise and Health, 17*(79). https://doi.org/10.4103/1463-1741.169698

Qi, Z.-Y., Zhao, J.-Y., Lin, F.-J., Zhou, W.-L., & Gan, R.-Y. (2021). Bioactive compounds, therapeutic activities, and applications of Ficus pumila L. *Agronomy*. https://doi.org/10.3390/agronomy11010089

Salehi, B., Prakash Mishra, A., Nigam, M., Karazhan, N., Shukla, I., Kiełtyka-Dadasiewicz, A., Sawicka, B., Głowacka, A., Abu-Darwish, M. S., Hussein Tarawneh, A., Gadetskaya, A. v., Cabral, C., Salgueiro, L., Victoriano, M., Martorell, M., Docea, A. O., Abdolshahi, A., Calina, D., & Sharifi-Rad, J. (2020). Ficus plants: State of the art from a phytochemical, pharmacological, and toxicological perspective. In *Phytotherapy Research*. https://doi.org/10.1002/ptr.6884

Seidler, A., Hegewald, J., Schubert, M., Popp, C., & Moebus, S. (2021). Effects of noise protection measures on annoyance, sleep disorders, and cardiovascular diseases: A model calculation. *Gesundheitswesen, 83*(5). https://doi.org/10.1055/a-1082-0809

Siriyong, T., Phungtammasan, S., Jansorn, S., Chonsongkram, N., Chanwanitsakul, S., Subhadhirasakul, S., & Voravuthikunchai, S. P. (2020). Traditional Thai herbal medicine as an alternative treatment for refractory chronic eczema. *Explore*. https://doi.org/10.1016/j.explore.2019.10.001

Skrzypek, M., Kowalska, M., Czech, E. M., Niewiadomska, E., & Zejda, J. E. (2017). Impact of road traffic noise on sleep disturbances and attention disorders amongst school children living in upper Silesian Industrial Zone, Poland. *International Journal of Occupational Medicine and Environmental Health, 30*(3). https://doi.org/10.13075/ijomeh.1896.00823

Stansfeld, S. A., & Shipley, M. (2015). Noise sensitivity and future risk of illness and mortality. *Science of the Total Environment, 520*. https://doi.org/10.1016/j.scitotenv.2015.03.053

Suzuki, K., Gonda, K., Kishimoto, Y., Katsumoto, Y., & Takenoshita, S. (2020). Potential curing and beneficial effects of Ooitabi (Ficus pumila L.) on hypertension and dyslipidaemia in Okinawa. *Journal of Human Nutrition and Dietetics*. https://doi.org/10.1111/jhn.12806

Sygna, K., Aasvang, G. M., Aamodt, G., Oftedal, B., & Krog, N. H. (2014). Road traffic noise, sleep and mental health. *Environmental Research, 131*. https://doi.org/10.1016/j.envres.2014.02.010

Thamapan, K., Laohakunjit, N., Kerdchoechuen, O., Vongsawasdi, P., & Mingvanish, W. (2020). Phytochemical evaluation and qualitative analysis as sweetness potentially of selected thai plants. *Chiang Mai University Journal of Natural Sciences*. https://doi.org/10.12982/CMUJNS.2020.0021

Thomazelli, R., Caetano, F. D. N., & Bertoli, S. R. (2017). *Acoustic Properties of Green Walls: Absorption and Insulation*. https://doi.org/10.1121/2.0000426

Trinh, P. T. N., An, N. H., An, P. N., Tri, M. D., van Du, C., Minh, P. N., le Thuy, N. T., Tuan, N. T., Thoa, V. T. K., Tuan, N. N., & Dung, L. T. (2018). A new benzofuran derivative from the leaves of Ficus pumila L. *Natural Product Research*. https://doi.org/10.1080/14786419.2017.1395427

Wang, H., Ding, Y., Liao, G., & Ai, C. (2016). Modeling and optimization of acoustic absorption for porous asphalt concrete. *Journal of Engineering Mechanics*. https://doi.org/10.1061/(asce)em.1943-7889.0001037

Wang, H., Luo, P., & Cai, M. (2018). Calculation of noise barrier insertion loss based on varied vehicle frequencies. *Applied Sciences (Switzerland)*. https://doi.org/10.3390/app8010100

Yu, Y. L., Chao, Y. T., Lee, L. C., Yen, J. Y., & Fan, Y. C. (2015). A novel soundproof ventilation plant design with high performance and no energy consumption. *Mathematical Problems in Engineering*. https://doi.org/10.1155/2015/496748

Zaiton Haron is an Associate Professor at School of Civil Engineering, Faculty Engineering, Universiti Teknologi Malaysia. She received PhD in Architecture and Building Engineering from University of Liverpool, England, United Kingdom. She is interested in research of noise control in the civil engineering area especially for reducing noise pollution in buildings due to traffic noise.

Khairulzan Yahya is a lecturer at School of Civil Engineering, Faculty Engineering, Universiti Teknologi Malaysia. He received PhD in Construction Management from University of Liverpool, England, United Kingdom. He is interested in research related to indoor environment quality in building.

Zanariah Jahya is PhD student at School of Civil Engineering, Faculty Engineering, Universiti Teknologi Malaysia. She is currently working in research related to sound absorption of porous concrete.

Nadirah Darus is a lecturer at School of Civil Engineering, Faculty Engineering, Universiti Teknologi Malaysia. She received PhD in Civil Engineering from Universiti Teknologi Malaysia, Malaysia. Her current research interests are in the area of noise pollution/traffic noise abatement.

Yap Zhen Shyong is PhD student at School of Civil Engineering, Faculty Engineering, Universiti Teknologi Malaysia (UTM). He is currently working in research related to sound absorption of porous concrete.

Herni Halim is a lecturer at the Department of Environment, School of Civil Engineering, Faculty Engineering, Universiti Sains Malaysia (USM). She is actively involved in research related to noise barriers for noise pollution reduction.

Chapter 16
Effects of Paclobutrazol and Potassium Nitrate in Improving the Flowering Performance of *Xanthostemon chrysanthus* (F. Muell.) Benth

Ahmad Nazarudin Mohd Roseli

Abstract Although *Xanthostemon chrysanthus* (F. Muell.) Benth. is not a local species, it is often preferred as an ornamental tree in Malaysian cities due to its unique yellow inflorescence. Previous studies have shown that *X. chrysanthus* was able to grow in sub-optimal urban soils. However, under local climate conditions, the flowering of the species is unpredictable and always less abundant than in its native range. Thus, an experiment to evaluate the potential of paclobutrazol (PBZ) and potassium nitrate (KNO_3) for improving the flowering of the species was conducted. Three dosages of PBZ (0, 0.125, and 0.25 g/L/tree) and KNO_3 (0, 100, and 200 g/tree) were combined and tested on *X. chrysanthus* aged approximately six years old after planting at Metropolitan Batu Park, Kuala Lumpur. PBZ application was made only once by soil drenching, in the early stages of the experiment in March 2012. The first application of KNO_3 was carried out concurrently with PBZ treatment, and the following application was repeated at three-month intervals, i.e. in June 2012, September 2021, and December 2012. Inflorescence size was significantly bigger with 0.125 g/L PBZ coupled with 100 g KNO_3 than that of other treatments. However, mean flower abundance was significantly improved with 0.25 g/L PBZ combined with 100 g KNO_3. Our previous studies also noted that tree height and leaf size were decreased with the presence of PBZ, however, no abnormal leaf formation was observed, suggesting that the PBZ dosages used were appropriate for the species. Treatment with 0.25 g/L PBZ coupled with 100 g KNO_3 potentially improves the flower abundance without detrimental effects, thus increasing the attractive appearance of *X. chrysanthus* trees.

Keywords Plant growth regulator · Flower abundance · Landscape tree · Urban tree management

A. N. Mohd Roseli (✉)
Forestry and Environment Division, Forest Research Institute Malaysia (FRIM), 52109 Kepong, Selangor, Malaysia
e-mail: nazarudin@frim.gov.my

S. Maruthaveeran et al. (eds.), *Urban Forestry and Arboriculture in Malaysia*,
https://doi.org/10.1007/978-981-19-5418-4_16

What Induces Flowering?

Flowering is an indicator of plant maturity and one of the crucial development phases in the whole cycle of plant growth. In general, plants will flower after reaching maturity, provided that all resource requirements are met. There are two factors namely endogenous and exogenous which stimulate the transformation of apical meristem from developing new shoots to the formation of flowers (Hanke et al., 2007). Endogenous factors such as gibberellic acids (GA) and photoassimilates (e.g. carbohydrates) are among the key elements involved in flowering. GA promotes the growth of vegetative buds but suppresses the growth of reproductive organs (Lenahan et al., 2006; Muñoz-Fambuena et al., 2012), whereas carbohydrates are essential sources of energy (Khandaker et al., 2011; Lebon et al., 2008; Pallardy, 2008).

Plants significantly rely on three major factors to regulate flowering, including photoperiod, temperature, and soil moisture (Neil & Wu, 2006; Srikanth & Schmid, 2011; Wilkie et al., 2008). Other factors like vegetative growth stages of the individual plant or specific plant parts and the involvement of plant growth regulators (PGRs) also give some effects on the initiation of flowering (Ahmad Nazarudin & Tsan, 2018; Wilkie et al., 2008). Vegetative growth stages involve plant ages and growth stages (phenological phases), while PGRs could be phytohormones (naturally produced) or externally applied synthetic compounds that have hormonal properties as those phytohormones. Thus, in a mature plant, the interaction between various factors determines the timing for the development of flower buds.

In the tropics, environmental factors are the keys that determine the formation of flowers as there is no significant season as in temperate regions. Trees that grow in the tropics are mostly dependent on exogenous factors for flowering, while flowering in the temperate regions is autonomous (Wilkie et al., 2008). Perennial tropical trees commonly bloom at any time of the year, causing inconsistency in flowering patterns between species and within the same growth range (Ahmad Nazarudin & Tsan, 2018; Bawa, 1983; Ramírez & Kallarackal, 2018). There are also species that flower in large quantities and can only be observed every few years. However, trees that produce flowers several times a year usually have less abundance of blooms with shorter periods than trees that flower once a year because there is competition for resources between the flowering and fruiting stages (Ahmad Nazarudin & Tsan, 2018; Bawa et al., 2003). In other words, vegetative and reproductive stages in certain species occur concomitantly and the flowers might be seen throughout the year regardless of the abundance rate.

Due to the relatively homogenous climate throughout the year, the landscape's atmosphere in the tropics is commonly monotonous except for certain periods when there are very significant weather changes. Therefore, the management of trees in urban areas is in dire need of alternative approaches to make the landscape more attractive and well appreciated. These approaches should be practically accepted for landscape tree management purposes.

Flowering Response to Paclobutrazol

Paclobutrazol (PBZ) is a synthesised PGR that has been widely used in managing plant growth. PBZ is a triazole compound, which contains three N atoms, chlorophenyl and carbon side chains (Fig. 16.1). PBZ inhibits the biosynthesis of gibberellic acids (GAs), reducing vegetative growth in plants. The application of PBZ resulted in various morpho-physiological responses in plants. It inhibits the growth of vegetative shoots, reduces internodes length, stem diameter and leaf size, modifies flowering pattern (Ahmad Nazarudin, 2012: Ahmad Nazarudin et al., 2012a; Mohammed et al., 2016), biochemical contents and the gas exchange of treated plants (Ju et al., 2019; Rodrigues et al., 2016; Xia et al., 2018). Furthermore, it helps increase plant tolerance levels to various stresses, for instance, extreme heat, prolonged drought, salinity, and disease attack (Ahmad Nazarudin & Mohd Farid, 2019; Matsoukis et al., 2014; Soumya et al., 2017).

PBZ is widely used in ornamental and commercial fruit industries. Previous studies have shown that PBZ profusely induced the flowering of commercial tropical fruit crops namely *Durio zibethinus* (Subhadrabandhu & Kaiviparkbunyay, 1998), *Garcinia mangostana* (Omran & Semiah, 2006), *Mangifera indica* (Rebolledo-Martínez et al., 2008) and *Psidium guajava* (Singh, 2010). The flowering performance of *M. indica* and *P. guajava* above also improved after treatment with a combination of PBZ and potassium nitrate (KNO_3). As for ornamental purposes, PBZ increased the blooming of various species such as *Bougainvillea glabra* (Karaguzel & Ortacesme, 2002), *Lantana camara* (Matsoukis et al., 2001), *Hibiscus rosa-sinensis* (Ahmad Nazarudin, 2012), and *Lagerstroemia indica* (Mohammed et al., 2016).

A previous PBZ study on *Consolida orientalis* found no significant effect on flower size, however, the colour and number of flowers were enhanced (Mansuroglu et al., 2009). In *Zinnia elegans*, flower size was not affected, though the plant height and branch length were declined (Pinto et al., 2005). Another PBZ study demonstrated that the number of flowers and flowering period in *Lilium* sp. remained unchanged (Francescangeli et al., 2007). In addition, the appearance of PBZ-treated *Dendrobium* orchid was normal (Te-Chato et al., 2009). These results show that PBZ can be used to improve flowering without detrimental effects on plants. However, appropriate dosages of the compound must be determined before being applied to target plants.

In general, PBZ reduces vegetative growth in plants by inhibiting GA biosynthesis. This reduction has indirect effects on the usage of relative sink sources in the treated plants, which eventually allows higher amounts of photoassimilates to reproductive

Fig. 16.1 The chemical structure of paclobutrazol

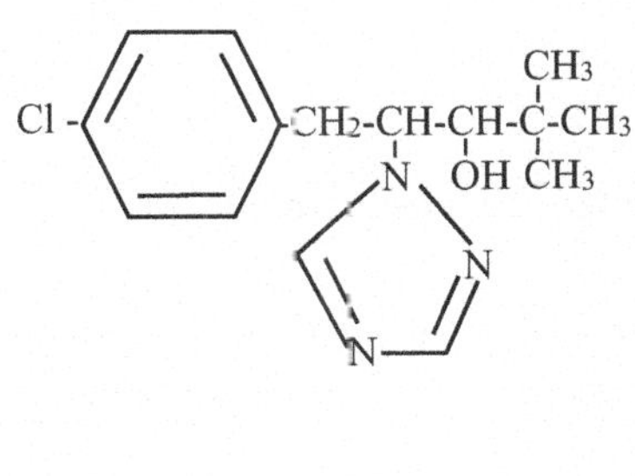

organs (Ahmad Nazarudin, 2015a). In other words, as the demand for energy to promote vegetative growth is reduced, the possibility to induce the reproductive parts is higher.

Study on Flowering of Urban Trees in Malaysia

In Malaysia, intensive studies on the flowering of urban trees are still lacking. As the flowering of tropical trees is mainly influenced by environmental cues which are generally uniform, an alternative method that can improve the flowering performance of urban trees is essential. The ability of PBZ and KNO_3 to increase the yield of various fruit crops was thought to be beneficial for improving the flowering performance of urban trees under local conditions. Thus, *Xanthostemon chrysanthus* (F. Muell.) Benth. was chosen for the experiment.

Xanthostemon chrysanthus (F. Muell.) Benth

Xanthostemon chrysanthus (F. Muell.) Benth. or Golden Penda (Myrtaceae) is native to tropical northern Australia, New Caledonia, New Guinea, Indonesia, and the Philippines. Naturally, the medium-sized species can reach approximately 20 m in height but are smaller in cultivation (ANPSA, 2020). The lance-shaped leaves are green and glossy, measuring about 15 × 5 cm in size (Ahmad Nazarudin, 2015b). The inflorescences are bright yellow in colour, with a diameter of about 8–10 cm and have many stamens. Nectar-feeding birds and insects are usually attracted to the blooms (Fig. 16.2). In Malaysia, *X. chrysanthus* is grown as a screen or as a specimen plant. This woody species is preferable to beautifying parks, roadsides, golf courses, pocket parks, and house gardens (Fig. 16.3). Golden Penda is very hardy, growing well in open areas, tolerant to acidic soil, and resistant to pollution and heat of the tropical urban areas. The previous investigation at selected locations in Kuala Lumpur exhibited that this species was able to withstand harsh urban soil, having high penetration resistance of 2.911–2.954 MPa with a low cation exchange capacity of 3.40–3.67 cmol kg^{-1} (Ahmad Nazarudin et al., 2014). Although *X. chrysanthus* is widely planted in urban areas, the flowering is not synchronized and is unpredictable. It is always less abundant and in some cases, the trees do not flower throughout the year.

Study Procedures

Investigation of flowering performance by the usage of PBZ and KNO_3 was conducted at one of the public parks in Kuala Lumpur, Metropolitan Batu Park

Fig. 16.2 *Xanthostemon chrysanthus* produces nectar for bees

Fig. 16.3 A group of *X. chrysanthus* planted in a pocket space between buildings in Putrajaya, Malaysia

(3.2137° N, 101.6776° E) (Fig. 16.4). A total of 81 existing *X. chrysanthus* approximately six years after planting were randomly selected (Fig. 16.5). Before PBZ and KNO_3 were applied, phenological observation was carried out by using the Biologische Bundesanstalt, Bundessortenamt and Chemical Industry (BBCH) scales as described by Meier (2001). The BBCH scales for *X. chrysanthus* were then established for further exploration of the flowering performance; this was also useful to determine the flowering occurrence as well as flowering period (Ahmad Nazarudin et al., 2012b, 2018).

As for the PBZ and KNO_3 experiment, Cultar® contains 250 g a.i. PBZ/L and Krista™ K Plus (KNO_3) (13.7:0:46.3) were used. A completely randomised design plot was established to arrange nine treatments, containing PBZ (0, 0.125, and 0.25 g/L/tree) and KNO_3 (0, 100, and 200 g/tree). Each dosage of PBZ was diluted

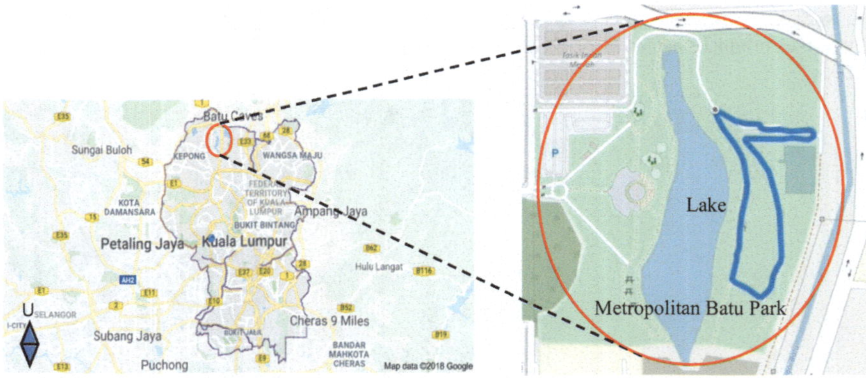

Fig. 16.4 The Metropolitan Batu Park, located in the northern part of Kuala Lumpur city (Google Maps)

Fig. 16.5 Some of the *X. chrysanthus* at the study site, Metropolitan Batu Park, Kuala Lumpur

with tap water up to 1 L volume prior to soil drenching near the root collar of each tree, while the control trees received the same volume of tap water. The treatment began in March 2012, at which point the PBZ was applied. PBZ application was not repeated throughout the experiment. The first application of KNO_3 was done concurrently with the first PBZ treatment, and the following application was repeated at three-month intervals, i.e. in June 2012, September 2021, and December 2012. The sub-soil method was used to apply KNO_3 under the tree canopy drip-line. Flower abundance (%) and inflorescence diameter (cm) were monitored for a year (March 2012 to February 2013). The flower abundance and inflorescence diameter were recorded as described by Ahmad Nazarudin et al. (2015a).

Study Findings

This study reveals that the flowering of *X. chrysanthus* occurred throughout the year though the abundance fluctuated. Our observations found that May 2012, September 2012, and January 2013 were the months in which profuse flowering occurred (Fig. 16.6) (Ahmad Nazarudin et al., 2020). The mean flower abundance was relatively similar in May and September 2012, and was higher in January 2013. In January 2013, trees treated with T8 (0.25 g/L PBZ + 100 g KNO_3) recorded the highest abundance (65%), suggesting that the combination of PBZ and KNO_3 at this rate successfully promoted more flowers than that of other treatments. In contrast, the lowest abundance of flowers (25%) was found in the control trees. Meanwhile, a lesser abundance of flowers was recorded from July to August 2012. This phenomenon is likely to result from the minimum monthly precipitation received in June and July 2012 that affected water availability in soil (Fig. 16.7). However, greater amounts of monthly precipitation received from October to December 2012 can be associated with profuse flowering in January 2013 (Figs. 16.6 and 16.7).

This study also shows that the availability of water in the soil due to higher amounts of rainfall received in previous months influenced the occurrence of flowering in the following months. In Malaysia, a greater volume of rainfall in the fourth quarter of each year usually occurs, increasing the abundance of flowers in the first quarter

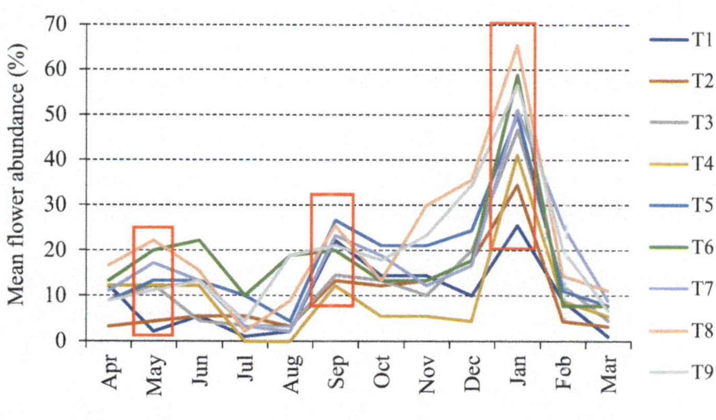

Fig. 16.6 Changes in mean monthly abundance of flowers in *X. chrysanthus* as influenced by PBZ and KNO_3 (April 2012 to March 2013). The red boxes show three distinct flowering occurrences. T1(Control); T2(100 g KNO_3); T3(200 g KNO_3); T4(0.125 gL^{-1} PBZ); T5(0.125 gL^{-1} PBZ + 100 g KNO_3); T6(0.125 gL^{-1} PBZ + 200 g KNO_3); T7(0.25 gL^{-1} PBZ); T8(0.25 gL^{-1} PBZ + 100 g KNO_3); T9(0.25 gL^{-1} PBZ + 200 g KNO_3)

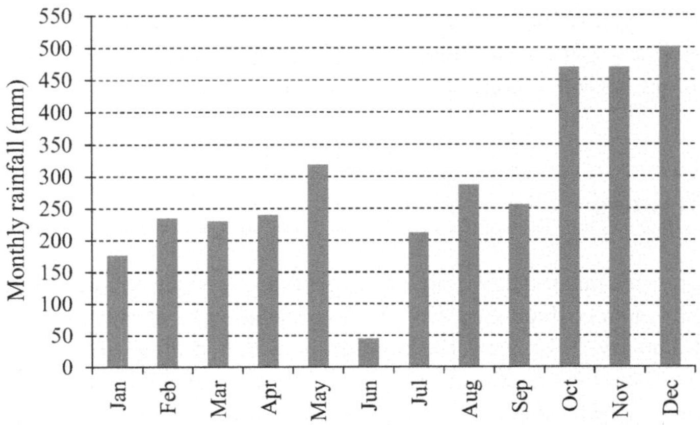

Fig. 16.7 Monthly rainfall during the study period (January–December 2012)

of the following year. The Northeast Monsoon occurs from November until March and yields more rainfall than the Southwest Monsoon which carries relatively drier weather from May until September (MMD, 2014). This result is in line with previous studies on *Jatropha curcas* (Jongschaap et al., 2007) and *Citrus limon* (Chaikiattiyos et al., 1994) that also found significant flowering following watering after being exposed to prolonged drought.

Previous studies indicated that the amount of blooms in *Lupinus varius* (Karaguzel et al., 2004) and *H. rosa-sinensis* (Ahmad Nazarudin, 2012) have increased with PBZ treatment. Another trial showed significant induction of inflorescences in *M. indica* as affected by PBZ and KNO_3 treatment (Rebolledo-Martínez et al., 2008). Potassium has resulted in better growth performance as it improves the translocation of carbohydrates, proteins, and photoassimilates in plants which indirectly produced more flowers (Akhtar et al., 2009). In addition to reducing GA production, PBZ also caused an increase in leaf water potential, abscisic acid and cytokinins (CK) contents in line with C:N ratio which stimulates reproductive growth in *M. indica* (Upreti et al., 2013).

In terms of flower quality, smaller inflorescences were measured in PBZ-treated trees compared to that of other treatments (Table 16.1). For example, 0.125 gL^{-1} PBZ resulted in 47.3% smaller inflorescence than that of 0.125 g/L PBZ + 100 g KNO_3 (T5). The same trend was also observed with 0.25 gL^{-1} PBZ, giving 34.7% difference compared to 0.25 g/L PBZ + 100 g KNO_3 (T8). T5-treated trees continuously yielded greater inflorescences measuring 17–20 cm for the entire study period. Meanwhile, trees treated with T8 have a relatively smaller inflorescence size of about 15 to 18 cm, but it is still acceptable in the landscape. T8 also gave a higher abundance of flowers in May and September 2012, and January 2013 compared to that of other treatments (Fig. 16.6). On the other hand, the mean monthly flower abundance of other trees was not consistent in those three distinct flowering occurrences (Fig. 16.6). These results

Table 16.1 Changes in the size of inflorescence in *X. chrysanthus* as influenced by PBZ and KNO_3 (Ahmad Nazarudin et al., 2020)

Treatment ID	Inflorescence size measured at two-month interval (cm)					
	First assessment	Second assessment	Third assessment	Fourth assessment	Fifth assessment	Sixth assessment
T1	10.00^{e}	11.16^{f}	10.50^{f}	10.40^{e}	10.00^{f}	9.70^{f}
T2	13.94^{d}	12.34^{e}	12.30^{e}	13.10^{d}	12.96^{e}	13.20^{e}
T3	13.96^{d}	13.86^{d}	14.16^{d}	13.44^{d}	14.10^{d}	13.82^{e}
T4	10.70^{e}	10.66^{f}	10.62^{f}	9.94^{e}	9.54^{fg}	9.78^{f}
T5	20.30^{a}	17.44^{a}	20.68^{a}	18.96^{a}	20.34^{a}	19.70^{a}
T6	17.42^{b}	16.16^{b}	17.80^{b}	18.54^{ab}	19.50^{b}	18.80^{b}
T7	10.44^{e}	9.98^{g}	9.94^{f}	8.76^{f}	9.00^{g}	9.02^{f}
T8	15.98^{c}	14.64^{c}	16.80^{c}	17.92^{b}	16.46^{c}	16.98^{c}
T9	14.02^{d}	14.34^{cd}	14.12^{d}	15.68^{c}	16.28^{c}	14.68^{d}

T1(Control); T2(100 g KNO_3); T3(200 g KNO_3); T4(0.125 gL^{-1} PBZ); T5(0.125 gL^{-1} PBZ + 100 g KNO_3); T6(0.125 gL^{-1} PBZ + 200 g KNO_3); T7(0.25 gL^{-1} PBZ); T8(0.25 gL^{-1} PBZ + 100 g KNO_3); T9(0.25 gL^{-1} PBZ + 200 g KNO_3); Means with different superscript letter(s) within column are significant at $p < 0.05$ by DMRT

suggest that T8 gave a higher abundance with an adequate size of flower compared to other treatments (Fig. 16.8).

This study suggests that PBZ coupled with KNO_3 enhanced the development of flower buds in *X. chrysanthus*. PBZ increased the number of axillary buds, while KNO_3 promoted the axillary buds into flowers when the environmental cues were appropriate (Bonnet-Masimbert & Webber, 1995). According to Yayat et al. (2013), PBZ increased CK content in roots and allocation of photosynthate utilisation. CK produced in the roots was then channelled to the above plant parts which eventually promoted the formation of axillary buds (Yayat et al., 2013). Depending on other endogenous and exogenous factors that have been mentioned earlier, these axillary buds may develop as leaves or flowers. In addition, a reduced rate of the vegetative part due to the inhibitory effect of PBZ was reported to increase the distribution of photosynthates into the reproductive part which encouraged the development of flower buds and hence improved the flowering (Abou-El-Khashab et al., 1997).

The flower production period of this species was extended due to PBZ and KNO_3 effects. It shows that flower senescence was delayed with PBZ, hence the attractive yellow flowers of the species can be displayed longer in the landscapes (Ahmad Nazarudin & Tsan, 2018). PBZ also reduced some of the vegetative growth such as leaf area and tree height of this species (Ahmad Nazarudin et al., 2014). However, no abnormal formation of the leaves occurred, implying that the PBZ dosages used in this study are appropriate. Other PBZ studies also reported similar results on the growth performance of woody species, *Acalypha siamensis, Ficus microcarpa,* and *Syzygium oleina* (Ahmad Nazarudin et al., 2003), *Syzygium myrtifolium* (Ahmad Nazarudin et al., 2014), and *Leonotis leonurus* (Teto et al., 2016). The inhibition

Fig. 16.8 Differences in inflorescence size and flower abundance of *X. chrysanthus* as influenced by PBZ alone (**A**) compared to combined effects of PBZ and KNO_3 (**B**) (Ahmad Nazarudin, 2015a)

growth effect by PBZ may be helpful in reducing the pruning activity of the treated trees, indirectly reducing the maintenance cost, while enhancing the flowering of the species.

Conclusion

In urban areas, the ability of trees to bear a significant abundance of flowers is crucial for the aesthetic enhancement of the landscape. However, under tropical climate conditions, flowering is influenced by multiple factors, resulting in unpredictable flowering habits. Thus, a proper selection of tree species, including the source of seedlings, and appropriate horticultural practices are among the key criteria that must be considered in planning and managing the urban trees. For instance, horticultural techniques which are practical and easily applied for flower improvement should be developed. A simple method, time-saving, and cost-effective is required for trees planted in cities.

Although the usage of PBZ and KNO_3 is not yet practiced in Malaysian cities, several studies carried out have confirmed many advantages of PBZ on tree growth and flowering. Application of PBZ and KNO_3 is thought of as one of the potential ways to improve the flowering performance of urban trees as well as to control the vigorous growth of tropical tree species, hence, reducing the pruning and trimming rotations. This study found improved flowering in *X. chrysanthus* with the presence of PBZ and KNO_3. A consistent inflorescence size was also measured throughout the study period following 0.125 g/L PBZ and 100 g KNO_3 (T5) treatment. However, the mean flower abundance was lower with T5 compared to the trees treated with T8 (0.25 g/L and 100 g KNO_3). In this case, T8 is preferred for the improvement of flowering and may be able to improve the appearance of the trees.

In general, PBZ inhibits vegetative growth, which indirectly reduces trimming activities and landscape operational costs. This will be beneficial for the landscape team as they can focus on other maintenance works. However, workers must be trained and competent with PBZ, and appropriate dosages must be identified prior to application. PBZ not only improves the flowering but it also reduces vigorous plant growth and increases plant tolerance to the harsh urban environment. Further studies are recommended to assess the possible frequency of PBZ and KNO_3 application, as well as more flowering tree species be tested.

Acknowledgements The authors would like to thank the Kuala Lumpur City Hall (DBKL) for permission to use the existing trees at the study site. Great appreciation also goes to the Ministry of Agriculture and Agro-based Industry Malaysia for the research funds (Science Fund 05-03-10-SF1030).

References

Abou-El-Khashab, A. M., El-Sammak, A. F., Elaidy, A. A., Salama, M. I., & Rieger, M. (1997). Paclobutrazol reduces some negative effects of salt stress in peach. *Journal of the American Society for Horticultural Science, 122*(1), 43–46.

Akhtar, M. E., Khan, M. Z., Ahmad, S., Ashraf, M., & Sardar, A. (2009). Effect of potassium on micromorphological and chemical composition of three cotton (Gossypium hirsutum L.) genotypes. *African Journal of Biotechnology, 8*(15), 3511–3518.

Ahmad Nazarudin, M. R. (2012). Plant growth retardants effect on growth and flowering of potted *Hibiscus rosa-sinensis* L. *Journal of Tropical Plant Physiology, 4*, 29–40.

Ahmad Nazarudin, M. R. (2015a). *Growth, physiological and flowering responses of Xanthostemon chrysanthus (F. Muell.) Benth. to paclobutrazol and potassium nitrate* (PhD Thesis), 322 pp. Universiti Teknologi MARA.

Ahmad Nazarudin, M. R. (2015b). *Xanthostemon chrysanthus* (F. Muell.) Benth : A new flowering tree for urban landscapes. *International Journal of Agriculture, Forestry and Plantation, 3*, 50–54.

Ahmad Nazarudin, M. R., & Mohd Farid, A. (2019). In vitro evaluation of paclobutrazol against selected pathogenic soil fungi. *Journal of Tropical Plant Physiology, 11*(2), 13–21.

Ahmad Nazarudin, M. R., Tsan F. Y., & Adnan, M. (2003). Maintaining landscape plants of Acalypha siamensis, Ficus microcarpa and Syzygium oleina by the application of paclobutrazol: a non-mechanical approach. *Transactions of the Malaysian Society of Plant Physiology, 12*, 231–233.

Ahmad Nazarudin, M. R., & Tsan, F. Y. (2018). Vegetative and reproductive growth behaviour of *Xanthostemon chrysanthus* (F. Muell.) Benth.—An ornamental tree in Malaysia. *Sains Malaysiana, 47*(2), 227–233.

Ahmad Nazarudin, M. R., Tsan, F. Y., & Mohd Fauzi, R. (2012a). Morphological and physiological response of *Syzygium myrtifolium* (Roxb.) Walp. to paclobutrazol. *Sains Malaysiana, 41*(10), 1187–1192.

Ahmad Nazarudin, M. R., Tsan, F. Y., Normaniza, O., & Adzmi, Y. (2012b). Phenological growth stages of the golden penda tree (*Xanthostemon chrysanthus*). *Annals of Applied Biology, 161*(1), 12–15.

Ahmad Nazarudin, M. R., Tsan, F. Y., & Normaniza, O. (2020). Induction of flower in *Xanthostemon chrysanthus* by the combined effects of paclobutrazol and potassium nitrate. *Transactions of the Malaysian Society of Plant Physiology, 27*, 13–17.

Ahmad Nazarudin, M. R., Tsan, F. Y., Normaniza, O., & Adzmi, Y. (2014). Growth performance and flowering of *Xanthostemon chrysanthus* at two urban sites in Kuala Lumpur, Malaysia. *Journal of Tropical Forest Science, 26*(3), 428–434.

Australian Native Plants Society (ANPSA). (2020). *Xanthostemon chrysanthus*. Retrieved July 15, 2020, from http://anpsa.org.au/x-chr.html

Bawa, K. S. (1983). Patterns of flowering in tropical plants. In C. E. Jones & R. J. Little (Eds.), *Handbook of experimental pollination biology* (pp. 394–410). van Nostrand Reinhold.

Bawa, K. S., Kang, H., & Grayum, M. H. (2003). Relationship among time, frequency, and duration of flowering in tropical rain forest trees. *American Journal of Botany, 90*(6), 877–887.

Bonnet-Masimbert, M., & Webber, J. E. (1995). From first flower induction to seed production in forest tree orchards. *Tree Physiology, 15*, 419–426.

Chaikiattiyos, S., Menzel, C. M., & Rasmussen, T. S. (1994). Floral induction in tropical fruit trees: Effects of temperature and water supply. *Journal of Horticultural Science, 69*, 397–415.

Francescangeli, N., Marinangeli, P., & Curvetto, N. (2007). Paclobutrazol for height control of two *Lilium* L.A. hybrids grown in pots. *Spanish Journal of Agricultural Research, 5*(3), 425–430.

Hanke, M. V., Flachowsky, H., Peil, A., & Hättasch, C. (2007). No flower no fruit-genetic potentials to trigger flowering in fruit trees. *Genes, Genomes and Genomics, 1*(1), 1–20.

Jongschaap, R. E. E., Corre, W. J., Bindraban, P. S., & Brandenburg, W. A. (2007). *Claims and facts on Jatropha curcas* L.: *Global Jatropha curcas evaluation, breeding and propagation programme* (Plant Research International B.V. 158). Wageningen, the Netherlands.

Ju, S., Xu, D., Zhan, C., Ji, L., Yin, T., Li, Z., & Lu, Z. (2019). Influence of paclobutrazol on the growth and photosynthesis of *Sequoia sempervirens* seedlings. *Journal of Horticultural Research, 27*(1), 21–30.

Karaguzel, O., & Ortacesme, V. (2002). Influence of paclobutrazol on the growth and flowering of *Bougainvillea glabra* 'Sanderiana.' *Ziraat Fakultesi Dergisi, Akdeniz Universitesi, 15*(1), 79–84.

Karaguzel, O., Baktir, I., Cakmakci, S., & Ortacesme, V. (2004). Growth and flowering responses of *Lupinus varius* L. to paclobutrazol. *HortScience, 39*(7), 1659–1663.

Khandaker, M. M., Boyce, A. N., Osman, N., & Hossain, A. S. (2011). Application of girdling for improved fruit retention, yield and fruit quality in *Syzygium samarangense* under field conditions. *International Journal of Agriculture and Biology, 13*(1), 18–24.

Lebon, G., Wojnarowiez, G., Holzapfel, B., Fontaine, F., Vaillant-Gaveau, N., & Clément, C. (2008). Sugars and flowering in the grapevine (*Vitis vinifera* L.). *Journal of Experimental Botany, 59*(10), 2565–2578.

Lenahan, O. M., Whiting, M. D., & Elfving, D. C. (2006). Gibberellic acid inhibits floral bud induction and improves 'Bing' sweet cherry fruit quality. *HortScience, 41*(3), 654–659.

Malaysian Meteorological Department (MMD). (2014). *General climate of Malaysia*. Retrieved August 1, 2014, from http://www.met.gov.my

Mansuroglu, S., Karaguzel, O., Ortacesme, V., & Sayan, M. S. (2009). Effect of paclobutrazol on flowering, leaf and flower colour of *Consolida orientalis*. *Pakistan Journal of Botany, 41*(5), 2323–2332.

Matsoukis, A. S., Chronopoulou-Sereli, A., Dimopoulus, I. D., & Kamoutsis, A. (2001). Responses of *Lantana camara* subsp. *camara* to paclobutrazol and shading. *Canadian Journal of Plant Science, 81*(4), 761–764.

Matsoukis, A., Gasparatos, D., & Chronopoulou-Sereli, A. (2014). Environmental conditions and drenched-applied paclobutrazol effects on lantana specific leaf area and N, P, K, and Mg content. *Chilean Journal of Agricultural Research, 74*(1), 117–122.

Meier, U. (2001). *Growth stages of mono- and dicotyledonous plants* (p. 204). BBCH Monograph.

Mohammed, N. T., Awang, Y., Ahmad, I., & Noori, R. S. (2016). Gas exchange, growth and flowering of *Lagerstroemia indica* treated with different concentration and application techniques of paclobutrazol. *Asian Journal of Plant Sciences, 16*(1), 37–44.

Muñoz-Fambuena, N., Mesejo, C., González-Mas, M. C., Iglesias, D. J., Primo-Millo, E., & Agustí, M. (2012). Gibberellic acid reduces flowering intensity in sweet orange [*Citrus sinensis* (L.) Osbeck] by repressing *CiFT* gene expression. *Journal of Plant Growth Regulation, 31*, 529–536.

Neil, K., & Wu, J. (2006). Effects of urbanization on plant flowering phenology: A review. *Urban Ecosystems, 9*, 243–257.

Omran, H., & Semiah, R. (2006). Effect of paclobutrazol application combined with potassium nitrate and bicomine spray on flowering and fruiting of mangosteen (*Garcinia mangostana* L.). *Acta Horticulturae, 727*, 151–156.

Pallardy, S. G. (2008). *Physiology of woody plants* (3rd ed.). Academic Press.

Pinto, A. C. R., Rodrigues, T. J. D., Leite, I. C., & Barbosa, J. C. (2005). Growth retardants on development and ornamental quality of potted 'lilliput' Zinnia elegans Jacq. *Scientia Agricola, 62*(4), 337–345.

Ramírez, F., & Kallarackal, J. (2018). Phenological growth stages of Feijoa [*Acca sellowiana* (O. Berg) Burret] according to the BBCH scale under tropical Andean conditions. *Scientia Horticulturae, 232*, 184–190.

Rebolledo-Martínez, A., Angel-Perez, A. L., & Moreno, J. R. (2008). Effects of paclobutrazol and KNO_3 over flowering and fruit quality in two cultivars of mango Manila. *Interciencia, 33*(7), 518–522.

Rodrigues and A., de Castro, E. M., Pereira, F. J., Maluleque, I. F., Barbosa, J. P. R. A. D., & da S. Rosado, S. C. , 2016 Rodrigues, L. C. de A., de Castro, E. M., Pereira, F. J., Maluleque, I. F., Barbosa, J. P. R. A. D., & da S. Rosado, S. C. (2016). Effects of paclobutrazol on leaf anatomy and gas exchange of *Toona ciliata* clones. *Australian Forestry, 79*(4), 241–247.

Singh, J. (2010). Influence of paclobutrazol and ethephon on vegetative growth of guava (*Psidium guajava* L.) plants at different spacing. *Notulae Scientia Biologicae, 2*(3), 110–113.

Srikanth, A., & Schmid, M. (2011). Regulation of flowering time: All roads lead to Rome. *Cellular and Molecular Life Sciences, 68*, 2013–2037.

Soumya, P. R., Kumar, P., & Pal, M. (2017). Paclobutrazol: A novel plant growth regulator and multi-stress ameliorant. *Indian Journal of Plant Physiology, 22*, 267–278.

Subhadrabandhu, S., & Kaiviparkbunyay, K. (1998). Effect of paclobutrazol on flowering, fruit setting and fruit quality of durian (*Durio zibethinus* Murr.) CV. Chanee. *Kasetsart Journal (Natural Science), 32*, 73–83.

Te-Chato, S., Nujeen, P., & Muangsorn, S. (2009). Paclobutrazol enhance budbreak and flowering of Friederick's dendrobium orchid in vitro. *Journal of Agricultural Technology, 5*(1), 157–165.

Teto, A. A., Laubscher, C. P., Ndakidemi, P. A., & Matimati, I. (2016). Paclobutrazol retards vegetative growth in hydroponically-cultured *Leonotis leonurus* (L.) R. Br. Lamiaceae for a multipurpose flowering potted plant. *South African Journal of Botany, 106*, 67–70.

Upreti, K. K., Reddy, Y. T. N., Shivu Prasad, S. R., Bindu, G. V., Jayaram, H. L., & Rajan, S. (2013). Hormonal changes in response to paclobutrazol induced early flowering in mango cv. Totapuri. *Scientia Horticulturae, 150*, 414–418.

Yayat, R. S., Ramadhani, R., & Jajang, S. H. (2013). Paclobutrazol application and shading levels effect to the growth and quality of Begonia (*Begonia rex-cultorum*) cultivar Marmaduke. *Asian Journal of Agriculture and Rural Development, 3*(8), 566–575.
Wilkie, J. D., Sedgley, M., & Olesen, T. (2008). Regulation of floral initiation in horticultural trees. *Journal of Experimental Botany, 59*(12), 3215–3228.
Xia, X., Tang, Y., Wei, M., & Zhao, D. (2018). Effect of paclobutrazol application on plant photosynthetic performance and leaf greenness of herbaceous peony. *Horticulturae, 4*(1), 5. https://doi.org/10.3390/horticulturae4010005

Ahmad Nazarudin Mohd Roseli is a Senior Research Officer at the Forest Research Institute Malaysia (FRIM). He obtained his doctorate in the field of plant ecophysiology from the Universiti Teknologi MARA (UiTM) in 2015. His research is focusing on plant ecophysiology, plant growth regulators, and urban horticulture. Currently, he is the President of the Malaysian Society of Plant Physiology (MSPP) which actively organises conferences and workshops related to the field of plant science. He also acts as the editorial board for the Journal of Tropical Plant Physiology (JTPP), the official journal of MSPP. He is a certified arborist (MY-0413A) under the International Society of Arboriculture (ISA), USA.

Chapter 17
Assessing Small Urban Parks as Habitats for Butterflies in Malaysia

Zanariah Jasmani, Hasanuddin Lamit, and Cecil C. Konijnendijk van den Bosch

Abstract Urban green spaces are getting smaller and more fragmented due to urbanisation. Thus their capability to serve as habitats for wildlife is threatened. However, there is a lack of in-depth discussion on biodiversity in small urban green spaces in tropical countries. This study analyses the attributes of biodiversity and ecological functions in small urban parks. An in-situ field survey that combines structured field observation and field measurements were conducted at nine small urban parks in Malaysia. Three categories of variables were measured, including physical characteristics, species richness and anthropogenic factors. Butterflies are used as the indicator because they are relatively easy to recognise, sensitive to environmental changes, have unique life cycles and biological demands that make them suitable for assessing biodiversity. Pearson correlations and multiple regressions are conducted to analyse the relationships between the variables and identify which variables significantly affect butterfly species richness and abundance. Findings show that butterfly species richness is positively influenced by the percentage of canopy cover, the total number of shrubs, the richness of shrub species, the presence of flowering shrubs, and the presence of exotic plants. Park size is noted to have a little significant impact on butterfly richness and abundance. Butterfly abundance is significantly (and positively) related to the percentage of canopy cover, overall quantity of vegetation, presence of native vegetation, shrub abundance, diversity of shrub species, total flowering plants, and number of palm species. Human-related factors and anthropogenic disturbance would not significantly impact the species richness and abundance, as butterflies are not affected by visitor density, activities and noise levels. The findings suggested two main considerations for small urban parks to enhance butterfly diversity: (1) development and management plans for small parks should emphasise

Z. Jasmani (✉) · H. Lamit
Faculty of Built Environment and Surveying, Universiti Teknologi Malaysia, UTM Skudai, 81310 Johor Bahru, Johor, Malaysia
e-mail: zanariahj@utm.my

C. C. K. van den Bosch
Department of Forest Resources and Management, Faculty of Forestry, University of British Columbia, Forest Sciences Building 2025, 2424 Main Mall, Vancouver, BC V6T 1Z4, Canada

S. Maruthaveeran et al. (eds.), *Urban Forestry and Arboriculture in Malaysia*,
https://doi.org/10.1007/978-981-19-5418-4_17

the spatial quality and incorporate social-ecological principles for achieving biodiversity conservation goals, and (2) newly developed parks should have a planting configuration design that can provide food, habitat and protection for butterflies.

Keywords Butterfly diversity · Human disturbance · Urban landscape ecology · Urban green space · Urban habitats

Introduction

The continuing expansion of urbanisation often results in smaller and more fragmented urban green spaces. Designing and establishing small parks in cities has the potential of improving the quality of life in many ways, especially for urban citizen's well-being, ecological networks and biodiversity conservation (Angold et al., 2006; Niemelä et al., 2011). Despite their small sizes, these urban parks provide recreational opportunities. Although small parks in cities can be regarded as mostly a type of intensively managed landscape, they may potentially provide unique natural habitats that are suitable for pollinators such as butterflies (Lizee et al., 2015; Matteson & Langellotto, 2010). Studies on butterfly communities in urban green spaces are becoming increasingly important because butterflies help to maintain ecosystem services and promote stewardship of biodiversity and natural resources in cities (New, 2015). Furthermore, butterflies are among the important ecological indicators for understanding the impact of urbanisation on biodiversity due to their sensitivity to habitat change/loss and through their visible presence (Bonebrake et al., 2010; Clark et al., 2007). Earlier studies suggest that butterfly populations in urban parks are closely linked with the availability of larval host plants and nectar sources for adult butterflies (Hogsden & Hutchinson, 2004; Lian & Sodhi, 2004; New, 2015), but their relations to direct disturbances or anthropogenic factors have received little attention.

Most studies on butterfly species richness and abundance in urban parks have been carried out in the Western world, while research in tropical countries has been limited to date (Bonebrake et al., 2010; Nielsen et al., 2014; Sing et al., 2016). Previous studies have demonstrated that large urban parks may host diverse butterfly assemblages (Lian & Sodhi, 2004), however, little is known about the potential value of small urban parks as butterfly habitats especially in the tropical Southeast Asian region (Sing et al., 2016; Tam & Bonebrake, 2015).

The purpose of the study is: (1) to evaluate the attributes of vegetation richness and structure, as well as anthropogenic factors on butterfly communities in small urban parks; and based on this, (2) to provide recommendations for planning and management of small urban parks aimed at enhancing butterfly diversity. It is hypothesized that (1) park characteristics such as size, land use and park maintenance can contribute to butterfly diversity, (2) vegetation diversity and structure are important factors for predicting butterfly species richness and abundance in small urban parks due to the availability of larval host plants and nectar sources, and (3) butterfly species richness and abundance will decline when human activities are intensive and surrounding disturbances increase (e.g. in terms of traffic volume and anthropogenic noise).

Method

Study Sites

The field survey was carried out in nine small urban parks in Petaling Jaya, Selangor, Malaysia, between August and September 2014. Petaling Jaya is Malaysia's first satellite new town established in 1952. The parks are located approximately 11 km from Kuala Lumpur city. In this study, small urban parks are defined as small-scale urban green spaces between 0.5 and 3.5 hectares, have a well-defined boundary and provide opportunities for public recreation. Figure 17.1 shows the location of all selected parks in this study. Apart from the information gathered from Petaling Jaya City Council, the satellite image generated from Selangor Planning Information System (SISMAPS) was used to identify the location and accurate size of the parks. All nine parks were chosen based on the criteria of size, the location of the urban area and function.

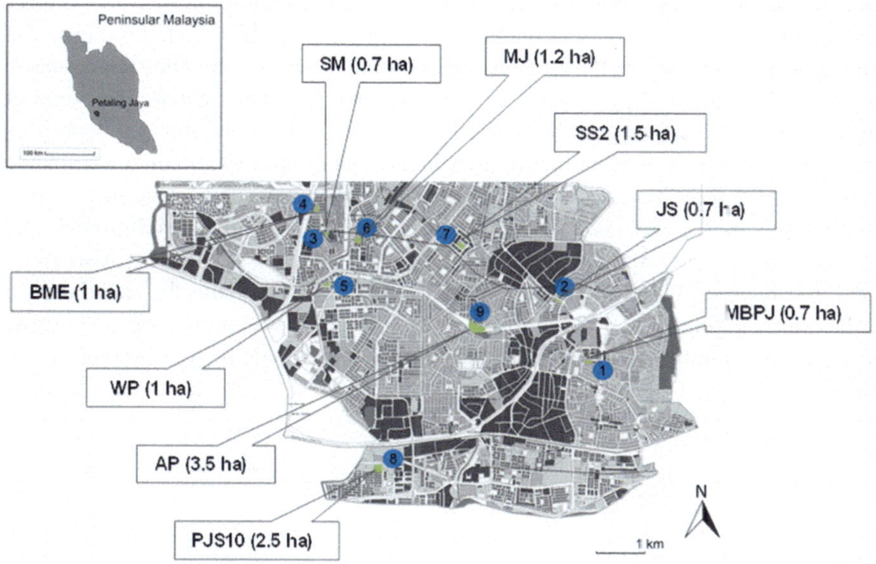

Fig. 17.1 Location of the nine parks in Petaling Jaya, Selangor Darul Ehsan. The map was created using data provided by the Petaling Jaya City Council and SISMAPS

Data Collection

In this study, the interest was mainly in investigating the influence of local landscape context and the quality of small urban parks on habitat provision for butterflies. Information was gathered on the surrounding land use and land cover using the Selangor Planning Information System (SISMAPS version 2). For the field survey, a combination of structured observation with field measurement method was used to investigate park characteristics, vegetation diversity and structure, butterfly species richness and abundance, as well as anthropogenic factors. *Structured observation* included observing human activities, butterfly species and soundscape. *Field measurement* involved the use of equipment in measuring sound levels, tree diameter, landscape features and ambient temperature. The field survey was carried out in three consecutive working days, and each park was observed for a total of six hours, spreading from the morning (7.30–9.30 a.m.), afternoon (12.30–2.30 p.m.) to evening (5.00–7.00 p.m.). Weekends were excluded for data consistency and comparability. Brief descriptions of all the studied parks are provided in Table 17.1.

Butterfly Surveys and Species Identification

Butterflies were identified using the transect walk and timed survey method. Since the studied parks were relatively small, both transect walk and timed survey allowed for detecting numbers of butterfly species and abundance (Kadlec et al., 2012; Lian & Sodhi, 2004). Using the transect method, we walked along the selected routes in the park and recorded any butterflies observed within 15 m width. The timed survey method did not use a fixed survey path. Instead, the researchers searched throughout the studied parks and especially in areas containing suitable habitats to identify as many butterfly species/individuals as possible. Butterfly species and their abundance were recorded three times a day following the survey periods as specified earlier. Photographs of the observed butterflies were taken with a high-resolution digital camera with 42-times optical zoom. Species identification was aided by treatises on local and regional butterfly collections (Kirton, 2014). For butterfly abundance, a scale value ranging from 1 (very low) to 10 (very high) was used. The scale value measured the number and frequency of butterfly individuals being observed in the parks.

Table 17.1 General descriptions of nine small urban parks studied (no. 1-9 according to the map in Fig. 17.1)

Study park	Characteristics
MBPJ (0.7 ha)	Located in the city centre, in front of a few government offices. Near a business area, shops, restaurants, banks and institutions are generally used for passive recreation. Vegetation is dominated by various ground covers and flowering shrubs, accompanied with small and medium-sized trees, palms and bamboos
Jalan Semangat (JS) 0.7 ha	Linear park located next to the main road junction and residential area. A pedestrian bridge connects the park with a secondary school across the road. Vegetation mostly comprises small and medium-sized trees and very few flowering shrubs. Park is intensely used during the school peak hours
Sunway Mas (SM) 0.7 ha	Triangular shaped, with an open lawn accompanied with small and medium-sized trees, a few palms and fewer shrubs. It is equipped with a small playground and basketball court. It is situated near offices, shopping lots and restaurants. The degree of cleanliness and maintenance is low
Bukit Mayang Emas (BME) 1 ha	It hosts an open lawn with small and medium-sized trees and few shrubs, equipped with outdoor fitness facilities, a playground, basketball court and badminton court. There are many stray dogs in the park, especially in the afternoons
Wawasan Park (WP) 1 ha	It is located very close to a private school, offices, shopping lots and residential areas. The vegetation is mostly small and medium-sized trees, together with a few tall trees as well. Facilities include a playground, shelters, badminton court and basketball court. The park is mostly used by the school kids and teenagers for sports activities
Mayang Jaya (MJ) 1.2 ha	The park has a dense canopy cover, dominated by trees, and hosts little shrubbery and ground covers. Facilities include a playground, futsal court, basketball court, and community hall. It is located close to an international college, shopping lots, restaurants and residential areas
SS2 Park (1.5 ha)	It is located at the business centre and the town's commercial area, food courts and restaurants. The vegetation mostly comprises small-medium trees with a few shrubs, as well as a few tall trees on the roadside close to the park. Many house crows (*Corvus splendens*) are present that feed on human food wastes
PJS10 Park (2.5 ha)	This park has a pond, well-equipped with outdoor fitness facilities, including a playground, basketball court, shelters and multipurpose area. The vegetation comprises trees in various sizes, some ground covers, flowering shrubs, palms and aquatic plants

(continued)

Table 17.1 (continued)

Study park	Characteristics
Aman Park (AP) 3.5 ha	The largest park in this study is located very close to a Light Rapid Transit (LRT) station. It has a small lake, open lawn, trees in different sizes, palms, various types of shrubs and ground cover, ferns and aquatic plants. It is well-maintained and equipped with various outdoor recreational facilities such as a playground, outdoor fitness and skateboard areas

Influencing Variables

The variables were categorised into three groups: physical characteristics, floral species richness and anthropogenic factors. Variables on the physical characteristics included surrounding land use, topography, size, shape, context (proximity to road and infrastructure), buildings and facilities, and the presence of water elements.

A series of vegetation variables were considered in this study. The presence of adult butterflies in urban parks is usually associated with the occurrence of larval host plants, nectar from plants and vegetation structure (Lian & Sodhi, 2004; Sing et al., 2016; Tam & Bonebrake, 2015). For some butterflies that have a limited mobility distance, they cannot disperse widely across fragmented landscapes (Bergerot et al., 2012). Therefore, accessibility to survival resources within their vicinity is essential to avoid local extinction. The availability of flowering plants can be an important determinant for butterflies in parks and gardens (Matteson & Langellotto, 2010; Öckinger et al., 2009). Research has also found that the planting of native vegetation can support more butterflies by increasing nectar sources and available niches (Giuliano et al., 2004). In this study, we calculated the percentage of canopy cover and percentage of open grass/ground. Vegetation types were classified into native or non-native (exotic) species. Other vegetation structures that were assessed include tree diameter at breast height (DBH), vegetation height, shrubs density, total flowering plants and flowering shrubs. Understanding the planting composition could provide information on the relation between vegetation characteristics and butterfly diversity.

Based on the classification by Tzoulas and James (2010), human activities in the park were grouped into four categories: utilitarian, recreation, sport and play. Besides these, we added one more group, namely special occasions to classify the activities that occasionally happen in the park such as weddings, team building and events. We observed activities in the surrounding area of the park and described these briefly in the survey form. The number of visitors was counted and divided into two groups; *park users* and *passers-by* (people who used the park as a pathway to other places).

Little is known about the impacts of anthropogenic noise related to roads on butterflies. A few previous studies have investigated the relation of road width, traffic volume and road verges on butterflies species richness and abundance (Clark et al., 2007; Munguira & Thomas, 1992; Skorka et al., 2013). However, none of these have related to traffic noise and its effects on butterflies. In this study, the noise levels were measured in decibel (dB) and the diversity of sound (types of sound) was identified. Noise levels were measured at three designated spots in the park. The spots were located as closely as possible to the sound sources and at a distance of approximately less than 100 m from each other. The observers recorded the noise levels for ten times at the intervals of three minutes at each spot. The diversity of sounds was categorised into two: natural, and anthropogenic or mechanical.

Data Analysis

All data recorded during the field surveys was compiled into Microsoft Excel in which the information was arranged and extracted from a larger set of data by using the pivot and matrix table function. Using the species matrix we computed the vegetation and butterfly species diversity based on the Shannon-Weiner diversity index (H) formula (Cornelis & Hermy, 2004; Jim & Chen, 2009). Schematic vegetation maps for each park were produced to calculate the coverage percentage, analyse the planting composition and species distribution.

Results

Park Characteristics

All studied parks are surrounded by commercial/ business, institutional, and residential areas. The topography of the parks is mostly flat except for AP, which has a rather undulating landscape. Park shape varies from irregular or square, to triangular or linear. Two parks (MBPJ and JS) are located next to a busy road (especially during peak hours) and noise levels (in dB) were high in these parks (Table 17.2). Only PJS10 and AP have a small lake. All parks are easily accessed by pedestrians and well equipped with facilities for outdoor recreation. Overall, park maintenance and cleanliness were satisfactory. Those parks located in important areas (e.g. near government offices and commercial lots) and exclusive residential areas in the city centre have been maintained more intensively, such as regular mowing.

Vegetation Structures and Butterfly Species Across the Parks

In total we recorded 89 different plant species (trees, palms and shrubs). For all studied parks, vegetation diversity was found to be moderate, with an average Shannon-Weiner index of $H = 2.27$. Of the nine parks, AP has the highest vegetation diversity ($H = 3.67$), while BME has the lowest at $H = 1.64$ (Table 17.2). The most

Table 17.2 Summary of vegetation characteristics, butterfly species richness and human factors

Park	MBPJ	JS	SM	BME	WP	MJ	SS2	PJS10	AP
Total vegetation (NOI)	424	90	74	103	59	82	54	354	381
Overall vegetation species	32	12	10	10	11	8	7	26	52
Number of shrubs (NOI)	346	52	44	25	28	25	20	266	139
Tree canopy covers (%)	76	77	34	53	48	90	68	53	67
Open grass/ground (%)	23	22	65	46	50	8	28	45	30
Shrubs species	15	4	3	2	2	4	2	13	10
Total flowering plants (NOI)	188	87	17	79	43	64	26	154	173
Flowering shrubs (NOI)	165	52	10	10	20	25	0	95	82
Area (hectare)	0.7	0.7	0.7	1	1	1.2	1.5	2.5	3.5
Native vegetation (NOI)	96	56	21	45	6	34	15	145	157
Exotic vegetation (NOI)	328	34	53	58	53	48	39	209	155
Native vegetation species	9	6	2	4	3	2	2	10	27
Exotic vegetation species	22	5	7	6	7	5	4	15	24
Vegetation diversity index	3.104	2.058	1.744	1.639	1.815	1.702	1.663	3.011	3.668
Butterfly species	13	7	4	4	4	14	6	9	8
Butterfly abundance	94	63	31	31	19	114	34	100	101
Butterfly diversity index	2.42	1.832	1.142	1.173	1.281	2.536	1.517	2.040	1.769
Average number of visitor (/day)	200	514	115	106	81	133	501	369	456
Average noise levels (dB/day)	77	81	72	63	60	58	71	66	68

Note NOI: Number of individuals. The vegetation and butterfly diversity index were based on Shannon-Weiner Diversity Index

popular tree species used as an ornamental plant is *Tabebuia rosea* (Trumpet Tree), and *Veitchia merrilli* (Christmas Palm) is the common palm species. *Hymenocallis speciosa* 'Variegata' (Spider Lily) is the most common ornamental shrub. Overall, exotic vegetation was more abundant than native species. For instance, Trumpet Tree and Spider Lily are very common. MBPJ has more shrubs than other parks, and it accommodates the highest number of exotic plants (Fig. 17.2 (a)). Although MBPJ is one of the smallest parks, it recorded a relatively high vegetation diversity (H = 3.1).

A total of 23 butterfly species from five families were recorded, with an average butterfly diversity index of $H = 1.746$ across the nine sampled parks. Although MJ has a low vegetation diversity ($H = 1.7$), it has the highest butterfly diversity ($H = 2.54$) (Table 17.2). *Hympolimnas bolina* (Great Egg-fly) was the most abundant butterfly species, followed by *Eurema hecabe* (Common Grass Yellow) and *Appias libythea* (Striped Albatross). The abundance of butterflies was higher during the afternoons than in the mornings and evenings (Fig. 17.2b). Individual butterfly characteristics,

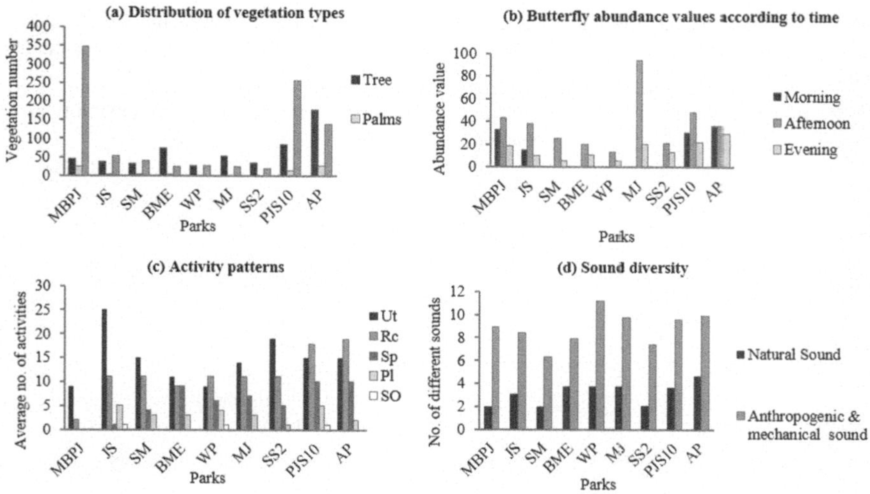

Fig. 17.2 **a** Distribution of vegetation types. **b** Butterfly abundance values according to time. **c** Activity patterns. *Ut-* Utilitarian, *Rc-* Recreation, *Sp-* Sports, *Pl-* Play, *SO-* Special occasion. **d** Sound diversity

their ecological traits and habitat zone were shown in Table 17.3. Most of the butterfly species were common in urban green areas and they could be specified as urban adapters. Moreover, the presence of forest butterfly species *Curetis santana* (Malayan Sunbeam) was also noted. We also observed *Lyssa zampa* (Tropical Swallowtail Moth or sometimes called Laos Brown Butterfly), a large, conspicuous moth species that resembles butterflies. In MBPJ, individuals of the butterfly species *Zizina otis* were seen flying around a clump of shrub *Phyllanthus myrtifolius* (Mousetail Plant) and the larvae (caterpillars) were observed attached to the stems. Meanwhile, in MJ several pupae were attached to the branches of shrub *Hibiscus rosa-sinensis* and more than five butterfly species were seen flying around the plants.

Butterflies found in the parks were mostly from the families of Nymphalidae and Pieridae. The distribution of different butterfly families in the studied parks and the locations where most butterflies were found were presented in Table 17.4. Most of the locations where butterflies were found were very close to the road and pathways. The schematic map in Table 17.4 showed the distribution of vegetated area and open grass/ground. Most butterflies were active in the hot sunny areas. We observed different types of butterfly behaviour, such as resting, mating, feeding on nectar or flying around the vegetation. Some of the butterflies are seen puddling in moist areas.

Table 17.3 Species and ecological traits of butterfly species found within the studied parks, including individual butterfly characteristics in relation to habitat requirements and adaptability in urban environments

No	Butterfly species	Relative abundance	Status	Habitat zone	Adaptability in urban environments
1	*Acraea terpsicore* (Tawny coster)	42	C	Villages, parks, gardens, bushes and wayside vegetation	Adapter
2	*Anthene emolus* (Ciliate Blue)	7	C	Villages, parks, gardens and forest in the lowlands and highlands	Adapter
3	*Appias libythea* (Striped Albatross)	82	C	Forest edge, secondary forest, mangrove edge, villages, parks, gardens, and waysides	Adapter
4	*Arhopala centurus* (Centaur Oakblue)	23	C	Villages, parks, gardens and lowland forest	Adapter
5	*Catopsilia pomona* (Lemon Emigrant)	47	C	Forest edge, villages, parks, gardens, open country and waysides	Adapter
6	*Catopsilia pyranthe* (Mottled Emigrant)	10	C	Forest edge, villages, parks, gardens, open country and waysides	Adapter
7	*Curetis santana* (Malayan Sunbeam)	4	R/CN	Inhabit forest up to the highlands	Adapter
8	*Elymnias hypermnestra* (Common Palmfly)	28	C	Forest edge, villages, gardens and oil palm plantations	Adapter
9	*Eurema hecabe* (Common Grass Yellow)	87	C	Almost every habitat at all elevations	Adapter

(continued)

Table 17.3 (continued)

No	Butterfly species	Relative abundance	Status	Habitat zone	Adaptability in urban environments
10	*Hympolimnas bolina* (Great Egg-fly)	95	C	Villages, wayside vegetation, parks and gardens	Adapter
11	*Hypolycaena erylus* (Common Tit)	27	C	Secondary forest, wayside vegetation, mangroves	Adapter
12	*Junonia atlites* (Grey Pansy)	33	C	Villages, parks, gardens, low vegetation, open grass	Adapter
13	*Junonia hedonia ida* (Chocolate Pansy)	9	C	Forest edges, villages, parks, garden	Adapter
14	*Junonia orithya* (Blue Pansy)	2	C	Villages, parks, gardens, low vegetation, open grass, coastal scrub	Adapter
15	*Lyssa zampa* (Tropical Swallowtail Moth/Laos Brown Butterfly)[a]	4	C	Forest edge to inland forest	Adapter/ exploiter
16	*Neptis hylas* (Common Sailer)	21	C	Villages, parks, gardens, open country, coastal and inland forest	Adapter
17	*Papilio demoleus* (Lime Butterfly)	10	C	Villages, parks and gardens	Adapter
18	*Papilio polytes romulus* (Common Mormon)	19	C	Villages, gardens and disturbed forest	Adapter
19	*Suastus gremius* (Indian Palm Bob)	7	C	Forest edge, gardens, palms cultivation area	Adapter
20	*Tajuria cippus* (Peacock Royal)	5	R/CN	Forest edge, villages, parks and gardens	Adapter

(continued)

Table 17.3 (continued)

No	Butterfly species	Relative abundance	Status	Habitat zone	Adaptability in urban environments
21	*Udaspes folus* (Grass Demon)	9	C	Forest edge, villages, parks and gardens	Adapter
22	*Ypthima baldus* (Common Five Ring)	10	C	Forest edge, villages, parks, gardens, wayside vegetation	Adapter
23	*Zizina otis* (Lesser Grass Blue)	6	C	Villages, parks, gardens, open grass, wayside vegetation, coastal scrub	Adapter

[a]Large and conspicuous moth species that resembles a butterfly (butterflies and moths are in the same order of Lepidoptera). R = Rare, C = Common, CN = Common but NOT easily seen

Human Factors

Small parks were used more for utilitarian purposes (42% of observed use, e.g. walking through and motorcyclist movement) than for recreation (32%, e.g., relaxing or family outings) or sports, play and special occasions (26%) (Fig. 17.2c). The highest number of visitors per day was recorded in JS with 514 visitors and the lowest in WP with only 81 visitors. Concerning noise sources, we experienced more anthropogenic sounds than natural sounds, and especially traffic noise (Fig. 17.2d). During the observations, we also experienced more unpleasant than pleasant smells from vehicles, animal manure, food waste, garbage and drains.

The Impacts of Explanatory Variables on Butterflies Species Richness and Abundance

With reference to Fig. 17.3, in order to identify the impact of the selected variables on butterfly abundance and diversity, we first conducted simple linear regressions between each predictor variable /explanatory variable (e.g., percent of canopy cover, park area and richness of shrub species) and butterfly species richness and abundance as dependent variables. A simple Pearson correlation matrix was used to check the relationships. Next, based on the simple linear regression and a correlation matrix, possible sets of predictors were selected to perform stepwise multiple regressions, using the criterion of Adjusted $R^2 > 0.5$ and $P < 0.05$ to add or remove variables. All statistical analyses were carried out in R version 3.1.2. and XLSTAT version 2015.

Table 17.4 Schematic map shows the composition of the vegetated and open area within the nine parks. The stacked bar shows the distribution of butterflies' abundance according to families. The yellow zone on the map represents the location where most butterflies were observed

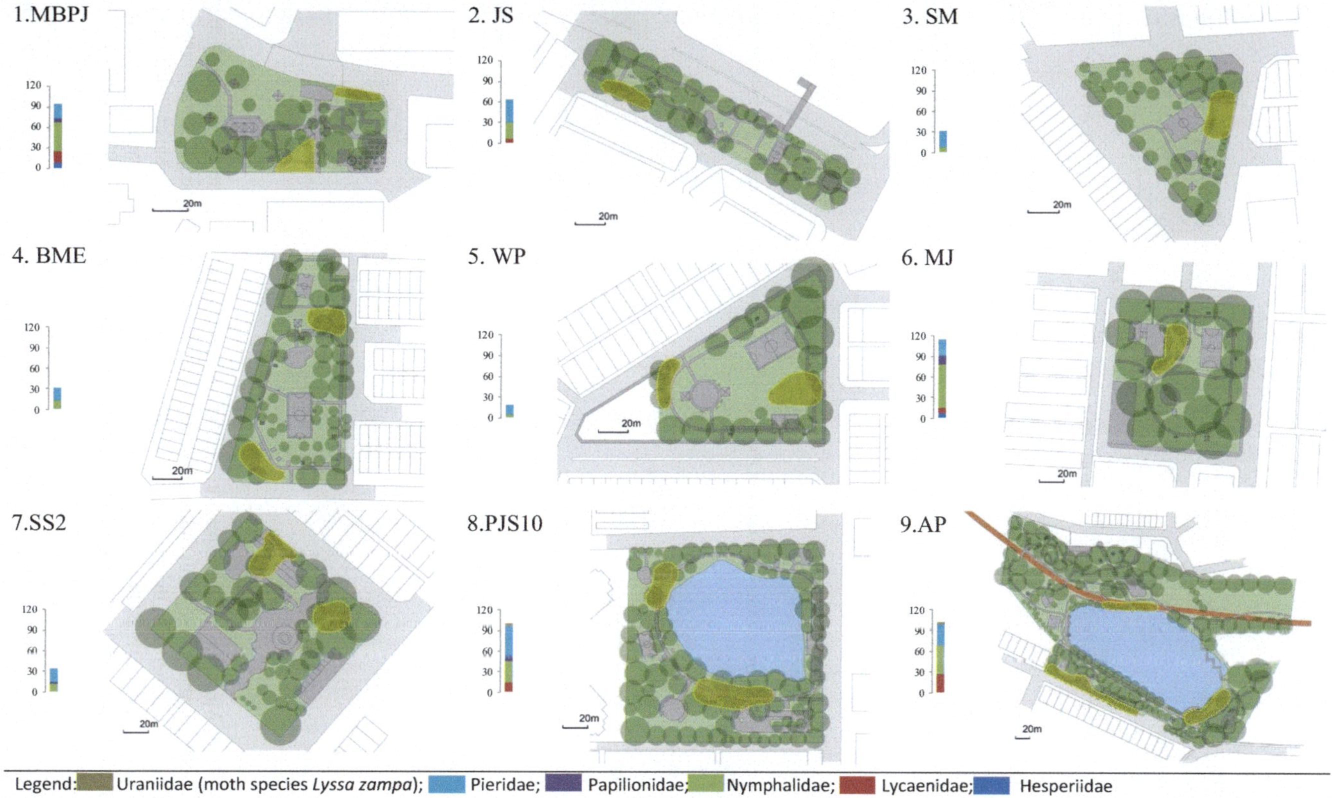

(a) Butterfly species richness

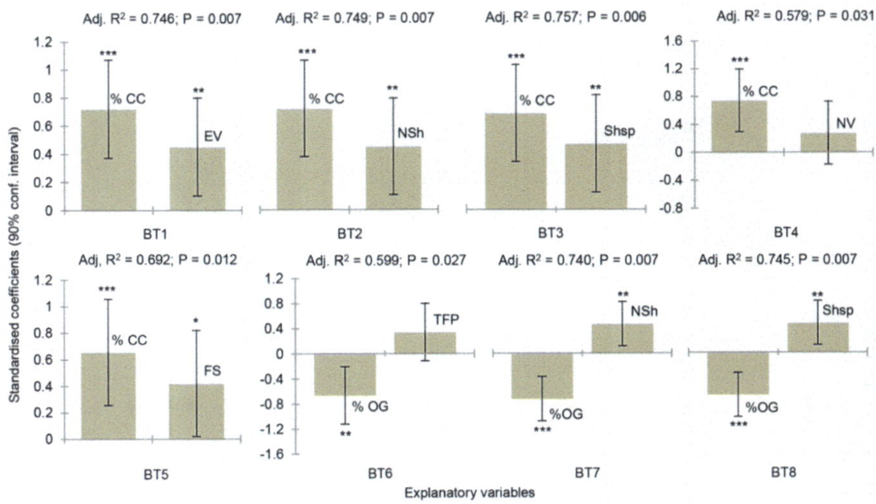

(b) Butterfly abundance

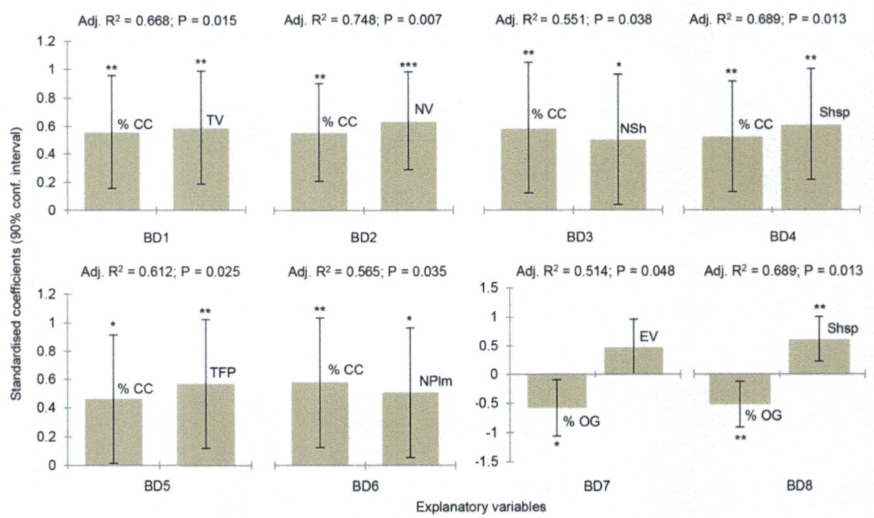

Fig. 17.3 Multiple linear regression coefficients charts for (**a**) butterfly species richness and (**b**) butterfly abundance. Only statistically significant results are retained. The bar graphs compare the relative influence of the explanatory variables on the dependent variable and their significance. BT1- BT8 and BD1- BD8 represent the name of the model. The positive and negative value of the bar graph indicates the direction of the influence. The error bars represent 90% confidence intervals and if closer to or includes zero (0), the variables are not significant. Note: The individual *X* variables significant at *** p value < 0.01; ** p value < 0.05; * p value < 0.1; %CC are: percentage of tree canopy cover; TV: total vegetation; EV: exotic vegetation; NV: native vegetation; NSh: number of shrubs; NPlm: number of palms; Shsp: shrubs species; %OG: percentage of open grass; TFP: Total flowering plants; FS: flowering shrubs; Ha: area in hectare

Pearson correlations (r) were firstly computed to explore the relationships between variables. The results showed that butterfly species richness was highly correlated with the percentage of canopy cover ($r = 0.78$), percentage of open grass/ground ($r = -0.77$), and moderately associated with number of shrub species ($r = 0.61$) as well as presence of flowering shrubs ($r = 0.62$). Meanwhile, butterfly abundance was highly correlated with presence of native vegetation ($r = 0.71$), shrub species diversity ($r = 0.71$), total flowering plants ($r = 0.72$), and moderately correlated with total vegetation ($r = 0.67$), percentage of canopy cover ($r = 0.65$), percentage of open grass/ground ($r = -0.64$), and presence of flowering shrubs ($r = 0.64$).

Figure 17.3 plotted the multiple linear regression coefficients charts for butterfly species richness (Fig. 17.3a) and butterfly abundance (Fig. 17.3b). All possible associations were tested with two different explanatory variables. Eight models for butterfly species richness and butterfly abundance were presented (BT1–BT8 and BD1–BD8), respectively. For each model, at least 50% of the variation in butterfly species richness and abundance could be explained (Adjusted $R^2 \geq 0.5$). The present paper only presents statistically significant results.

Results revealed that the percentage of canopy cover had a high significant positive influence on butterfly species richness (BT1, BT2, BT3, BT4 and BT5 with %CC p value < 0.01) and butterfly abundance (BD1, BD2, BD3, BD4 and BD6 with %CC p value < 0.05). In contrast, the percentage of open grass/ground was found to have a significant negative influence on butterfly species richness and abundance (BT7, BT8 with %OG p value < 0.01 and BT6, BD8 with %OG p value < 0.05). Pearson's correlation indicated that both native and exotic vegetation had a positive association with butterfly species richness ($r = 0.38$ for native vegetation, and $r = 0.55$ for exotic vegetation) and butterfly abundance ($r = 0.71$ for native vegetation and $r = 0.55$ for exotic vegetation). In the regression model BT1 (Adjusted $R^2 = 0.746$; $P = 0.007$), particularly exotic vegetation had a significant positive effect on butterfly species richness (for EV, p value < 0.05), while model BD2 (Adjusted $R^2 = 0.748$; $P = 0.007$) demonstrated that only native vegetation had a significant positive influence on butterfly abundance. The total number of vegetation was found to have a significant positive effect on butterfly abundance (BD1: TV p value < 0.05), but with insignificant impact on butterfly species richness.

The total number of shrubs (NSh) and the number of shrub species (Shsp) could significantly and positively influence the butterfly species richness and abundance (BT2, BT3, BT7, BT8, BD4, and BD8). The presence of flowering shrubs (FS) also had a significant effect on butterfly species richness (BT5: FS p value 0.012), and for butterfly abundance the total flowering plants was the significant variable (BD5: TFP p value 0.025). The abundance of palms (NPlm) also had a significant positive influence on butterfly abundance (BD6). Pearson correlation indicated that park size was positively correlated with butterfly species richness ($r = 0.10$) and butterfly abundance ($r = 0.49$), but the association was weak. Moreover, regression analyses revealed that park size did not have a significant effect on butterfly species richness and abundance. The results from correlation showed that human-related factors and anthropogenic disturbances had no significant impact on butterfly species diversity and abundance.

Discussion

The Influence of Park Area and Local Landscape Characteristics

This study demonstrated that although small urban parks were commonly surrounded by urbanised areas and harboured large shares of cultivated vegetation, they can still provide habitats for various butterflies. Of the 60 butterfly species recorded in 10 large parks (with the size ranging from 10 to 101.1 ha) in Kuala Lumpur (Sing et al., 2016), 23 were found in the studied parks, corresponding to nearly 30% of the total butterfly species. Due to the similarity in microclimate and environmental factors, butterfly species found in Petaling Jaya and Kuala Lumpur are very similar. Our findings are consistent with an empirical study conducted by Shwartz et al. (2013) who found that butterfly species in 36 small public gardens (0.5–2.0 ha) in Paris accounted for more than half of overall butterfly species in the region. In the present research, although only a small range of park sizes (0.5–3.5 ha) was considered, we still observed various butterfly species from different families.

Empirical studies suggested that park size can be one of the decisive determinants of butterfly diversity (Di Mauro et al., 2007; Giuliano et al., 2004; Soga & Koike, 2013). However, in the present study, Pearson correlation results showed that park size had only a weak positive association with butterfly species richness and abundance. And regression analysis (Fig. 17.3) revealed that park size was not a significant predictor. This result contradicted our expectation but agreed with previous studies by Lizée et al. (2011) and Sing et al. (2016). A possible reason was that all studied parks were relatively small, and habitat quality was more important than park size. We observed more butterfly species in MBPJ compared to AP. Although it was the smallest park in our study, it had various dense shrubs that probably favoured butterflies. However, butterfly individuals can disperse more widely and scattered in larger parks such AP and therefore, species richness and abundance could be—higher than observed.

Previous studies suggested that land use patterns, levels of urbanisation and proximity to busy roads might lower butterfly species richness and abundance (Casner et al., 2014; Clark et al., 2007; Di Mauro et al., 2007; Lee et al., 2015). Increases in road lane density may also negatively affect butterfly species richness (Chong et al., 2014). In this study, we found no significant effect of the urban surroundings on butterfly species richness and abundance. This result was in accordance with empirical findings from the study conducted by Öckinger et al. (2009). The schematic diagram of all parks in. In reference to Table 17.4 it can be suggested that potential human disturbance from adjacent roads had no negative effects on butterfly communities. This finding can be true for butterflies with the presence of suitable habitat variables that provide the opportunity for nectar feeding and protection from wind or predators.

Land use intensity and configuration such as building density were not measured in this study. However, previous research by Lizée et al. (2011) suggested that butterfly species diversity would decline if the built matrix surrounding parks was aggregated in convoluted shapes. In addition, the shape of individual urban parks can also play a role in supporting urban biodiversity and ecological functions. As described by Forman (1995) and Beck (2013), several green patch shapes, such as circular, squares and lobes, tended to have more interior habitat and less edge habitat. Yamaura et al. (2008) revealed that circular shape patches significantly increased abundance of forest butterfly species. Meanwhile, more irregular green patch shapes increase edge effects and disturbances within species and can therefore decrease insect species density (Su et al., 2015). In this study, the relation between park shape and butterflies were found to be insignificant.

One common butterfly behaviour is puddling, which means sipping water from the moist ground and wet sandy soil or damp surface (Tam & Bonebrake, 2015; Yücel, 2013). According to the observations made in this study, there were no significant differences between butterfly species found in small parks with and without the presence of water bodies (ponds and small lakes). One possible explanation for this is the lack of dependency of butterflies on open freshwater. Adult butterflies only take in mineral-rich water (particularly sodium) for the mating reason. and compared to wet soil freshwater contains low sodium concentrations (Beck et al., 1999).

The Influence of Vegetation Diversity and Structure

The results of this study show that vegetation variables are the most important predictor for butterfly species richness and abundance in small urban parks (Fig. 17.3). As we anticipated, the percentage of canopy cover has a significant positive influence on butterfly richness and abundance. Our findings in MJ indicate that the park with the highest percentage of canopy cover (90%) recorded the highest butterfly diversity index (H = 2.536). This is in line with a recent study conducted by Tam and Bonebrake (2015). In contrast, the percentage of open grass or open ground was found to have a significant negative impact on butterfly species richness and abundance. Although butterflies require a sunny spot for basking and becoming active (Yücel, 2013), our results are somewhat contradictory to the findings. Less butterflies were seen when large open grass without flowering plants existed. Shwartz et al. (2013) highlighted that the appearance of butterflies is correlated with the existence of lawn with wildflowers. While Matteson and Langellotto (2010, 2012) explained that butterflies are highly mobile, and thus unlikely to remain in green space for long periods especially in open grass as they may become prey for insectivorous birds. Therefore, a right proportion of vegetative cover and open grass is needed for higher butterfly species and abundance in small parks.

As expected, shrub's characteristics are significant determinants for butterfly richness and abundance, which is consistent with the former study by Shwartz et al. (2013) who explained that bush cover has a positive impact on butterfly presence. Shrub species and abundance significantly influence butterfly species richness and abundance. Shrubs with ornamental flowers make a significant contribution to species richness and the total number of flowering plants has a significantly positive impact in butterfly abundance, which is in agreement with several earlier studies (Clark et al., 2007; Matteson & Langellotto, 2010, 2012). This would imply that urban butterflies prefer plants with flowers and flowering shrubs may provide nectars for adult butterflies. Although we did not specifically measure host plant diversity, the existence of larvae and pupae on shrubs was observed.

Previous studies have demonstrated that butterfly species richness and abundance can be enhanced with the planting of more native plants than exotic ones (Burghardt et al., 2009; Tallamy & Shropshire, 2009). However, exotic plants have been widely used in urban parks as ornamental plants due to their aesthetic value and contribution to floristic diversity (Hitchmough, 2008; Nielsen et al., 2014). Urban butterflies may exploit several exotic plant species. For example, in Hong Kong's urban parks, *Widelia trilobata* was commonly used as a nectar source (Tam & Bonebrake, 2015), and in India *Lantana camara* was utilised by butterflies as surrogate host plant (Mukherjee et al., 2015). In this study, we found that both native and exotic vegetation was positively correlated to butterfly species richness and abundance. However, only exotic vegetation had a significant (positive) impact on butterfly species richness (Fig. 17.3: BT1), while native plants significantly affected butterfly abundance. In the absence of native vegetation, butterflies will use exotic species as alternatives (Giuliano et al., 2004) and in particular cases butterflies may utilise exotic flowers more frequently than native flowers (Matteson & Langellotto, 2011). For example, in urban gardens in Davis, California, USA butterflies were found to mostly breed on alien plant species instead of native ones (Shapiro, 2002). We suggest that if small urban parks should provide habitats for butterflies, appropriate mixtures of native and exotic vegetation could support a wider range of butterfly species. However, appropriate exotic plants must be carefully selected because invasive species can be problematic for biodiversity. The advanced technology of hybridisation between native and exotic species may produce the plants that can be used safely in urban green spaces (Shwartz et al., 2013). Small urban green space may be used mostly by generalist butterfly species, hence, butterflies' ability in adapting to vegetation change is fairly high (Matteson & Langellotto, 2010).

Apart from shrubs, the abundance of palms has a significant positive relation with butterfly abundance. Palms are highly favourable host plants of some butterfly species from the families *Hesperiidae* and *Nymphalidae* (Frank, 2002; Kirton, 2014). In this study, the butterfly species associated with palms were found to be *Elymnias hypermnestra* and *Suastus gremius*. Seven species of palms recorded in this study are commonly used by butterflies. These species are *Archontophoenix alexandrae* (Alexandra Palm), *Areca catechu* (Betel Palm), *Cocos nucifera* (Coconut), *Elaeis guineensis* (Oil Palm), *Licuala grandis* (Fan Palm), *Cyrthostachys renda* (Lipstick Palm) and *Nypa fruticans* (Nipah Palm). This suggests that the planting of palms in small urban parks might be useful to enhance habitat diversity and offers additional food sources for butterflies.

Human Factors and Implications for Park Management

Based on the regression analysis, butterflies were thus not impacted by the visitor density, human activities and noise levels. Butterfly species richness and abundance may be affected by human activities and disturbances from the adjacent areas. Earlier research by Clark et al. (2007) revealed that the number of people in small urban green spaces was negatively correlated with butterfly richness. Another study conducted in the Washington DC metropolitan area found that an increase of human population negatively affected butterfly diversity in urban gardens (Di Mauro et al., 2007). Meanwhile, Lee et al. (2015) explained that high coverage of built-up areas and roads may negatively influence butterfly diversity. In contrast, we found that neither the number of people using small parks nor human activities could affect species richness and abundance of the butterflies. A possible reason is that the density of people and intensity of their activities in the studied parks were not heavy and thus only moderate disturbance might be introduced. Our finding is in line with the suggestion that butterfly species richness is at its peak when intermediate or mild disturbances occur due to the opportunities of new resources that decrease when activities are extensive (Blair & Launer, 1997; Matteson et al., 2013). These new resources may include the introduction of non-native plant species, changes in vegetation configuration and existence of wet soils due to trampling that can be used as puddling areas for butterflies.

Anthropogenic noise from the surrounding urban environment may be detrimental to butterflies by disturbing their capacity to detect danger or predators (Morley et al., 2014). However, our hypothesis regarding the negative impact of noise cannot be validated, as no significant impact of anthropogenic noise to butterflies was found in this study. This might indicate that butterflies have the capability to cope with anthropogenic noise in small parks and/or that the noise level was not high enough to disturb butterflies. Several butterfly species can be very sensitive to low frequency sound (Morley et al., 2014), but since the frequency was not measured, we were unable to draw any conclusions on this point in the present study.

The management of small urban parks may influence whether these parks can serve as habitats for butterflies. Appropriate management strategies and actions can help to protect or establish habitats, increase species diversity, improve vegetation composition and even reduce maintenance costs (Konvicka & Kadlec, 2011; Muratet & Fontaine, 2015; Shwartz et al., 2013). Weber et al. (2008) demonstrated that butterfly individuals decreased significantly after mowing due to the sudden reduction of nectar sources, but the abundance recovered after a few weeks. Leaving the tall grass adjacent to the mowed grass or reducing the mowing frequency may support the growth of wildflowers which contribute to unmanaged microhabitats for butterflies (Chong et al., 2014; Sing et al., 2016). The application of pesticides in small urban parks can decrease the species richness and abundance through habitat loss and host plant destruction (Di Mauro et al., 2007). The impact of pesticides on butterflies depends on the types of pesticide and quantity applied and may vary according to the landscape. For example, in MBPJ, warning signs of pesticides usage on certain groups of plants were noticeable, and fewer butterflies were seen flying or basking around these plants. On and around other vegetation where no sign of pesticides was found, butterfly abundance was higher and we were able to record the presence of larvae. Although pesticides are used in MBPJ, the butterfly diversity is relatively high. Perhaps the type of pesticides used is organically safe (Muratet & Fontaine, 2015) or butterflies are able to utilise other habitat resources that are not treated with pesticides.

Conclusion

This paper highlights some of the important characteristics of small urban parks that should be considered in their planning, design and management specifically aimed at enhancing butterfly abundance and diversity. The findings support the suggestion made by Lizee et al. (2015) that several small parks can accumulate higher total butterfly species than one large park. In general, our hypotheses at the outset of this study were tested and validated. And some of the results were interesting, e.g., the limited impact of park size on butterfly populations. Based on the overall findings, we provide two general recommendations for small urban park management to support diverse butterfly communities. First, the development and management plans for small urban parks should emphasise the internal quality of the parks (e.g., by mowing and pruning practices, the proportion of vegetated area and open grass). Incorporating social-ecological principles in setting up the goals for biodiversity conservation in small urban parks is essential, which can help enhance both benefits of parks to humans and wildlife. Although direct impacts of human disturbances on butterfly communities were not significant—in this study, we encourage further investigation into the interaction between human and biodiversity in small urban green spaces. It may be difficult to control the impact from the adjacent urban matrix entirely, but minimising this impact could benefit butterfly diversity. Second, newly established small urban parks should have a planting design which can provide- habitat and food for butterflies. The availability of larval host plants and nectar sources for adult butterflies is important for their survival. Appropriate inclusion of both native and exotic shrubs creates floral varieties that would attract butterflies.

This study shows that small urban parks in Malaysia do provide habitat for diverse butterfly populations, but limited to only generalist species. In the absence of large parks, small parks may also function as the surrogates and urban habitat refuges.

Acknowledgements The authors gratefully acknowledge the Ministry of Higher Education, Malaysia and Universiti Teknologi Malaysia (UTM) for funding this study. We also would like to thank all research assistants for their help in conducting the field survey in Malaysia. Thank you also to Petaling Jaya City Council for their cooperation in providing all necessary input for the study.

References

Angold, P. G., Sadler, J. P., Hill, M. O., Pullin, A., Rushton, S., Austin, K., Small, E., Wood, B., Wadsworth, R., Sanderson, R., & Thompson, K. (2006). Biodiversity in urban habitat patches. *Science of the Total Environment, 360*, 196–204. https://doi.org/10.1016/j.scitotenv.2005.08.035

Beck, J., Mühlenberg, E., & Fiedler, K. (1999). Mud-puddling behavior in tropical butterflies: In search of proteins or minerals? *Oecologia, 119*, 140–148. https://doi.org/10.1007/s004420050770

Beck, T. (2013). Principles of ecological landscape design. *Island Press*. https://doi.org/10.1007/s13398-014-0173-7.2

Bergerot, B., Merckx, T., Van Dyck, H., & Baguette, M. (2012). Habitat fragmentation impacts mobility in a common and widespread woodland butterfly: Do sexes respond differently? *BMC Ecology, 12*, 5. https://doi.org/10.1186/1472-6785-12-5

Blair, R. B., & Launer, A. E. (1997). Butterfly diversity and human land use: Species assemblages along an urban gradient. *Biological Conservation, 80*, 113–125.

Bonebrake, T. C., Ponisio, L. C., Boggs, C. L., & Ehrlich, P. R. (2010). More than just indicators: A review of tropical butterfly ecology and conservation. *Biological Conservation, 143*, 1831–1841. https://doi.org/10.1016/j.biocon.2010.04.044

Burghardt, K. T., Tallamy, D. W., & Shriver, W. G. (2009). Impact of native plants on bird and butterfly biodiversity in suburban landscapes. *Conservation Biology, 23*, 219–224. https://doi.org/10.1111/j.1523-1739.2008.01076.x

Casner, K. L., Forister, M. L., O'Brien, J. M., Thorne, J., Waetjen, D., & Shapiro, A. M. (2014). Contribution of urban expansion and a changing climate to decline of a butterfly fauna. *Conservation Biology, 28*, 773–782. https://doi.org/10.1111/cobi.12241

Chong, K. Y., Teo, S., Kurukulasuriya, B., Chung, Y. F., Rajathurai, S., & Tan, H. T. W. (2014). Not all green is as good: Different effects of the natural and cultivated components of urban vegetation on bird and butterfly diversity. *Biological Conservation, 171*, 299–309. https://doi.org/10.1016/j.biocon.2014.01.037

Clark, P. J., Reed, J. M., & Chew, F. S. (2007). Effects of urbanization on butterfly species richness, guild structure, and rarity. *Urban Ecosystem, 10*, 321–337. https://doi.org/10.1007/s11252-007-0029-4

Cornelis, J., & Hermy, M. (2004). Biodiversity relationships in urban and suburban parks in Flanders. *Landscape and Urban Planning, 69*, 385–401. https://doi.org/10.1016/j.landurbplan 2003.10.038

Di Mauro, D., Dietz, T., & Rockwood, L. (2007). Determining the effect of urbanization on generalist butterfly species diversity in butterfly gardens. *Urban Ecosystem, 10*, 427–439. https://doi.org/10.1007/s11252-007-0039-2

Forman, R. T. T. (1995). Some general principles of landscape and regional ecology. *Landscape Ecology, 10*, 133–142. https://doi.org/10.1007/BF00133027

Frank, J. H. (2002). *Insects on Palms, Florida entomologist*. https://doi.org/10.1653/0015-4040(2002)085[0402:IOP]2.0.CO;2

Giuliano, W. M., Accamando, A. K., & Mcadams, E. J. (2004). Lepidoptera-habitat relationships in urban parks. *Urban Ecosystem, 7*, 361–370. https://doi.org/10.1007/s11252-005-6835-7

Hitchmough, J. (2008). New approaches to ecologically based, designed urban plant communities in Britain: Do these have any relevance in the United States? *Urban Ecology, 1*, 1–15.

Hogsden, K. L., & Hutchinson, T. C. (2004). Butterfly assemblages along a human disturbance gradient in Ontario, Cananda. *Canadian Journal of Zoology, 82*, 739–748. https://doi.org/10.1139/z04-048

Jim, C. Y., & Chen, W. Y. (2009). Diversity and distribution of landscape trees in the compact Asian city of Taipei. *Applied Geography, 29*, 577–587. https://doi.org/10.1016/j.apgeog.2009.01.002

Kadlec, T., Tropek, R., & Konvicka, M. (2012). Timed surveys and transect walks as comparable methods for monitoring butterflies in small plots. *Journal of Insect Conservation, 16*, 275–280. https://doi.org/10.1007/s10841-011-9414-7

Kirton, L. G. (2014). *Butterflies of Peninsular Malaysia, Singapore and Thailand*. John Beaufoy, Malaysia.

Konvicka, M., & Kadlec, T. (2011). How to increase the value of urban areas for butterfly conservation? A lesson from Prague nature reserves and parks. *European Journal of Entomology, 108*, 219–229.

Lee, C. M., Park, J. W., Kwon, T.-S., Kim, S.-S., Ryu, J. W., Jung, S. J., & Lee, S. K. (2015). Diversity and density of butterfly communities in urban green areas: An analytical approach using GIS. *Zoological Studies, 54*, 4. https://doi.org/10.1186/s40555-014-0090-7

Lian, P. K., & Sodhi, N. S. (2004). Importance of reserves, fragments, and parks for butterfly conservation in a tropical urban landscape. *Ecological Applications, 14*, 1695–1708.

Lizée, M.-H., Manel, S., Mauffrey, J.-F., Tatoni, T., & Deschamps-Cottin, M. (2011). Matrix configuration and patch isolation influences override the species–area relationship for urban butterfly communities. *Landscape Ecology, 27*, 159–169. https://doi.org/10.1007/s10980-011-9651-x

Lizee, M. H., Tatoni, T., Deschamps-Cottin, M. (2015). Nested patterns in urban butterfly species assemblages: Respective roles of plot management, park layout and landscape features. *Urban Ecosystem*, 205–224. https://doi.org/10.1007/s11252-015-0501-5

Matteson, K. C., Grace, J. B., & Minor, E. S. (2013). Direct and indirect effects of land use on floral resources and flower-visiting insects across an urban landscape. *Oikos, 122*, 682–694. https://doi.org/10.1111/j.1600-0706.2012.20229.x

Matteson, K. C., & Langellotto, G. (2012). Evaluating Community Gardens as Habitat for an Urban Butterfly. Cities Environ. 5, Article 10.

Matteson, K. C., & Langellotto, G. A. (2011). Small scale additions of native plants fail to increase beneficial insect richness in urban gardens. *Insect Conservation and Diversity, 4*, 89–98. https://doi.org/10.1111/j.1752-4598.2010.00103.x

Matteson, K. C., & Langellotto, G. A. (2010). Determinates of inner city butterfly and bee species richness. *Urban Ecosystem, 13*, 333–347. https://doi.org/10.1007/s11252-010-0122-y

Morley, E. L., Jones, G., Radford, A. N. (2014). The importance of invertebrates when considering the impacts of anthropogenic noise. *Proceedings of the Royal Society B: Biological Sciences, 281*, 20132683. https://doi.org/10.1098/rspb.2013.2683

Mukherjee, S., Banerjee, S., Basu, P., Saha, G. K., & Aditya, G. (2015). Lantana camara and butterfly abundance in an urban landscape: Benefits for conservation or species invasion. *Ekologia Bratislava, 34*, 309–328. https://doi.org/10.1515/eko-2015-0029

Munguira, M. L., & Thomas, J. A. (1992). Use of road verges by butterfly and burnet populations, and the effect of roads on adult dispersal and mortality. *Journal of Applied Ecology, 29*, 316–329. https://doi.org/10.1016/0006-3207(93)90403-N

Muratet, A., & Fontaine, B. (2015). Contrasting impacts of pesticides on butterflies and bumblebees in private gardens in France. *Biological Conservation, 182*, 148–154. https://doi.org/10.1016/j.biocon.2014.11.045

New, T. R. (2015). Insect conservation and urban environments. *Springer Internal Publishing Switzerland*. https://doi.org/10.1007/978-3-319-21224-1

Nielsen, A. B., van den Bosch, M., Maruthaveeran, S., & van den Bosch, C. K. (2014). Species richness in urban parks and its drivers: A review of empirical evidence. *Urban Ecosystem, 17*, 305–327. https://doi.org/10.1007/s11252-013-0316-1

Niemelä, J., Breuste, J. H., Elmqvist, T., Guntenspergen, G., James, P., & Mcintyre, N. E. (2011). *Urban ecology: Patterns, processes, and applications*. Oxford University Press.

Öckinger, E., Dannestam, Å., & Smith, H. G. (2009). The importance of fragmentation and habitat quality of urban grasslands for butterfly diversity. *Landscape and Urban Planning, 93*, 31–37. https://doi.org/10.1016/j.landurbplan.2009.05.021

Shapiro, A. M. (2002). The Californian urban butterfly fauna is dependent on alien plants. *Diversity and Distributions, 8*, 31–40. https://doi.org/10.1046/j.1366-9516.2001.00120.x

Shwartz, A., Muratet, A., Simon, L., & Julliard, R. (2013). Local and management variables outweigh landscape effects in enhancing the diversity of different taxa in a big metropolis. *Biological Conservation, 157*, 285–292. https://doi.org/10.1016/j.biocon.2012.09.009

Sing, K.-W., Jusoh, W. F. A., Hashim, N. R., & Wilson, J.-J. (2016). Urban parks: Refuges for tropical butterflies in Southeast Asia? *Urban Ecosyst*. https://doi.org/10.1007/s11252-016-0542-4
Skorka, P., Lenda, M., Moron, D., Kalarus, K., & Tryjanowski, P. (2013). Factors affecting road mortality and the suitability of road verges for butterflies. *Biological Conservation, 159*, 148–157. https://doi.org/10.1016/j.biocon.2012.12.028
Soga, M., & Koike, S. (2013). Patch isolation only matters for specialist butterflies but patch area affects both specialist and generalist species. *Journal of Forest Research, 18*, 270–278. https://doi.org/10.1007/s10310-012-0349-y
Su, Z., Li, X., Zhou, W., & Ouyang, Z. (2015). Effect of landscape pattern on insect species density within urban green spaces in Beijing, China. *PLoS ONE, 10*, 1–13. https://doi.org/10.1371/journal.pone.0119276
Tallamy, D. W., & Shropshire, K. J. (2009). Ranking lepidopteran use of native versus introduced plants. *Conservation Biology, 23*, 941–947. https://doi.org/10.1111/j.1523-1739.2009.01202.x
Tam, K. C., & Bonebrake, T. C. (2015). Butterfly diversity, habitat and vegetation usage in Hong Kong urban parks. *Urban Ecosystem*. https://doi.org/10.1007/s11252-015-0484-2
Tzoulas, K., & James, P. (2010). Peoples' use of, and concerns about, green space networks: A case study of Birchwood, Warrington New Town, UK. *Urban Forestry & Urban Greening, 9*, 121–128. https://doi.org/10.1016/j.ufug.2009.12.001
Weber, P. G., Preston, S., Dlugos, M. J., & Nelson, A. P. (2008). The effects of field mowing on adult butterfly assemblages in Central New York State. *Natural Areas Journal, 28*, 130–143. https://doi.org/10.3375/0885-8608(2008)28[130:teofmo]2.0.co;2
Yamaura, Y., Kawahara, T., Iida, S., & Ozaki, K. (2008). Relative importance of the area and shape of patches to the diversity of multiple taxa. *Conservation Biology, 22*, 1513–1522. https://doi.org/10.1111/j.1523-1739.2008.01024.x
Yücel, G. F. (2013). Integrating ecosytem landscapes in cityscape: Birds and butterflies. In: M. Özyavuz (Ed.), *Advances in landscape architecture*. InTech. https://doi.org/10.5772/55753

Zanariah Jasmani is a senior lecturer of landscape architecture at the Faculty of Built Environment and Surveying, Universiti Teknologi Malaysia (UTM). She received her Master in Land Resource Management from Universiti Putra Malaysia in 2011 and completed her PhD in Landscape Architecture and Planning at the University of Copenhagen in 2016. She involves in research related to urban landscape ecology, vertical greeneries and green space management. Currently, she is the Head of the Heritage Landscape Research Group in UTM. She also holds the position as coordinator for undergraduate final year projects and postgraduate coordinator for the landscape architecture program. She is a corporate member of the Institute of Landscape Architects Malaysia (ILAM) and a certified landscape architect (LAr).

Hasanuddin Lamit is a professor of landscape architecture at the Faculty of Built Environment and Surveying, Universiti Teknologi Malaysia (UTM). He received his Master and PhD from the University of Sheffield, UK. His research area is on urban design and environmental psychology, and currently supervising undergraduates and postgraduates research in the field of the built environment.

Cecil C. Konijnendijk van den Bosch is a professor of urban forestry at the University of British Columbia, Canada and Program Director for the UBC Master of Urban Forestry Leadership. He is also the Director at Nature-Based Solutions Institute (NBSI). He is actively involved in research on various aspects of nature-based solutions and urban forestry and urban greening; green space governance and policy.

Chapter 18
Malaysian Roadside Tree Species Selection Model in Urban Areas

Ramly Hasan and Noriah Othman

Abstract This chapter explains a Malaysian roadside tree species selection model for reference to local authorities and assists landscape practitioners to improve their practices in selecting roadside tree species. The objectives are to determine additional attributes in urban roadside plant species selection and to recommend additional attributes for the Malaysian roadside tree species selection model. The in-depth interview aims to describe the current practices in the selection of roadside tree species by selected Malaysian local authorities. The respondents of the survey questionnaire were 764 landscape practitioners who registered with the Institute of Landscape Architect Malaysia (ILAM) session 2016/2018. Additional attributes (namely location, space, trend, personal preference, theme, expertise, landscape policies, framework and plan, knowledge, experience and skill) were added to this model, making it fit for use in the Malaysian context. The recommended additional attributes for the Malaysian roadside tree species selection model were agreed upon by all 12 experts including senior landscape practitioners, foresters, horticulturists and arborists. This study expands the existing body of knowledge in the selection of relevant roadside plant species in urban areas perceived to be crucial for Malaysian landscape practitioners.

Keywords Roadside tree species · Model · Landscape practitioners · Attributes · Experts

R. Hasan (✉)
Department of Landscape Architecture, Faculty of Architecture and Ekistics, Universiti Malaysia Kelantan (UMK), 16300 Bachok, Kelantan Darul Naim, Malaysia
e-mail: ramly.h@umk.edu.my

N. Othman
Department of Landscape Architecture, Faculty of Architecture Planning and Surveying, Universiti Teknologi MARA (UiTM), 42300 Bandar Puncak Alam, Selangor Darul Ehsan, Malaysia

S. Maruthaveeran et al. (eds.), *Urban Forestry and Arboriculture in Malaysia*,
https://doi.org/10.1007/978-981-19-5418-4_18

Introduction

Trees originate in several land-living areas, extending from private parks to public gardens as well as roadside, commercial, residential and institutional areas. Roadside tree planting is one of the most popular strategies in environmental design for improving the outdoor climate in urbanized areas. Landscape practitioners might explain tree planting and the selection of plant species by listing ecological, communal, cultural and economic benefits. These profit through the expanse of urban tree planting and thus areas an indirect influence on tree species selection. In Malaysia, the highest statistics of public complaints pertaining to tree problems were recorded from 2014 to 2016; this has left the local authorities worried. The effects of the cumulative amount of communal complaints have forced local authorities to spend more than hundreds of thousands of Ringgit Malaysia in paying compensation (Yaman et al., 2011).

Refer to statistics on the complaints on the roadside tree from the public four chosen local authorities namely Kuala Lumpur City Hall (DBKL), Petaling Jaya City Council (MBPJ), Subang Jaya Municipal Subang Jaya Municipal Council (MPSJ) and Selayang Municipal Council (MPS) showed nine prominent roadside plants issues of fallen trees, brittle branches, visibility obstruction, dead leaves and debris on the road, heavy branches obstructing traffic, old and dead trees, debris obstructing drainage system, tree location near to houses and leaning tree trunk.

Certain roadside trees implanted in town areas are species with energetic growing patterns, brittle branches, and frail bifurcation and are vulnerable to disease. These problems occur due to some influences such as location, economic and communal factors, which are not considered during the practice in the selection of tree species (Miller, 1997). According to Nor Azah (2015), two noticeable problems in the practices of selecting roadside tree species among local authorities in Malaysia are the lack of expertise and experience to make an appropriate selection. Because of this, the total number of public complaints on roadside trees is increasing every year. However, the goal of this paper is to establish a Malaysian roadside tree species selection model as a reference for the local authorities. This model is to assist landscape practitioners to improve their practices in selecting roadside tree species. The objectives are to discover additional attributes in urban roadside tree species selection and to recommend additional attributes for the Malaysian roadside tree species selection model.

Roadside Tree Management in Urban Settings

Roadside planting management is about planning (Laurian, 2019); planting and maintaining, protecting and taking care of trees, forests (Olawuyi, 2019); green space (Bedi & Tripathi, 2020); the related resource for economic, environmental, social and public safety (Bettinger et al., 2016; Jones & Davies, 2017; Rafiuddin, 2011;

Shams et al., 2020; Wolf & Robbins, 2015). Roadside planting management includes the expansion of resident participation and cares for funds in the long-term management of tree planting, protection and maintenance programmes. Roadside planting management such as maintenance and conservation of unsafe trees can be realized efficiently through the improvement and operation of strategic tree management (Hong Kong Conservancy Association, 2009). This conservancy proposes a strategy to manage urban tree resources comprising planning and tree selection, tree planting, tree maintenance, tree assessment and risk management, and tree protection to make sure trees remain in a healthy condition. Management of trees falls under the authority of local authorities (Hasan et al., 2016; Tomlinson et al., 2015; National Landscape Guidelines, 2011). In organisations, these local authorities have departments or divisions to plan, implement and develop, maintain and regularly control the landscape in town areas (Hisyam, 2014). The Malaysia Plan 11 is Malaysia's five-year development plan towards realizing the aim of vision 2020. These local authorities are required by Malaysia Plan 11 to plant half a million of trees in urban areas. However, tree planting is only one of several programmes. Trees planted previously should not be neglected and left unattended as they should be looked after and maintained to reach maturity while posing minimum potential to be hazardous to public safety and its properties. Identifying hazard trees and taking corrective measures can protect property and save human lives (Fig. 18.1).

Fig. 18.1 *Swietenia macrophylla* (Broad-leaved Mahogany) are felled at the Perak Road, Georgetown, Penang, due to strong winds

Tree Species Selection Model

The five main studies cited in the evolution of tree selection; Street Tree Planning Framework (Amir & Misgav, 1990), Tree Selection Model (Miller, 1997), Tree Selection Process, Decision Support System (Kirnbauer et al., 2009), TreeSelect Modules (Friederike, 2011) and Tree Species Selection Model (Roy, 2014). From the lens of this model, the researcher focused on the three important factors (1) site factors namely environmental constraint and cultural constraint; (2) economic factors, and (3) social factors that influence the selection of tree species. This is to establish their correlation with the process of selecting roadside tree species. The selection is based on the attributes suggested by Roy's (2014) Street Tree Species Selection Model that are deemed suitable and appropriate for the Malaysian context. Roy (2014) in his study revised and adjusted the model according to the situation in Australia. To come up with the appropriate tree species selection in Australia, additional factors of institutional factors, tree characteristics and existing tree population, were brought into Roy's (2014) model. However, since Australia is not a tropical country like Malaysia, therefore the process of selecting tree species will be different in terms of the type of tree species, location, space, education and culture of the people.

Important Factors Influencing Roadside Tree Species Selection

Site factors can be divided into two constraints of (1a) environmental constraints and (1b) cultural constraints. The word environmental is used to define plant selection attributes that are related to the interaction of the plants with their physical surroundings. Environmental conditions were generally explained in the literature as environmental constraints or environmental limitations, producing the list of appropriate tree species (Chantalakhana & Skunmun, 2015; Cunningham et al., 2016; Guneroglu et al., 2019; Swoczyna et al., 2015). The term *culture* is used here to explain tree selection attributes that are related to the interactions of trees with the elements affecting tree species selection. Cultural attributes were mostly explained in the literature as cultural constraints or cultural limitations assuming the list of appropriate tree species (Jim, 2013; Lo & Jim, 2015; McPherson et al., 2011). The costs of tree planting, tree maintenance and tree removal are the further considerations in tree selection. As explained by Miller (1997), tree planting rates encompass several economic factors for consideration. For example, tree species for planting may become expensive in the long run because of high care costs. Tree planting can also be more costly if matured trees are to be planted in industrial areas or to avoid early tree mortality through vandalism. Social factors are things that affect human lifestyle and property. Social factors in tree selection have been addressed in many works of literature (Avolio et al., 2015; Bourne & Conway, 2014; Ko et al., 2015;

Koeser et al., 2014). According to Miller (1997), social attributes refer to aesthetic or design, negative externalities and functional utility (Table 18.1).

Method

The in-depth interview aims to describe the current practices in the choice of roadside plants species at selected Malaysian local authorities. A qualitative method was employed to collect in-depth knowledge of the process. In the context of the present research, social science refers to the decision-making process among practitioners at local authorities upon selecting roadside trees for urban environments. Any selection practices are regarded as choice-making and understood as problem-solving activities. The interview which was conducted face-to-face to allow personal communication makes it possible to gather more information on the process in the range of roadside plant species. According to Bourhill and Rancourt (2020) and Kumar (2015), interview questions can either be open or close-ended. While the answers for close-ended are chosen from those given by the interviewer depending on which are deemed most appropriate, open-ended questions are answered freely by respondents. This study specifically incorporated open-ended questions to investigate the method of the application of different attributes in roadside plant selection. The respondents of the survey questionnaire were landscape practitioners who registered with the Institute of Landscape Architect Malaysia (ILAM). There were 764 landscape architects in the directory of members of the Institute of Landscape Architects Malaysia (ILAM) batch 2016/2018. The record of the landscape practitioners in the ILAM directory was categorized according to the sequence of membership numbers. The respondents of the appraisal were carefully chosen based on the sequence in the list. The survey used simple random sampling techniques, starting with the first landscape practitioners in the list, followed by the second landscape practitioners and so on. In the case where the researcher failed to reach the landscape practitioners in the sequence, the choice thus enthused the next respondents on the list. The survey questionnaire was distributed by hand (468 sets), mail (149 sets) and email (147 sets). Respondents' addresses were confirmed through phone calls (twice a week). However, six months after the distribution of the questionnaire, only 375 respondents responded. Due to time constraints, the researcher had to proceed with data analysis, which took three months. In total, only 49% of the respondents participated in this survey questionnaire (Fig. 18.2).

Results

Table 18.2 indicates the four additional factors and eleven additional attributes mentioned in the interview. Four factors along with their respective attributes were mentioned including limiting constraints (location and space) as well as social

Table 18.1 Existing factors and attributes influencing roadside tree species selection

Factors	Attribute	Description
1. Site		
1a. Environmental constraints	Physiographic Climatic Biologic aspects Edaphic	Landform or topography Weather condition Pests and diseases Soil type
1b. Cultural constraints	Utilities Structures Surface cover Pollutions	Infrastructure Urban structures Hardscape material Pollution from chemicals material
2. Economic	Tree planting Tree maintenance Tree removal Management cost Infrastructure repair cost Insurance payout	Tree, planting preparation, transportation, establishing costs, worker costs Tree cutting, pest controlling, irrigation, leaf litter deduction, worker costs Transport, labour costs Labour cost, related to management cost Road, facilities, utilities, infrastructure repair cost Compensation
3. Social	Aesthetic/Design Aspects Negative externalities Functional utility	Leaves, flower, seed stands Tree form, tree size, tree arrangement, Hazard potential and vandalism potential Crime prevention through planting design

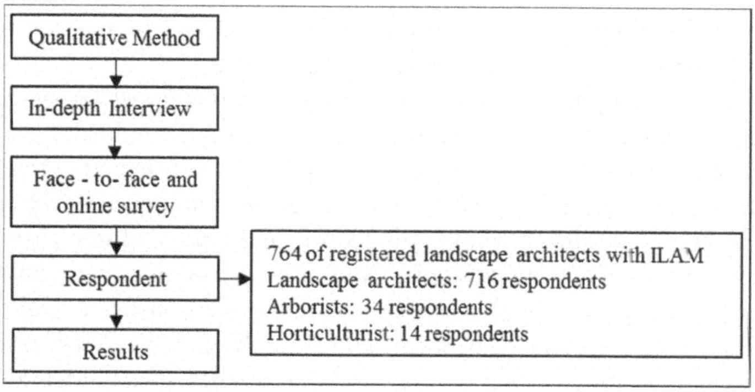

Fig. 18.2 Flowchart of research method

(trending, personal preference, themes and expertise), institutional (landscape policies, framework and plan) and education factors (knowledge, experience and skill). Currently, the development of buildings and construction of highways (e.g. road widening, and the construction of facilities and utilities) have resulted in limited location and space for roadside tree planting to become a barrier to the selection of roadside trees. Hence, the selection of tree species to be planted in a limited space along the roadside usually includes small trees with moderate branches so drivers' views will not be obstructed. The location (near building and other facilities, for instance) of a tree planting is also one attribute to be considered when selecting roadside tree species. Four additional attributes (including trend, personal preference, theme and expertise) of social factors were also mentioned in the interview. The current trend in the choice of roadside plant species generally depends on the local authorities themselves. A city with a beautiful landscape and flowering trees which add beauty to the roadside basically sets a good example for other cities. This explains why many landscape architects choose to plant flowering tree species along the roadside so a good ambiance to reduce stress can be created for road users. In fact, one of the missions of the National Landscape Department requires local authorities to include gardens in the city- this is to create a sense of place. Thus, trends in the selection of roadside tree species are not only important to beautify, but also to instill more value in the roadside. Apart from these factors and their related attributes, personal preference, expertise and themes were mentioned by the interviewees to be additional attributes involved in roadside tree species selection. Upon decision-making in roadside tree selection, senior officers and the top management, other than landscape architects and arborists, are also the responsible authority. Nonetheless, experts such as certified arborists are noted to be able to provide relevant advice and recommendations during a decision-making process. The themes of different roads (including primary, secondary, tertiary, highway and roundabout) were also noted to influence the decision-making of inappropriate tree species selection. Roadside tree species to be selected most importantly is not to obstruct drivers' view, in any way.

Also, the benefits of roadside trees to the public and the theme of each road were observed to determine the choice of plants to be planted.

In the interview, additional attributes making up institutional factors were also mentioned. According to the interviewees, it is necessary for landscape policies, framework and landscape plan to be taken into account when selecting roadside tree species. While landscape policies suggest the selection of appropriate trees requiring low maintenance, framework and plan (including guidelines and standard operation plan) aim to provide a better landscape, especially in roadside areas. In making a decision pertaining to the selection of roadside trees, decision-makers are to follow the framework and plan produced by local authorities. Nevertheless, it is to be pointed out that certain local authorities were observed to have their own guidelines to guide tree species selection. Apart from institutional factors, the interviewees also claimed the education factor is important when selecting roadside tree species. The role of personal preference in a decision-making process should include the three important attributes of knowledge, experience and skill: persons in charge of the choice of plant species should not only be knowledgeable in tree species as well as roadside tree maintenance and management but also have considerable experience and skill in taking care of trees.

Discussion

Framework and plan are the basic supporting structures or rules of the selection of roadside trees (Steenberg et al., 2017). Tree species selection in local authorities is guided by a tree management framework and plan including the guidelines provided by the National Landscape Department. Other than that, several books were also found to be used as a reference by local authorities upon selecting roadside trees. Certain local authorities however have their own specific guideline and plan to facilitate the selection of roadside trees. Potential hazardous trees such as *Pterocarpus indicus* (Angsana) and *Khaya senegalensis* (Khaya) are not encouraged to be planted at the roadside. This is because the trees have a wide-spreading crown and little branches that are quickly broken. The lateral roots of *Pterocarpus indicus* (Angsana) potentially damage sidewalks and cause accidents to people (Syahbudin et al., 2018). The provision of landscape in local and structure plans in contrast are published by each local authority to guide tree management in specific areas. The selection of flowering trees and trees with beautiful shapes are two current trends in many landscape practices. This is because the public nowadays is more concerned with creating an attractive environment in roadside areas. As indicated by Cole (2014), flowering trees along the roadside provide many environmental and psychological advantages to drivers. In fact, past research has shown beautiful scenery to be able to effectively minimise crash rates and reduce stress and frustration among drivers. In the Malaysian context, the trend is one of the major attributes that affect the selection of roadside plant species. The planting of flowering trees at many roadsides in Malaysia is one instance proving the role of trends in tree species

Table 18.2 Additional factors and attributes reported in the interview

Factors	Attributes	Description	Number of respondents vouching
Limiting constraint	Location and space	Conditions that limit the roadside tree growth. Example: location near buildings, limited space	10 persons
Social	Trending, personal preference, themes and expertise	Change in the selection of tree species based on current trendy tree characteristics; Persons with a say in decision making; The need for the features of a street to be matched with suitable trees species; A qualified/professional	12 persons
Institutional	Landscape policies, framework and plan	Principle of action by National Landscape Department such as blueprint, approach and guidelines of the proposal of the roadside tree; Basic supporting structures or rules incorporated upon selecting roadside tree	11 persons
Education	Knowledge, experience and skill	Understanding of the facts and specific information pertaining to the choice of tree species; Substantial familiarity with the selection of plants species and tree problems; Skills involved in the choice of plant species and expertise in tree maintenance	12 persons

selection. According to Elaine (2017), the trend in tree planting among house developers involves quite a number of *Peltophorum pterocarpum* (Yellow flame) trees in roadside areas. Every year, from March until April, flowers from the trees can be observed along roads. Its golden-yellow flowers that congregate at the tree top like a beautifully crafted royal crown attract home buyers and increase the value of houses. Most of the people's love for flowering trees is because of the beauty they add to the roadside, creating a sense of place and enlivening the areas. This is also seen in other flowering trees, *Tabebuai rosea* (Tecoma) planted across the roadside in Shah Alam, for instance, which was indicated by Othman et al. (2015) to also receive positive feedback from the respondents who assert the light flowers to enhance the surrounding. The community was found to beyond enjoy their beauty as being near trees gives them a feeling of calm. According to Antonia (2015), the trend of planting flowering trees *Tabebuai rosea* (Tecoma) at the roadside is also seen along the roads in Kuching city including Tan Sri Datuk William Tan Road, Simpang Tiga Road, Central Road, Airport Road, Canna Road, Foochow Road, Stutong Road and Rock Road. The trend of planting flowering trees along the roadside is one way of promoting Kuching as a beautiful city with natural profusion. Rahman (2013) on contrary argues landscape practitioners to more consciously select appropriate species to be planted at the roadside. These species are usually fast-growing trees, offering more money and a high wood value, while also being environment-friendly and multi-purpose. In making a decision related to the selection of roadside tree species, landscape policies play an important role in ensuring a successful landscape development in the urban fabric. The National Landscape Policy (NLP) is generated to produce a beautiful, comfortable and safe living environment as a prerequisite to increasing the comprehensive quality of human life (National Landscape Policy, 2011). The introduction of the National Landscape Policy (NLP) is evidence of the country's intent to create Malaysia as a Beautiful Garden Nation by 2020. Thus, it is important for roadside landscapes in national development to be balanced with green, well-maintained and attractive surroundings through the creation of special places. In selecting tree species, local authorities in Malaysia have taken into account three National strategies to fulfill NLP vision. The three strategies are (1) managing the Nation towards a higher income wealth (2) ensuring holistic and sustainable growth, and (3) focusing on the quality of human well-being. The selection of roadside tree species should be in line with the production of a conducive landscape environment and the creation of the identity of the respective landscape in order to benefit road users. In addition, the beauty of flowering trees added to the roadside can amplify road use among the locals and foreigners. The selection of roadside tree species is also associated with National Urbanisation Policy (NUP) and Environmental Policy (EP). The National Urbanisation Policy (NUP) stresses the development of identity and town image that is suitable for community responsibility and this can be interpreted through landscape establishment. Proposed landscape planting involving suitable tree species along the roadside will boost city's image and create a 'special' identity to represent Malaysia. Therefore, urban landscape design goes hand in hand with construction developments. The Environmental Policy (EP) on the contrary aims to achieve clean, rich conditions, secure and healthy for the present and future populations. It was prepared to realize the aims of sustainable development through optimal and successful means. Associated with the selection of roadside tree species by local authorities, tree stock should be in a safe and healthy condition to ensure sustainable urban environments for future generations. The National Landscape Department has published the second edition of Landscape Planning Guidelines (LPG) (2008) to guide landscape architects, especially those working with local authorities. The guidelines (Chap. 3, pages 22–34) apart from listing suitable tree species for the roadside also discuss other criteria such as planting distance and space requirements. The approvals of all landscape planting plans for landscape development are to follow the Landscape Planning Guidelines (LPG). At the local authority level, they have their own guidelines to administer the proposal of planting design for roadside areas. Tree species including *Acacia sp.* (Acacia), *Pterocarpus indicus* (Angsana), *Khaya senegalensis* (Khaya) and *Hura crepitans* (Sandbox Tree) are prohibited to be planted at the roadside, due to their inappropriate characteristics of fast growth, little branches, thorny and gummy. The suggested roadside trees such as *Hopea odorata* (Merawan Siput Jantan) and *Cinnamomum iners* (Wild Cinnamon) all have strong resilience, moderate branches and medium growth speed.

Personal preference is also one of the attributes influencing the selection of roadside tree species in Malaysian local authorities. Landscape architects and the top management of local authorities are authorised in making decisions regarding the approval of landscape planting plans. In addition, these persons are also ordered to make decisions pertaining to roadside tree species selection. Personal preference in tree species was observed to be able to affect tree growth in roadside areas. The consideration of the notion of 'right trees at the right place' is hence important when selecting roadside trees. To guarantee the selection of appropriate trees for the right location, decision-making is therefore to be conducted by individuals with considerable experience and expertise in tree management. Interference of individuals with a non-landscape background upon the choice of plant species can negatively affect plant growth and consequently threaten the public and its property. In the Malaysian context, the amount of compensation exceeding RM 5,000 a year was reported by officers from the Kuala Lumpur City Hall (DBKL) to include the compensation for human injuries, transport damages, roof damage and damage to other facilities. Themes are another important attribute that affects the selection of tree species. There are various types of trees that are appropriate for different road types including primary road, protocol road and secondary road. These different roads have different widths. The spaces for tree planting however vary according to the type of roads.

Developing Malaysian Urban Roadside Tree Species Selection Model

The development of the model concerns two important processes. Firstly, define the roadside tree species problems based on public complaints. Secondly, determine the similarity and additional attributes in urban roadside tree species selection. Next is developed a Malaysian roadside tree species selection model based on additional factors and attributes. A Malaysian Roadside Tree Species Selection Model was developed to provide a comprehensive reference to Malaysian local authorities and landscape practitioners. Additional attributes (namely location, space, trend, personal preference, theme, expertise, landscape policies, framework and plan, knowledge, experience and skill) were added to this model, making it fit for use in the Malaysian context. The recommended additional attributes for the Malaysian roadside tree species selection model were agreed upon by all 12 experts. The 12 experts include senior landscape practitioners, foresters, horticulturists and arborists. In this research, an expert or specialist is defined as an individual with an active experience of being employed in the area of attentiveness for some years and holding a position of responsibility. This model can be used to guide landscape practitioners in Malaysia in selecting tree species for roadside tree planting. The person who makes the decision should be knowledgeable, proficient in tree species, experienced and has attended relevant courses such as conferences and workshops. In addition, they also need to pass the tests administered by ILAM and the arborist association. Other than that, social, economic, and institutional factors as well as tree characteristics should be considered when selecting tree species (Fig. 18.3).

Conclusion

In this study, which expands the present body of knowledge in the area of the choice of proper and suitable roadside plants species in urban areas, the goal is to establish a Malaysian roadside tree species selection model, perceived to be crucial for the Malaysian landscape practitioners (such as landscape consultants, landscape contractors, landscape developers and local authorities). This is because the developed model is able to guide landscape practitioners to the right selection of roadside trees. The proposed model reinforces 11 additional attributes of location, space, trend, personal preference, themes, expertise, landscape policies, framework and plan, knowledge, experience and skill that should be taken into account in any decision-making in roadside tree species selection. The proposed roadside tree species selection model is expected to offer the following benefits to the listed landscape practitioners. Landscape practitioners in Malaysia can use the model as a guide for proposed planting design and making a proper tree selection. In addition, students and the public also can use the model in selecting tree species.

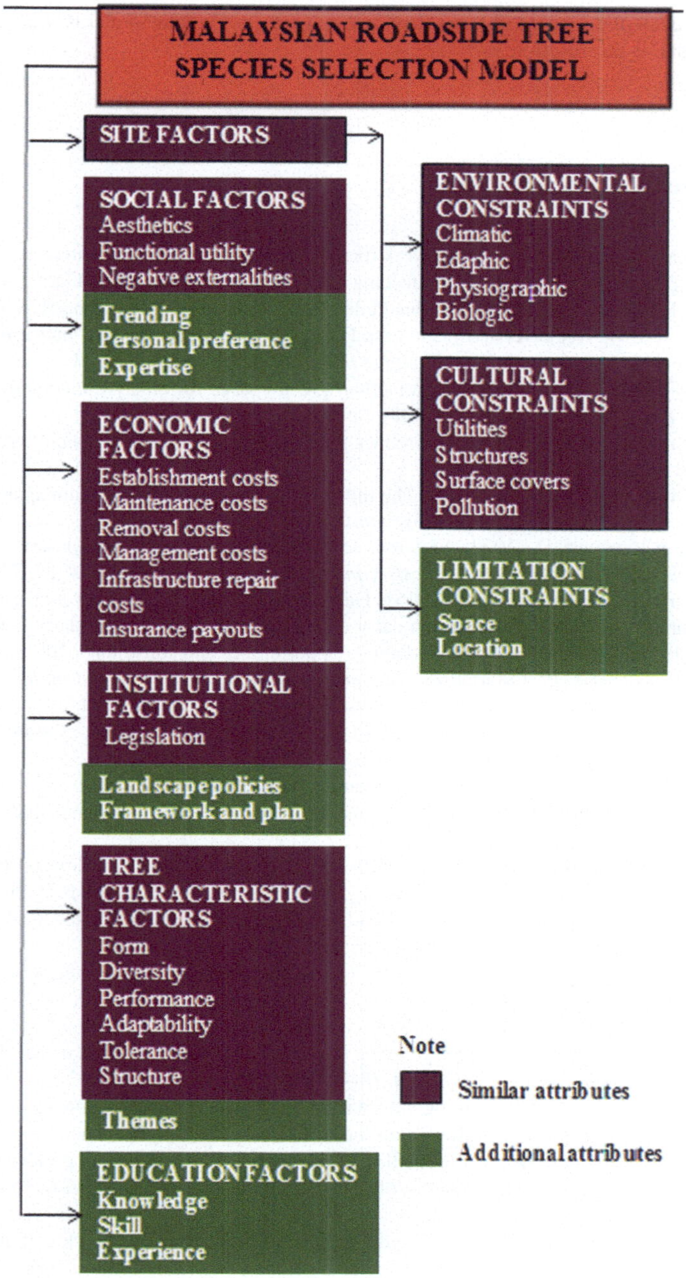

Fig. 18.3 Model for roadside tree species selection in Malaysia

Acknowledgements I would like to extend my gratitude to Mybrain15 and Research Incentive Grant for funding this research

References

Amir, S., & Misgav, A. (1990). A framework for street tree planning in urban areas in.

Antonia, C. (2015). Proliferation of pink tecoma colours the city. Borneo Post Online Aronson.

Avolio, M., Pataki, D. E., Gillespie, T., Jenerette, G. D., McCarthy, H. R., Pincetl, S., & Weller-Clarke, L. (2015). Tree diversity in southern California's urban forest: The interacting roles of social and environmental variables. *Frontiers in Ecology and Evolution, 3*, 73.

Bedi, P., & Tripathi, N. G. (2020). Smart urban green spaces for smart Chandigarh. In *Smart environment for smart cities* (pp. 111–148). Springer.

Bettinger, P., Boston, K., Siry, J. P., & Grebner, D. L. (2016). *Forest management and planning*. Academic press.

Bourne, K. S., & Conway, T. M. (2014). The influence of land use type and municipal context on urban tree species diversity. *Urban Ecosystems, 17*(1), 329–348.

Bourhill, T., & Rancourt, D. (2020). Vicarious learning through Informational Interviews: A new experiential learning tool. *Papers on Postsecondary Learning and Teaching, 4*, 32–38.

Chantalakhana, C., & Skunmun, P. (2015). Dairy development in Thailand and a case study onenvironmental impacts of peri-urban dairy colonies. Part I. Smallholder dairy development. Smallholder dairy production and marketing—Opportunities and constraints, 105.

Cole, D. F. (2014). *The effect of roadside elements on driver behavior and run-off the-road crash severity* (Unpublished Master Thesis). University of Massachusetts Amherst.

Cunningham, H. R., Rissler, L. J., Buckley, L. B., & Urban, M. C. (2016). Abiotic and biotic constraints across reptile and amphibian ranges. *Ecography, 39*(1), 1–8.

Elaine, Y. (2017). Add a 'royal touch' to your garden with Yellow flame tree.

Executive Management Team. (2010). Tree management guidelines for council managed land. Version 1.2.

Guneroglu, N., Bekar, M., & Sahin, E. K. (2019). Plant selection for roadside design:"The view of landscape architects." *Environmental Science and Pollution Research, 26*(33), 34430–34439.

Hasan, R., Othman, N., & Ismail, F. (2016). Roadside tree management in selected local authorities for public safety. *Procedia-Social and Behavioral Sciences, 234*, 218–227.

Hong Kong Conservancy Association. (2009). Urban tree management in Hong Kong problems and recommendations (pp. 1–21).

Hisyam, K. (2014). Introduction to sustainable community based rural tourism.

Jim, C. Y. (2013). Sustainable urban greening strategies for compact cities in developing and developed economies. *Urban Ecosystems, 16*(4), 741–761.

Jones, N., & Davies, C. (2017). Linking the environmental, social and economic aspects of urban forestry and green infrastructure. In *The urban forest* (pp. 305–313). Springer.

Kirnbauer, M. C., Kenney, W. A., Churchill, C. J., & Baetz, B. W. (2009). A prototype decision support system for sustainable urban tree planting programs. *Urban Forestry & Urban Greening, 8*, 3–19.

Ko, Y., Lee, J. H., McPherson, E. G., & Roman, L. A. (2015). Factors affecting longterm mortality of residential shade trees: Evidence from Sacramento. *California. Urban Forestry & Urban Greening, 14*(3), 500–507.

Koeser, A. K., Gilman, E. F., Paz, M., & Harchick, C. (2014). Factors influencing urban tree planting program growth and survival in Florida, United States. *Urban Forestry & Urban Greening, 13*(4), 655–661.

Kumar, R. (2015). The forest resources of Malaysia: Their economics and development.

Laurian, L. (2019). Planning for street trees and human–nature relations: Lessons from 600 years of street tree planting in Paris. *Journal of Planning History, 18*(4), 282–310.
Lo, A. Y., & Jim, C. Y. (2015). Protest response and willingness to pay for culturally significant urban trees: Implications for Contingent Valuation Method. *Ecological Economics, 114*, 58–66.
McPherson, E. G., Simpson, J. R., Xiao, Q., & Wu, C. (2011). Million trees Los Angeles canopy cover and benefit assessment. *Landscape and Urban Planning, 99*(1), 40–50.
Miller, R. W. (1997). *Urban forestry: Planning and managing urban greenspaces* (2nd ed.). Prentice Hall.
Nor Azah, A. A. (2015). Ringkasan keperluan latihan pelanggan. institut latihan kesejahteraan bandar, perumahan dan kerajaan tempatan (IKPKT).
National Landscape Policy. (2011). Jabatan Landskap Negara (JLN). Kementerian Perumahan dan Kerajaan Tempatan.
Othman, N., Isa, M. M., Mohamed, N., & Hasan, R. (2015). Street planting compositions: The public and expert perspectives. *Procedia-Social and Behavioral Sciences, 170*, 350–358.
Olawuyi, E. B. (2019). Urban tree planting: Implication on sustainable forest management in Idi-Shin, Oyo State. *International Journal of Environment, Agriculture and Biotechnology, 4*(4).
Rafiuddin, R. (2011). *Urban tree management: Towards best practices and application* (Unpublished Master of Science thesis). Case study: Kuala Lumpur.
Rahman, P. M. M. (2013). Species composition trend of roadside tree plantation program: A case study of Rangpur district of Bangladesh.
Roy, S. (2014). *Factor influencing Australian local governments' street tree species selection* (Doctorate dissertation). Griffith University, Brisbane, Australia.
Syahbudin, A., Syaufina, R. L., Yudhistira, R., & Sadono, R. (2018). Tree architecture models, canopy maintenance, and associated root problems of angsana (Pterocarpus indicus Willd.) in the urban trees of Yogyakarta. In IOP Conference Series: Earth and Environmental Science (Vol. 203, No. 1, p. 012010). IOP Publishing.
Steenberg, J. W., Millward, A. A., Nowak, D. J., & Robinson, P. J. (2017). A conceptual framework of urban forest ecosystem vulnerability. *Environmental Reviews, 25*(1), 115–126.
Swoczyna, T., Kalaji, H. M., Pietkiewicz, S., & Borowski, J. (2015). Ability of various tree species to acclimation in urban environments probed with the JIP test. *Urban Forestry & Urban Greening, 14*(3), 544–553.
Shams, Z. I., Shahid, M., Nadeem, Z., Naz, S., Raheel, D., Aftab, D., & Roomi, M. S. (2020). Town socio-economic status and road width determine street tree density and diversity in Karachi, Pakistan. *Urban Forestry & Urban Greening, 47*, 126473.
Tomlinson, I., Potter, C., & Bayliss, H. (2015). Managing tree pests and diseases in urban settings: The case of Oak Processionary Moth in London, 2006–2012. *Urban Forestry & Urban Greening, 14*(2), 286–292.
Wolf, K. L., & Robbins, A. S. (2015). Metro nature, environmental health, and economic value. *Environmental Health Perspectives, 123*(5), 390.
Yaman, A. R., Jamil, M. Z., & Yaakob, R. (2011). Manual pengurusan risiko pokok.

Ramly Hasan completed his Ph.D. in Built Environment at the Faculty of Architecture, Planning and Surveying, Universiti Teknologi MARA (UiTM) in August 2019. He received his MSc. in Heritage and Conservation Management at the Faculty of Architecture, Planning and Surveying, UiTM. Currently, he is a senior lecturer at the Department of Landscape Architecture, Faculty of Architecture and Ekistics, Universiti Malaysia Kelantan (UMK). Ramly is actively involved in research related to the built environment focusing on roadside tree species selection and tree management. He is a certified landscape architect (LAr) and registered with the Institute of Landscape Architects Malaysia (ILAM).

Noriah Othman is an Associate Professor at the Centre of Studies for Landscape Architecture, Faculty of Architecture, Planning and Surveying at Universiti Teknologi MARA (UiTM), Puncak Alam, Selangor Darul Ehsan and has been lecturing at UiTM since 1990. She obtained her Bachelor's and Master's degrees from Universiti Pertanian Malaysia (UPM) and her doctorate degree in Landscape Horticulture at UiTM. Noriah is actively involved in research in the field of environment-behaviour studies and has presented papers at various conferences and contributed several articles in research journals. Being a member of the working committee for plant standards in SIRIM, she has been regularly invited to speak on topics related to landscape horticulture by local authorities and other agencies and is occasionally involved as a horticulturist in several landscape master plans by local authorities and the National Landscape Department. She is a certified arborist (CA) with a Tree Risk Assessment Qualification (TRAQ). At present, she is also holding the position of Head of the Quality Unit at FSPU and Lead Auditor of ISO 9001:2015 at UiTM.

Chapter 19
Can Urban Forests Help to Manage Academic Stress Among Undergraduate University Students?

Keeren Sundara Rajoo, Daljit Singh Karam, and Arifin Abdu

Abstract Academic stress is a major concern affecting students worldwide. This phenomenon is especially serious in a conservative society like Malaysia, where mental health issues are a taboo topic due to superstitious beliefs and cultural insensitivities. Although the Ministry of Health Malaysia reported an exponential increase in mental health problems among Malaysian students, there is still a lack of effort in addressing this issue. Urban forests can serve as a solution to this growing problem since forest therapy has been found to be able to reduce stress-related diseases such as anxiety and elevated blood pressure. Despite its growing popularity across the globe, forest therapy remains a relatively novel concept in Malaysia. A survey found that up to 95% of Malaysian university students in public and private universities across Malaysia were unaware of forest therapy. However, after learning of the benefits, the majority of these students were eager to participate in a forest therapy program organised by Universiti Putra Malaysia. To determine whether forest therapy had the potential to alleviate academic stress among students, a forest therapy program was developed based on the expectations of the students. A total of 29 students participated in a half-day forest therapy program at Ayer Hitam Forest Reserve in Selangor; It was discovered that there was a sustained reduction in the blood pressure of students after having undergone forest therapy, and the effects were maintained for up to 5 days. The students also confirmed this result, as they revealed that they did feel less stressed after the forest therapy program. In conclusion, forest therapy

K. S. Rajoo (✉)
Department of Forestry Science, Universiti Putra Malaysia Bintulu Sarawak Campus (UPMKB), Bintulu, Sarawak, Malaysia
e-mail: keeren.rajoo@upm.edu.my

D. S. Karam
Department of Land Management, Universiti Putra Malaysia (UPM), Serdang, Selangor Darul Ehsan, Malaysia

A. Abdu
Department of Forestry Science and Biodiversity, Universiti Putra Malaysia (UPM), Serdang, Selangor Darul Ehsan, Malaysia

K. S. Rajoo
Institute of Ecosystem Science Borneo, Universiti Putra Malaysia Bintulu Sarawak Campus (UPMKB), Bintulu, Malaysia

S. Maruthaveeran et al. (eds.), *Urban Forestry and Arboriculture in Malaysia*,
https://doi.org/10.1007/978-981-19-5418-4_19

has the capability to be a form of preventive medicine in reducing chronic stress, especially among students.

Keywords Forest therapy · Academic stress · Preventive medicine · Mental health · Blood pressure

Academic Stress in Malaysia

Present-day students have been found to be more prone to experiencing academic stress and as a result, are susceptible to stress-related diseases, due to exceedingly high academic expectations (Leppink et al., 2016). Stress is a major cause of cardiovascular-related diseases such as elevated blood pressure. Moreover, stress also leads to mental health issues, namely anxiety and depression (Song et al., 2017). Academic stress is beyond a mere "student problem", but rather a national concern. This is due to the issue affecting an entire generation, whereby having an entire generation plagued with physical and mental illnesses would negatively impact the societal development of the nation (Nur et al., 2017).

This phenomenon is especially serious in a conservative society like Malaysia, where mental health issues are a taboo topic due to superstitious beliefs and cultural insensitivities (Boon, 2016). In Malaysia, mental diseases are viewed by some Malaysians as either a chastisement from God, instigated by supernatural beings or attention-seeking behaviour (Bernama, 2016). Thus, it is not surprising that most Malaysians hide their mental health issues in order to avoid senseless persecution from society. The Deputy President of the Malaysian Mental Health Association (MMHA) perfectly sums up the problem, as he states that people suffering from mental illnesses, "would get judged, ridiculed or criticized" (Boon, 2016).

Nevertheless, Malaysians are experiencing severe mental health deterioration. An international research found that up to 70% of Malaysian employees were experiencing stress-related diseases (FMT, 2017). This issue is also being experienced by students, as the Malaysian Health Ministry discovered a significant increase in mental issues among students, from 1 in 10 students in 2011 to 1 in 5 in just four years later (Bernama, 2016). There is also evidence that these young adults are having elevated blood pressure which is highly unusual for individuals in their age group (Teh et al., 2015).

Students often fail to identify and address their chronic stress. Some even perceive it to be normal, while others fear societal stigma (Ramli et al., 2018). Thus, there needs to be a form of self-care for these students to rely on, that can be efficiently implemented in conservative societies like Malaysia. Forest therapy has the potential to be used in this regard, as it has been discovered that it has the capability to reduce elevated blood pressure and anxiety (Song et al., 2017).

Forest Therapy in Malaysia

Forests have always played an important part in Malaysia, serving as the life support of the country and its people. Not only does it ensure environmental stability and ecological balance, but it is also a very important source of income. Moreover, forests are spiritually vital to some Malaysian societies, especially the indigenous people of Malaysia. There are no Malaysians who have a deeper connection with forests than the indigenous people or locally known as the *Orang Asli*. Historically, the first contact of the *Orang Asli* was with Indian traders in the first millennium B.C., where they traded forest products for salt, cloth and iron tools. To this day, there are *Orang Asli* communities that still depend on forests as a source of livelihood. Accounting for only 0.5% of the Malaysian population, *Orang Asli* has an extremely strong spiritual bond with forests, still forming the basis for the culture and religion of some tribes to this day.

The *Orang Asli* community has for generations believed in the practice of *Mandi Embun*, which loosely translates to "bathing in the forest dew". They believe that walking barefooted in forests would lead to better physical and spiritual health (Ayob, 2018). *Mandi Embun* will usually be done during the break of dawn, as the "spiritual energy" of the forest would be transferred to the forest dew via evaporation, and this "spiritual energy" would be transferred to anyone who comes in contact with the forest dew (Ayob, 2018). In other words, the "energy" of the forest would be transferred to individuals practicing *Mandi Embun*. It goes without saying that this belief does not hold any scientific merit. However, it is safe to assume that the Orang Asli community was in actuality practicing a form of Forest Therapy, just heavily hidden under the veil of superstition and animistic beliefs. One could even say that *Mandi Embum* is a form of alternative medicine.

Today, *Mandi Embun* is practiced by some Malaysians as a leisure activity. Some forest guides even promote the practice by stating that it would improve the immune system (Irshad, 2017), although as of today, no scientific study has been conducted on the subject in Malaysia. Moreover, forest therapy is still a very novel concept in Malaysia. Nevertheless, studies have shown that forest therapy is effective in addressing stress-related issues, thus it has the potential to be an excellent self-care tool for Malaysian students who are suffering from chronic stress.

Students' Perception of Forest Therapy and Their Stress Levels

A survey involving 412 students aged 19–24 years (21.73 ± 2.33 years) was conducted to determine the perception of Malaysian university students regarding forest therapy. There are over seven hundred thousand Malaysian university students (Nur et al., 2017). Thus, to obtain a margin error of 5% at 95% confidence level, the sample size should be 384. Therefore, the number of respondents for this survey is

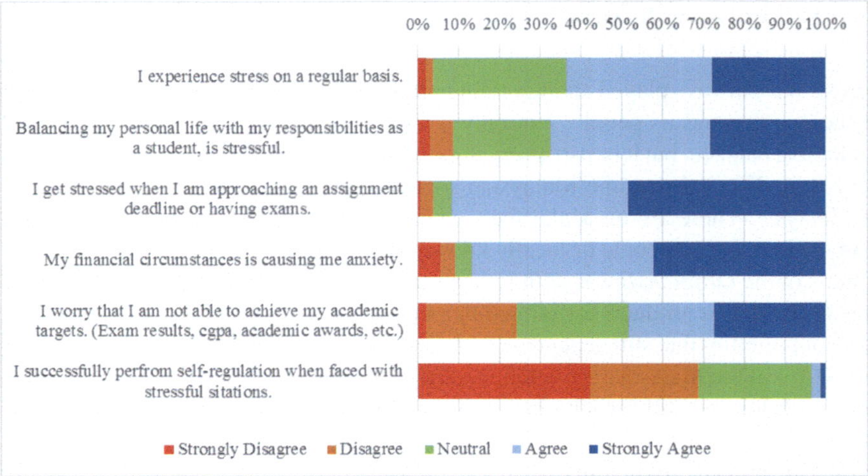

Fig. 19.1 Students' academic stress self-assessment ($N = 412$)

sufficient to be representative of the population. Besides evaluating their perception regarding forest therapy, the students' views on their personal stress levels were also evaluated. The respondents for the survey were Malaysian university students from universities across Malaysia.

As expected, the data revealed that most students were experiencing stress (Fig. 19.1). The students stated that their stress levels increased depending on the situation, for instance, 91.7% of the respondents admitted to feeling overwhelmed when deadlines for tests or assignments were imminent. Another interesting finding was that most students (68.7%) did not perform self-regulation when experiencing stressful situations. Therefore, the students need to be exposed to an effective self-regulation method to manage their academic stress.

As expected, a large majority of students did not know about forest therapy (95.2%) but were interested to participate once they knew of its benefits (Fig. 19.2A and 19.2B). Nevertheless, almost half of the respondents (47.3%) were disinterested in participating in forest therapy, for various reasons such as religious beliefs, financial constraints and societal stigma (Fig. 19.2C). Therefore, these concerns need to be addressed when propagating forest therapy to university students. Students were most interested in a half-day weekend program, and were interested in sensory enjoyment, sleeping in hammocks, river soaking and forest strolling.

Effects of Forest Therapy on Academic Stress Levels

To determine whether forest therapy had the potential to alleviate academic stress among students, a forest therapy program was developed based on the survey

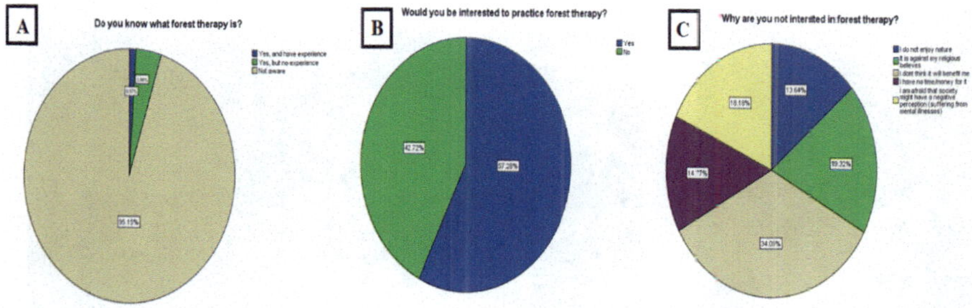

Fig. 19.2 Students' perception regarding forest therapy ($N = 412$), **A**: Students' awareness regarding forest therapy, **B**: Students' interest to practice forest therapy, **C**: Reasons for students' lack of interest

responses of the students. A total of 29 students from Universiti Putra Malaysia who believed that they were suffering from stress, participated in the forest therapy program. Fig. 19.3 shows a collage of the forest therapy in action. The program's purpose was to evaluate the sustained effects a forest therapy program had on students' blood pressure. Systolic (SBP) and diastolic (DBP) blood pressure were used as measurement indices and were measured in fixed intervals before and after the forest therapy program. Measurements were recorded three times before breakfast (7:15 a.m.), before lunch (12:00 p.m.) and before dinner (5:30 p.m.) on each assessment day. This study was conducted at the Ayer Hitam Forest Reserve, the only secondary forest reserve located in the Klang Valley.

The baseline values for this study were the blood pressure values taken 2 days prior to the programme, since it reflected the participant's "normal" cardiovascular health (Table 19.1). The baseline mean SBP value before breakfast (119.48 ± 1.27 mmHg) did not exhibit any significant decrement on the day of the forest therapy program (120.34 ± 1.1 mmHg), but significantly decreased 3 days after (113.14 ± 1.33 mmHG), 5 days after (114.69 ± 0.68). The SBP readings 7 days after (122.66 ± 1.03) did not have any significant difference with the baseline values. When compared with the SBP baseline values before lunch (127.17 ± 1.16 mmHg), the SBP values were significantly lower during forest therapy (117.28 ± 1.32 mmHg), 3 days after (114.03 ± 1.37 mmHg) and 5 days after (117.45 ± 1.04 mmHg). The same trend was observed for the SBP readings taken before dinner, whereby the baseline values (126.1 ± 1.46 mmHg) were significantly higher when compared to the readings during forest therapy (117.34 ± 1.38 mmHg), 3 days after (115.62 ± 1.16 mmHg) and 5 days after (119.9 ± 0.914 mmHg). All the SBP readings taken 7 days after were not significantly different from the baseline values.

The DBP baseline value before breakfast (80.79 ± 0.75 mmHg) was significantly higher when compared to the values during forest therapy (79.03 ± 0.7 mmHg), 3 days after (76.45 ± 0.55 mmHg) and 5 days after (78.24 ± 0.47 mmHg) (Table 19.1). The same trend was observed for the DBP readings before lunch and dinner, whereby the baseline value before lunch (83.31 ± 0.938 mmHg) was significantly higher when compared to the readings during forest therapy (77.55 ± 0.762 mmHg),

Fig. 19.3 Activities performed during forest therapy

Table 19.1 Mean physiological measurements according to different assessment days and measurement times

Measurement time	Day	SBP (mmHg)		DBP (mmHg)		Pulse rate (bpm)	
		Mean ± SE	*p* value	Mean ± SE	*p* value	Mean ± SE	*p* value
Before breakfast	2 days before	119.48 ± 1.27	–	80.79 ± 0.75	–	85.48 ± 0.86	–
	Forest therapy day	120.34 ± 1.1	0.981	79.03 ± 0.7	0.311	86.1 ± 1.29	0.99
	3 days after	113.14 ± 1.33	0.001*	76.45 ± 0.55	0.0*	87.82 ± 1.29	0.71
	5 days after	114.69 ± 0.68	0.021*	78.24 ± 0.47	0.047*	87.48 ± 1.25	0.76
	7 days after	122.66 ± 1.03	0.253	80.86 ± 0.73	1.0	88.21 ± 1.23	0.48
Before lunch	2 days before	127.17 ± 1.16	–	83.31 ± 0.938	–	86.76 ± 1.34	–
	Forest therapy day	117.28 ± 1.32	0.0*	77.55 ± 0.762	0.0*	86.62 ± 1.43	1.0
	3 days after	114.03 ± 1.37	0.0*	78.1 ± 0.56	0.0*	85.86 ± 1.23	0.994
	5 days after	117.45 ± 1.04	0.0*	79.48 ± 0.38	0.003*	85.83 ± 1.27	0.993
	7 days after	126.97 ± 1.32	1.0	84.45 ± 0.86	0.805	85.83 ± 1.37	0.993
Before dinner	2 days before	126.1 ± 1.46	–	82.52 ± 0.812	–	87.45 ± 1.226	–
	Forest therapy day	117.34 ± 1.38	0.0*	77.52 ± 0.783	0.0*	85.59 ± 1.14	0.826
	3 days after	115.62 ± 1.16	0.0*	77.9 ± 0.453	0.0*	87.1 ± 1.426	1.0
	5 days after	119.9 ± 0.914	0.007*	79.89 ± 0.405	0.096	87.45 ± 1.209	1.0
	7 days after	127.79 ± 1.4	0.884	81.69 ± 1.082	0.936	87.66 ± 6.399	1.0
Mean	2 days before	124.36 ± 0.771	–	81.92 ± 0.501	–	86.40 ± 0.669	–
	Forest therapy day	117.49 ± 0.78	0.0*	77.78 ± 0.432	0.0*	86.66 ± 0.748	0.999
	3 days after	114.45 ± 0.720	0.0*	77.79 ± 0.312	0.0*	86.75 ± 0.772	0.997
	5 days after	118.13 ± 0.576	0.0*	79.39 ± 0.272	0.0*	86.68 ± 0.714	0.999
	7 days after	125.66 ± 0.757	0.72	82.07 ± 0.544	0.999	87.23 ± 0.733	0.929

SBP: systolic blood pressure, DBP: diastolic blood pressure, SE: Standard error
$N = 30$, mean ± SE, * $p < 0.05$ by ANOVA with Tukey post-hoc test

3 days after (78.1 ± 0.56 mmHg) and 5 days after (79.48 ± 0.38 mmHg). The DBP values before dinner were also significantly lower when compared to the baseline value (82.52 ± 0.812 mmHg), that is during forest therapy (77.52 ± 0.783 mmHg), 3 days after (77.9 ± 0.453 mmHg) and 5 days after (79.89 ± 0.405 mmHg). All DBP readings taken 7 days after the forest therapy program didn't exhibit any significant difference from the baseline values taken 2 days prior.

Moreover, the exit survey revealed that the students believed that the program reduced their stress levels. The majority (82.8%) also said that they plan to continue the practice in the future, some even said they would encourage their family and friends to participate.

The findings of this study were consistent with past studies (Park et al., 2007, 2009; Song et al., 2017; Tsunetsugu et al., 2013), but most of these studies tested forest therapy programs that were for several days and participated by working adults. Therefore, the findings of this study have shown that even a half-day weekend program can yield positive results, and the benefits can be experienced by students too.

Conclusion

University students are experiencing chronic levels of academic stress, but they are either unable or unwilling to manage it. This has led to severe deterioration in their physical and mental wellbeing. Therefore, a self-care method to manage the stress that is applicable in conservative societies is needed. This study revealed that forest therapy has the potential to be one such method. The study found that a half-day weekend program was sufficient in reducing elevated blood pressure for up to 5 days. The participants also agreed that they felt reduced stress once undergoing forest therapy. However, there is a large number of students who expressed reluctance in practicing forest therapy, mainly due to preconceived societal and religious beliefs. Therefore, these concerns need to be addressed when propagating forest therapy to university students.

References

Abdullah S., & Hassan S. (2004). Freedom of religion, apostasy and Islam (illustrated ed). Ashgate Publishing, Ltd. pp. 3, 124, 126, 132. ISBN 0–7546–3083–8.

Ayob, H. (2018). Amalan Mandi Embun. Retrieved from http://siasahpolitik.blogspot.com/2018/04/amalan-mandi-embun-pahlawan-melayu.html

Bernama. (2016). Mental health of Malaysian students' cause of worry: Health Ministry. Retrieved from https://www.nst.com.my/news/2016/09/172683/mental-health-malaysian-students-cause-worry-health-ministry

Boon, S. W. (2016) The Mental State of Malaysia. Retrieved from https://www.esquire.my/lifestyle/health/feature/The-Mental-State-Of-Malaysia

Centers for Disease Control and Prevention—CDC (2018). Blood Pressure Levels. Retrieved from www.cdc.gov/bloodpressure/measure.htm

Daljit, S. K. (2017). Terapi hutan untuk pesakit psikiatrik. Retried from http://www.upm.edu.my/article/forest_therapy_for_psychiatric_patients-32787

DeLuca, J. S., & Yanos, P. T. (2016). Managing the terror of a dangerous world: Political attitudes as predictors of mental health stigma. *International Journal of Social Psychiatry, 62*(1), 21–30.

FreeMalaysiaToday (FMT). (2017). Poor health among M'sians due to work stress. Retrieved from http://www.freemalaysiatoday.com/category/nation/2017/11/18/poor-health-among-msians-due-to-work-stress-says-survey/

Hansen, M. M., Jones, R., & Tocchini, K. (2017). Shinrin-Yoku (Forest Bathing) and Nature Therapy: A State-of-the-Art Review. *International Journal of Environmental Research and Public Health., 14*(8), 851. https://doi.org/10.3390/ijerph14080851.PMID28788101

Harvey, G., & Wallis, R. J. (2007). *Historical Dictionary of Shamanism.* Scarecrow Press.

Hayward, D. (2016). Political conservatives more likely to stigmatize mentally ill people, study finds. Retrieved from psypost.org/2016/03/conservative-people-tended-hold-stigmatized-beliefs-mentally-ill-41921.

Irshad, M. (2017). Mandi Embun at The Datai. Accessed https://www.thedatai.com/nature/the-datai-diaries/mandi-embun-or-foresttherapy

Laskowski, E. R. (2018). Heart Rate FAQs. Retrieved from www.mayoclinic.org/healthy-lifestyle/fitnes/expert-answers/heart-rate/faq-20057979

Lal, K. (2014). Academic Stress among Adolescents in Relation to Intelligence and Demographic Factors. Am. *Int. J. Res. Humanity. Arts Soc. Sci., 5*, 123–129.

Leppink, E., Odlaug, B., Lust, K., Christenson, G., & Grant, J. (2016). The Young and the Stressed. *The Journal of Nervous and Mental Disease, 204*, 931–938.

Lum, M. (2018). Malaysia is Asia's fattest country. Retrieved from http://www.star2.com.myhealth/2018/08/14/fat-state-of-affairs

Malaysian Digest (2015). Working to Death: Recent Study shows more Malaysians getting stress-related illnesses. Retrieved from http://www.malaysiandigest.com/opinion/557139-working-to-death-recent-study-shows-malaysians-getting-stressed-related-illnesses.html

Nur, N., Kibik, A., Kilic, E., & Sumer, H. (2017). Health-related Quality of Life and Associated Factors Among Undergraduate University Students. *Oman Medical Journal, 32*(4), 329–334.

Panahi, S.; Yunus, A.; & Panahi, M. (2016, May 10–12). Influence of Cognitive Emotion Regulation on Psychological Well-being of Malaysian Graduates. In Proceedings of the 4th International Congress on Clinical and Counselling Psychology.

Park, B. J., Tsunetsugu, Y., Kasetani, T., Hirano, H., Kagawa, T., Sato, M., & Miyazaki, Y. (2007). Physiological effects of Shinrin-Yoku (taking in the atmosphere of the for-est)–using salivary cortisol and cerebral activity as indicators. *Journal of Physiological Anthropology, 26*, 123–128.

Park, B. J., Tsunetsugu, Y., Kasetani, T., Morikawa, T., Kagawa, T., & Miyazaki, Y. (2009). Physiological effects of forest recreation in a young conifer forest in Hinokage Town, Japan. *Silva Fennica, 43*, 291–301.

Ramli, N. H. H., Alavi, M., Mehrinezhad, S. A., & Ahmadi, A. (2018). Academic stress and self-regulation among university students in Malaysia: Mediator role of mindfulness. *Behavioural Science, 8*(12), 1–9.

Shin, W. S., Shin, C. S., Yeoun, P. S., & Kim, J. J. (2011). The influence of interaction with forest on cognitive function. *Scandinavian Journal of Forest Research, 26*, 595–598.

Song, C., Ikei, H., & Miyazaki, Y. (2017). Sustained effects of a forest therapy program on the blood pressure of office workers. *Urban Forestry & Urban Greening, 27*, 246–252.

SPRINT (The SPRINT Research Group). (2015). A randomized trial of intensive versus standard blood-pressure control. *New England Journal of Medicine, 373*, 2103–2116.

Star (2017). Survey: Sedentary lifestyle puts one at risk of obesity. Retrieved from http://www.thestar.com.my/news/nation/2016/05/20/survey-sedentary-lifestyle-puts-one-at-risk-of-obesity/

Teh, C., Ngo, C., Zulkifli, R., Vellasamy, R., & Suresh, K. (2015). Depression, Anxiety and Stress among Undergraduate Students: A Cross Sectional Study. *Open J. Epidemiol., 5*, 260–268.

Tsunetsugu, Y., Lee, J., Park, B. J., Tyrväinen, L., Kagawa, T., & Miyazaki, Y. (2013). Physiological and psychological effects of viewing urban forest landscapes assessed by multiple measurements. *Landscape Urban Plann., 113*, 90–93.

Keeren Sundara Rajoo is currently a senior lecturer at the Department of Forestry Science, Faculty of Agricultural Science and Forestry, Universiti Putra Malaysia (UPM) Bintulu Campus. He completed his Ph.D. in the field of Ecology at the Department of Biology, Faculty of Science, Universiti Putra Malaysia, Serdang. He did his Ph.D. research attachment at the Tokyo University of Agriculture, Japan, funded by SEARCA. His research focus is soil science, environmental waste management, ecology and forest therapy.

Daljit Singh Karam is a senior lecturer at the Department of Land Management, Faculty of Agriculture, Universiti Putra Malaysia (UPM) since 2013. He received his Bachelor's in Forestry Science in 2009 and Ph.D. in Forestry (Soil Science) in 2012 from the Faculty of Forestry, UPM. He pursued his post-doctoral studies at the University of Liverpool, the UK from 2014 to 2015. His research interests include soil science, forest therapy and soil quality management.

Arifin Abdu is a full-time professor and the current Deputy Vice-Chancellor (Student Affairs and Alumni), Universiti Putra Malaysia (UPM). He received his Bachelor of Forestry Science in 2002 and his Master of Science in 2005, both from the Faculty of Forestry, UPM. He was awarded a scholarship to pursue his Ph.D. in Life Environmental Science at the Ehime University, Japan and graduated in 2008. His research interests include forest rehabilitation, soil fertility, forest therapy and forest social science.

Chapter 20
Urban Green Space, Green Exercise and Health Outcomes: Evidence from Kuala Lumpur, Malaysia

Tapan Kumar Nath

Abstract Urban green spaces (UGS) are freely accessible private, communal or publicly owned areas in urban settings covered with vegetation and wetlands generally used for recreation. Physical activities (e.g. walking, jogging), leisure, and social gatherings in UGS bring positive health and well-being benefits. In this chapter, drawing evidence from two studies in Kuala Lumpur, Malaysia, illustrates respondents' perception of the well-being outcomes of UGS. The respondents reported that the UGS (e. g. parks) provided a place for jogging, cycling, coming close to nature, meeting friends, and children's opportunity for physical exercise. They believed that physical activities in UGS had a positive contribution to overall well-being through stress reduction, mindfulness, physical fitness, and body weight control. In light of the study results, we suggest that Malaysia can look forward to the promotion of and support for UGS as a crucial public health program.

Keywords Parks · Urban forests · Physical activities · Social interaction · Health benefits

Introduction

Urban green spaces (UGS) are generally located within urban areas, and own and managed by private, communal or public entities. They are freely accessible and usually used for recreation/leisure and light physical activities (Cilliers, 2015; Zupancic et al., 2015; Zhu et al., 2017; van Hecke et al., 2018; Koohsari et al., 2018; Carmona, 2019). UGS can be of varying sizes, have plants, open spaces, and wetlands, and may have small equipment for children's play and physical exercises. These UGS can be parks, gardens, children's playgrounds, mountain trails, and other open natural areas (Cilliers, 2015; Lee & Kim, 2015; Lee et al., 2015; Yung et al., 2016), which are considered important places for urban dwellers for restoration and amelioration of psychological stress (Braubach et al., 2017; Lee et al., 2015; Zhang

T. K. Nath (✉)
University of Nottingham Malaysia, Semenyih, Selangor, Malaysia
e-mail: Tapan.Nath@nottingham.edu.my

S. Maruthaveeran et al. (eds.), *Urban Forestry and Arboriculture in Malaysia*,
https://doi.org/10.1007/978-981-19-5418-4_20

et al., 2015). Urban green spaces within neighbourhoods confer several well-being benefits and contribute to urban sustainability (Koohsari et al., 2015, 2018; Villanueva et al., 2015).

Access to UGS could boost "green exercise", which covers exercises such as walking and cycling, and offers social interactions among users to form social capital and cohesion thus help to achieve positive well-being benefits (Barton & Pretty, 2010; Mackay & Neill, 2010; Sugiyama et al., 2010; Groenewegen et al., 2012; Jackson et al., 2013; Bancroft et al., 2015; Pretty et al., 2011, 2016; Braubach et al., 2017; Pretty, 2017; Nath et al., 2018; Jennings & Bamkole, 2019). Urban green spaces encourage neighbouring people to participate in diverse outdoor activities which can promote regular social interactions, and create a sense of community (Arnberger, 2012; Dinnie et al., 2013; Jennings et al., 2016; Nath et al., 2018). A wide range of positive effects derived from UGS has been reported. Some of these are reduced hypertension, lowering the risk of non-communicable diseases, and enhanced psycho-physiological well-being (Bratman et al., 2012; Horiuchi et al., 2013; Jennings et al., 2016; Jiang et al., 2018; Lamb et al., 2019; Lee et al., 2011; Silva et al., 2018; Ward-Thompson et al., 2012).

Another key role of urban green spaces is to serve as venues for socio-cultural functions (Braubach et al., 2017; Lee et al., 2015). Residents visiting green spaces do more social interactions which create a sense of cohesiveness, reduces feelings of loneliness and enhances social support for individual resilience and well-being (Arnberger, 2012; Coley et al., 1997; Dinnie et al., 2013; Haq, 2011; Jennings et al., 2016; Prezza et al., 2001). Researchers stated that patients in hospitals showed quicker recovery when they were exposed to green spaces than that of built environments (Gladwell et al., 2012; Project Evergreen, 2017).

Green exercise, social cohesion and seeing picturesque sceneries in UGS help to improve physical and mental well-being (Nath et al., 2018). In appreciation of such health and mental well-being benefits, urban planners and public health researchers put particular attention to the sustainability of UGS (Koohsari et al., 2015, 2018; Subramanian & Jana, 2018; Villanueva et al., 2015; Wood et al., 2017). The association between UGS and human well-being has been widely researched in the global north (Hooper et al., 2018; Koohsari et al., 2018; Labib et al., 2019; Lamb et al., 2019; Pretty, 2017; Wood et al., 2017), however, there is limited research in the global south (Florindo et al., 2017). Notable studies in Malaysia on UGS include Nor Akmar et al., 2011, 2012, Aida et al. (2016) and Karuppannan et al. (2014). They studied planning and management of UGS, health and well-being, and biodiversity in green spaces. There are few studies across Asia, which investigated the relationship between UGS and well-being. These are Zhu et al. (2017), Paul and Nagendra (2017), Chang and Chen (2015), and Pham et al. in Beijing, Delhi, Taiwan and Vietnam, respectively.

This chapter, based on two studies from Kuala Lumpur, Malaysia shows visitors' perception on the association between UGS and health outcomes. This chapter is based on the author's earlier works, personal experience and related literature reviews.

Case Study 1: Urban Parks, Uses and Health Outcomes

This study was based on randomly selected 310 structured interviews (250 visitors and 60 neighbourhoods) in three parks within Kuala Lumpur (KL) (Fig. 20.1). These three parks are being managed by Kuala Lumpur City Hall (DBKL). Lembah Kiara park has dense greeneries, and several walkways for hiking which are connected to adjacent Bukit Kiara Forest. This has water bodies and playgrounds for children.

Rimba Kiara park has jogging trails and seating areas. There are plenty of shaded areas for resting and a few small ponds at the centre. Zaaba Park has basketball courts and playgrounds as well as walkways for jogging and cycling. Walkways are covered with trees.

Frequency of Parks' Visit and Purposes

Both residents from neighbouring residential areas and outside visitors use these parks. Neighbourhood residents (33%) responded that they visited parks once a week (Fig. 20.2). They visited usually on the weekend because of their jobs and family activities. Park visitors (35%) commented that they visit parks 2–3 days per week while homemakers and retired persons visited parks every day because they largely have more time and convenient access to neighbouring parks. In addition to studying parks, park visitors and neighbourhood residents commented that they also go to other parks. However, neighbourhood residents reported that they liked to visit neighbouring parks instead of visiting distant ones. Residents who have easy entry to nearby UGS are more likely to visit them (Lee & Maheswaran, 2010).

We identified four principal reasons for visiting parks (Fig. 20.3). For 68% of respondents, the purposes were jogging and light gym using available equipment. There is a positive relationship between UGS and physical activities (Mytton et al., 2012). On the other hand, about 22% of the respondents said that they visit parks

Fig. 20.1 Location of three sampled parks within Kuala Lumpur, Malaysia. Map in the inset is Peninsula Malaysia. *Source* Nath et al. (2018)

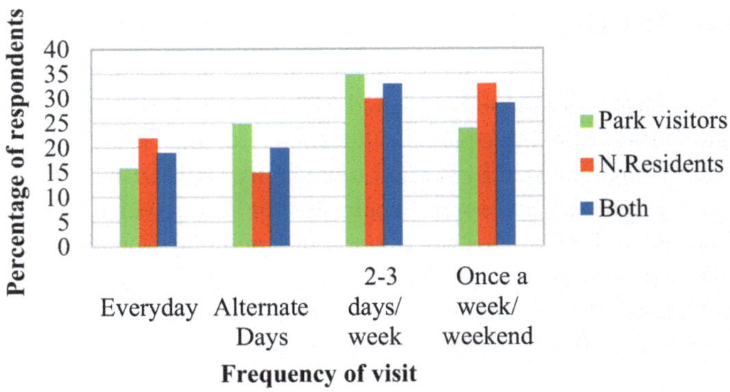

Fig. 20.2 Frequency of green space visits by respondents. *Source* Nath et al. (2018)

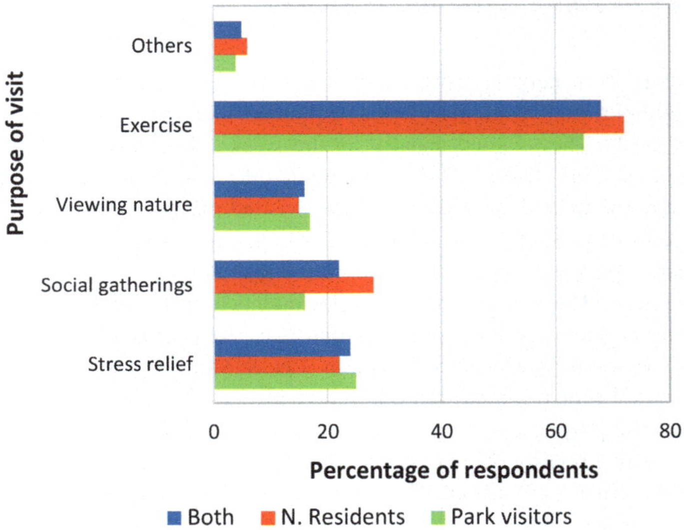

Fig. 20.3 Respondents' opinions on purposes of park visit. *Source* Nath et al. (2018)

for social gatherings, 24% for getting relief from stress and 16% to view natural scenery. Ward-Thompson et al. (2012) It was reported that UGS has been related to several health benefits along with a reduction of mental stress. With regards to a social gathering, the responses between park visitors and neighbourhood residents were significantly different (Chi-squared = 4.61; $P < 0.03$). This reason is that neighbouring people visit the same UGS frequently and so have more social interactions with others. Peters et al. (2010) reported that UGS enhance social capital among the users. Social capital can affect a number of factors which are connected to physical and mental well-being (Jennings et al., 2016).

Park Visit, and Health and Well-being Outcomes

About 90% of park visitors and neighbourhood residents reported that peaceful areas of parks were relaxing, stress-relieving and improved mindfulness (Table 20.1). They claimed that light physical exercises in UGS helped to sustain body fitness and weight. Light exercises in UGS can be related to mental and physical health benefits especially for stressed urban residents (Hamer & Chida, 2008; Regeneris Consulting, 2009; Thompson Coon et al., 2011; Richardson et al., 2013; White et al., 2013; White et al., 2016). Regular exercise (e. g. walking) in parks has a positive link with health outcomes (Mytton et al., 2012). Respondents' opinions on healthy feelings, physically sounds and sustaining body mass were significantly different (Table 20.1), which are because of regular and more frequent visitation of parks by neighbourhood residents and most probably they had understood the health benefits of physical activities. Parks' visitation opened up the avenues for children to meet and interact, which might have effects on their healthy growth (Table 20.1). A significant difference with regards to the opinion of positive impact on wellbeing was found between park visitors and neighbourhood residents (Table 20.1). Douglas et al. (2017). Reported that living adjacent to UGS is usually related to increased physical activities, a healthy lifestyle, and enhanced well-being.

Case Study 2: Urban Community Forest and Peoples' Well-being

Population growth and development pressure in KL, Malaysia has caused a decline in urban forest cover from 13 m^2 per capita in 2010 to 8.5m^2 in 2014 (Kanniah, 2017; Maryanti et al., 2017). Such a significant loss of urban greeneries has motivated urban residents to conserve the remnant urban forests, particularly those living in the surrounding vicinity. A number of local associations/NGOs have come forward to collaborate informally with forestry/local government agencies for the protection and conservation of these forests as urban community forests (UCF). The Kota Damansara Community Forest Reserve (KDCFR) is one of the pioneer UCFs which has been informally managed and used by local groups for the last 10 years. This study was based on 80 structured interviews with KDCFR users.

The KDCFR (Fig. 20.4) is a remnant logged over secondary lowland mixed dipterocarp rainforest of the once extensive Sungai Buloh Forest Reserve. This oldest Sungai Buloh Forest reserve has been seriously threatened and affected by development activities and hence the area had been reduced from 6590 to 2903 ha in 1972, then further reduced to the current 321.7 ha (Shaharuddin et al., 2012). Much of the Sungai Buloh Forest Reserve had been degazetted and developed for housing, roads, highways, shopping malls and other land uses in the last decades (Latiff, 2012). The KDCFR has also been under serious pressure from the government for housing development (Vaz & Lim, 2012). Relentless campaigning and lobbying

Table 20.1 Perception of respondents on health and well-being outcomes in Kuala Lumpur

Variable	Percentage of respondents		
	Park visitors ($n = 250$)	*N*. residents ($n = 60$)	Both ($N = 310$)
Park is a place for relaxation and reduction of stress levels			
Yes	90	87	89
No	8	10	9
No answer	2	3	2
Park visitation improves mindfulness			
Yes	95	90	93
No	–	–	–
No answer	5	10	7
Parks' uses improve health condition (Chi-squared = 11.65, $P < 0.01$)			
Yes	68	90	79
No	20	8	14
No answer	12	2	7
Parks' uses improve general physical fitness (Chi-squared = 10.54, $P < 0.01$)			
Yes	72	92	82
No	10	-	5
No answer	18	8	13
Activities in parks help to maintain body weight (Chi-squared = 12.49, $P < 0.01$)			
Yes	62	86	74
No	28	10	19
No answer	10	4	7
Parks' visit has a positive impact on children growth and development			
Yes	85	77	81
No	13	20	16
No answer	2	3	2
Parks' visit has a positive impact on well-being (Chi-squared = 24.72, $P < 0.01$)			
Yes	69	100	84
No	3	–	2
No answer	28	–	14

Source Nath et al. (2018)

have been conducted by a coalition of resident groups and the Malaysian Nature Society (an NGO) since 2003. In the end, the authority (Selangor State Government) was convinced to recognize the values of the forest for conservation, education, research and residents' uses and it was officially re-gazetted as KDCFR in early 2010 (Sulaiman et al., 2012; Vaz & Lim, 2012). Since then, the residents' groups in

Fig. 20.4 Location of KDCFR (green portion within the yellow square) in Kuala Lumpur with a map of Peninsular Malaysia (inset). *Source* Nath and Magendran (2020)

collaboration with the Selangor State Forestry Department (SSFD), several corporate agencies and other like-minded organizations have been managing the forest and built a network of forest trails for public recreational uses. The KDCFR is located in Petaling Jaya district of Selangor, on the western outskirts of KL (Fig. 20.4).

Purposes of Visit to KDCFR

The visitors to KDCFR had mentioned 10 purposes of their visit (Fig. 20.5). More than half of the visitors indicated forest walks and closeness to nature (51%) as their purposes to visit KDCFR, while 42% reported recreation. Meanwhile, responses related to health issues, stress reduction, and practice cycling were recorded at 33%, 30% and 29%, respectively. Nath et al. (2018) in their study in KL found that people use urban green spaces for several purposes including stress reduction, physical activities and social interactions. Meeting friends as part of social interaction was recorded at 21%. This data showcases the importance of overall health and mental wellbeing for most of the visitors of KDCFR. Researchers (Duinker et al., 2015; Atmis, 2016; Ordóñez et al., 2016) reported that people visit urban forests because they provide many health and social benefits (e. g. social interactions). Nearly two-thirds of the visitors (more than 30 years old) believe that they had achieved their objectives of visiting KDCFR and the remaining 31% of the visitors felt it to some extent which is still a positive note. Denga et al. (2017) and Gashu et al. (2019) observed that older age respondents had a tendency to take a walk-in forest and were more positive in response to their satisfaction with visiting urban forests because they possess more experience in interacting with nature.

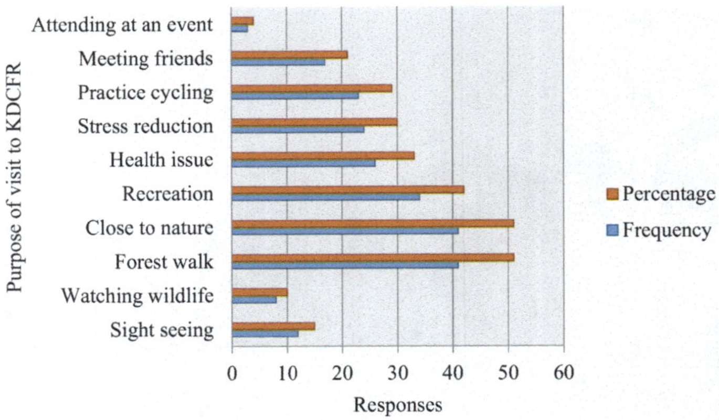

Fig. 20.5 Purposes of visitors visiting KDCFR in Kuala Lumpur, Malaysia. *Source* Nath and Magendran (2020)

We found a significant difference (Chi-square = 39.69, df = 9, P < 0.001) between male and female visitors with respect to their purposes of visiting KDCFR (Table 20.2). To more than 80% of female visitors, the main driver of KDCFR visit was forest walks and closer to nature while for the majority of male (58%) visitors the main purpose was recreation. The majority of the visitors (94%) were positive when being asked about their opinion on contribution to KDCFR conservation. Only 8% of male visitors did not want to contribute (Table 20.2). Meanwhile, a notable percentage of visitors would like to align their contribution towards trail management (45%) and forest protection (34%). Other contributions included communication support and cutting and removing fallen trees to clear the trails for use. There was a significant difference between male and female visitors with respect to their contribution (Chi-square = 11.46, df = 3, P < 0.01) (Table 20.2).

Conclusion

Urban green spaces offer users numerous opportunities for green exercise, social gatherings and recreation. Parks and urban forests' environments were conducive to users' physical and mental health, and their overall well-being. Findings of these two studies were in line with research conducted in the global north. Studies suggest that green exercise in UGS has ameliorative effects on both mental and physical health (NEA, 2011; Saraev, 2012). These health benefits can be translated into monetary values, which may include a decrease in public health expenditures and enhanced financial returns because of a decrease in illness and absence from jobs (Regeneris Consulting, 2009). A study reported that if 2.12 million persons walk half an hour a

Table 20.2 Male and female visitors' responses to selected survey questions in KDCFR, Kuala Lumpur, Malaysia. Values are frequencies followed by percentages in parentheses

Variable	Male ($n = 62$)	Female ($n = 18$)
Objectives of visit* (Chi-square statistic = 39.69, df = 9, $P < 0.001$)		
Sight seeing	10(16)	2(13)
Watching wildlife	8(13)	2(11)
Forest walk	35(56)	16(87)
Close to nature	35(56)	14(80)
Recreation	36(58)	6(33)
Health issue	20(32)	6(33)
Stress reduction	12(19)	5(27)
Practice cycling	20(32)	1(7)
Meeting friends	14(22)	6(33)
Attending at an event	2(4)	-
Do you like to contribute to KDCFR conservation?		
Yes	57(92)	18(100)
No	5(8)	-
If yes, what kind contribution are you willing to offer?* (Chi-square statistic = 11.46, df = 3, $P < 0.01$)		
Trail management	27(44)	8(47)
Forest protection	20(32)	6(33)
Donation	37(60)	7(40)
Others	1(2)	2(13)
How much do you like to donate in a year? ** (Chi-square statistic = 7.02, df = 3, $P < 0.07$)		
RM 20	2(6)	2(17)
RM 20–50	11(35)	5(41)
RM 50–100	17(53)	3(25)
> RM 100	2(6)	2(17)

Note *Multiple responses were counted. **$n = 44$ (Male 32, Female 12)
Source Nath and Magendran (2020)

day for three days per week, this can generate a financial benefit of £1.75 billion per year (White et al., 2016).

In recent years, developing countries experience an increasing rate of urbanization and migration of rural populace into urban areas. Malaysia is no exception: the urban population increased from 34.2% in 1980 to 77.2% in 2020 (World Data Atlas-Malaysia 2021) and it was estimated that 80% of the Malaysian population would live in urban areas by 2030 (FMT, 2017). In the greater KL, the population size was 7.2 million in 2016, which might increase to 10 million by 2020 (KL Calling, 2016). The increasing population and associated development would put pressure on UGS. In light of the perceived benefits of UGS for human health and well-being, the Malaysian government may take an intervention towards conservation, motivation, and uses of UGS by people, which would have positive effects on society, the environment and the overall economy of the country.

References

Aida, N., Sasidhran, S., Kamarudin, N., Aziz, N., Puan, C. L., & Azhar, B. (2016). Woody trees, green space and park size improve avian biodiversity in urban landscapes of Peninsular Malaysia. *Ecological Indicators, 69*, 176–183.

Atmis, E. (2016). Development of urban forest governance in Turkey. *Urban Forestry & Urban Greening, 19*, 158–166.

Arnberger, A. (2012). Urban densification and recreational quality of public urban green spaces—A Viennese case study. *Sustainability, 4*(4), 703–720.

Bancroft, C., Joshi, S., Rundle, A., Hutson, M., Chong, C., Weiss, C. C., Genkinger, J., Neckerman, K., & Lovasi, G. (2015). Association of proximity and density of parks and objectively measured physical activity in the United States: A systematic review. *Social Science and Medicine, 138*, 22–30.

Barton, J., & Pretty, J. (2010). What is the best dose of nature and green exercise for improving mental health? A multi-study analysis. *Environmental Science and Technology, 44*, 3947–3955.

Bratman, G. N., Hamilton, J. P., & Daily, G. C. (2012). The impacts of nature experience on human cognitive function and mental health. *Annals of NY Academy of Sciences, 1249*, 118–136.

Braubach, M., Egorov, A., Mudu, P., Wolf, T., Thompson, C. W., & Martuzzi, M. (2017). Effects of urban green space on environmental health, equity and resilience. *Nature-based solutions to climate change adaptation in urban areas,* (pp. 187–205). https://doi.org/10.1007/978-3-319-560 91-5, https://doi.org/10.1007/978-3-319-56091-5_11. Accessed 12 January 2020.

Carmona, M. (2019). Place value: Place quality and its impact on health, social, economic and environmental outcomes. *Journal of Urban Design, 24*(1), 1–48.

Chang, H.-S., & Chen, T.-L. (2015). Decision Making on Allocating Urban Green Spaces Based upon Spatially-Varying Relationships between Urban Green Spaces and Urban Compaction Degree. *Sustainability, 7*, 13399–13415.

Cilliers, E. J. (2015). A Framework for Planning Green Spaces in Rural South Africa. *Agriculture, Forestry and Fisheries, 4*, 80–86.

Coley, R. L., Kuo, F. E., & Sullivan, W. C. (1997). Where does community grow? The social context created by nature in urban public housing. *Environmental Behaviour, 29*, 468–494.

Denga, J., & Ilb, R.A., & Pierskalla, C. (2017). Visitors' and residents' perceptions of urban forests for leisure in Washington. *Urban Forestry & Urban Greening, 28*, 1–11.

Dinnie, E., Brown, K. M., & Morris, S. (2013). Community, cooperation and conflict: Negotiating the social well-being benefits of urban greenspace experiences. *Landscape Urban Planning, 112*, 1–9.

Douglas, O., Lennon, M., & Scott, M. (2017). Green space benefits for health and well-being: A life-course approach for urban planning, design and management. *Cities, 66*, 53–62.

Duinker, P.N., Ordóñez, C., Steenberg, J.W.N., Miller, K.H., Toni, S.A., & Nitoslawski, S.A. (2015). Trees in Canadian cities: indispensable life form for urban sustainability. *Sustainability,* 7, 7379–7396.

Florindo, A. A., Barrozo, L. V., Cabral-Miranda, W., Rodrigues, E. Q., Turrell, G., Goldbaum, M., Cesar, C. L. G., & Giles-Corti, B. (2017). Public open spaces and leisure-time walking in Brazilian adults. *International Journal of Environmental Research and Public Health, 14*, 553.

FMT. (2017). Report: 80% of Malaysia will be urban by 2030. Free Malaysia Today. Retrieved July 14, 2019 from https://www.freemalaysiatoday.com/category/nation/2017/01/19/report-80-of-malaysia-will-be-urban-by-2030/

Gashu, K., Gebre-Egziabher, T., & Wubneh, M. (2019). Local communities' perceptions and use of urban green infrastructure in two Ethiopian cities: Bahir Dar and Hawassa. *Journal of Environmental Planning & Management, 63*(2), 287–316.

Gladwell, V. F., Brown, D. K., Barton, J. L., Tarvainen, M. P., Kucppa, P., Pretty, J., Suddaby, J. M., & Sandercock, G. R. H. (2012). The effects of views of nature on autonomic control. *European Journal of Applied Physiology, 112*, 3379–3386.

Groenewegen, P. P., et al. (2012). Is a green residential environment better for health? If so, why? *Annals of the Association of American Geography, 102*(5), 996–1003.

Hamer, M., & Chida, Y. (2008). Walking and primary prevention: A meta-analysis of prospective cohort studies. *British Journal of Sports Medicine, 42*, 238–243.

Haq, S. M. (2011). Urban green spaces and an integrative approach to sustainable environment. *Journal of Environmental Protection, 2*, 601–608.

Hooper, P., Boruff, B., Beesley, B., Badland, H., & Giles-Corti, B. (2018). Testing spatial measures of public open space planning standards with walking and physical activity health outcomes: Findings from the Australian national liveability study. *Landscape and Urban Planning, 171*, 57–67.

Horiuchi, M., Endo, J., Akatsuka, S., Uno, T., Hasegawa, T., & Seko, Y. (2013). Influence of forest walking on blood pressure, profile of mood states and stress markers from the viewpoint of aging. *Journal of Aging & Gerontology, 1*, 9–17.

Jackson, L. E., Daniel, J., McCorkle, B., Sears, A., & Bush, K. F. (2013). Linking ecosystem services and human health: The Eco-health relationship browser. *International Journal of Public Health, 58*, 747–755.

Jennings, V., Larson, L., & Yun, J. (2016). Advancing sustainability through urban green space: Cultural ecosystem services, equity, and social determinants of health. *International Journal of Environmental Research and Public Health, 13*, 196. https://doi.org/10.3390/ijerph13020196

Jennings, V., & Bamkole, O. (2019). The relationship between social cohesion and urban green space: An avenue for health promotion. *International Journal of Environmental Research and Public Health* 16, 452; doi:https://doi.org/10.3390/ijerph16030452.

Jiang, B., Schmillen, R., & Sullivan, W. C. (2018). How to waste a break: Using portable electronic devices substantially counteracts attention enhancement effects of green spaces. *Environmental Behaviour, 0013916518788603*,. https://doi.org/10.1177/0013916518788603

Kanniah, K. D. (2017). Quantifying green cover change for sustainable urban planning: A case of Kuala Lumpur, Malaysia. Malaysia Sustainable Cities Program, Working Paper Series. https://scienceimpact.mit.edu/sites/default/files/documents/Kanniah.pdf

Karuppannan, S., Baharuddin, Z., Sivam, A., & Daniels, C. (2014). Urban Green Space and Urban Biodiversity: Kuala Lumpur, Malaysia. *Journal of Sustainable Development, 7*, 1–16.

KL Calling. (2016). KL calling investor's guide. http://cdn.ey.com/echannel/my/EY-kl-calling-an-investor-guide-2016.pdf.

Koohsari, M. J., Badland, H., Mavoa, S., Villanueva, K., Francis, J., Hooper, P., Owen, N., & Giles-Corti, B. (2018). Are public open space attributes associated with walking and depression? *Cities, 74*, 119–125.

Koohsari, M. J., Mavoa, S., Villanueva, K., Sugiyama, K., Badland, H., Kaczynski, A. T., Owen, N., & Giles-Corti, B. (2015). Public open space, physical activity, urban design and public health: Concepts, methods and research agenda. *Health & Place, 33*, 75–82.

Labib, S. M., Lindley, S., & Huck, J. J. (2019). Spatial dimensions of the influence of urban green-blue spaces on human health: A systematic review. *Environmental Research*. https://doi.org/10.1016/j.envres.2019.108869

Lachowycz, K., & Jones, A. P. (2011). Greenspace and obesity: A systematic review of the evidence. *Obesity Reviews, 12*, e183–e189.

Lamb, K. E., Mavoa, S., Coffee, N. T., Parker, K., Richardson, E. A., & Thornton, L. E. (2019). Public open space exposure measures in Australian health research: A critical review of the literature. *Geographical Research, 57*(1), 67–83.

Latiff, A. (2012). Role and functions of Kota Damansara Forest Reserve, Selangor as a community forest in biodiversity education, research and conservation in an urban setting. In: Prosiding ekspedisi saintifik kepelbagaian biology- Hutan Simpan Kota Damansara, Selangor (pp 42–51). Selangor State Forestry Department, Malaysia.

Lee, A., & Maheswaran, R. (2010). The health benefits of urban green spaces: A review of the evidence. *Journal of Public Health, 33*, 212–222.

Lee, J., Park, B. J., Tsunetsugu, Y., Ohira, T., Kagawa, T., & Miyazaki, Y. (2011). Effect of forest bathing on physiological and psychological responses in young Japanese male subjects. *Public Health, 125*, 93–100.

Lee, A. C. K., Jordan, H. C., & Horsley, J. (2015). Value of urban green spaces in promoting healthy living and wellbeing: Prospects for planning. *Risk Management and Healthcare Policy, 8*, 131–137.

Lee, Y., & Kim, K. (2015). Attitudes of Citizens towards Urban Parks and Green Spaces for Urban Sustainability: The Case of Gyeongsan City, Republic of Korea. *Sustainability, 7*, 8240–8254.

Mackay, M. J., & Neill, J. T. (2010). The effect of "green exercise" on state anxiety and the role of exercise duration, intensity, and greenness: A quasi-experimental study. *Psychology of Sport and Exercise, 11*, 238–245.

Maryanti, M. R., Khadijah, H., Uzair, A. M., & Ghazali, M.A.R.M.M. (2017). The urban green space provision using the standards approach: issues and challenges of its implementation in Malaysia. In: Brebbia, C. A., Zubir, S. S., Hassan, A.S., (eds) WIT Transactions on ecology and the environment, WIT Press, UK. 210, 369–379.

Mytton, O. T., Nick, T., & Foster, H. C. (2012). Green space and physical activity: An observational study using Health Survey for England data. *Health & Place, 18*, 1034–1041.

Nath, T. K., S. S. Zhe Han, & Lechner, A. M. (2018). Urban green space and well-being in Kuala Lumpur, Malaysia. *Urban Forestry & Urban Greening* 36, 34–41.

Nath, T. K., & M. Magendran (2020): Urban Community Forest in Kuala Lumpur, Malaysia: Current Management, Public Uses and Willingness Toward Conservation. *Journal of Sustainable Forestry*. https://doi.org/10.1080/10549811.2020.1796709.

NEA (National Ecosystem Assessment). (2011). UK National Ecosystem Assessment. UNEP-WCMC, Cambridge. http://uknea.unep-wcmc.org/LinkClick.aspx?fileticket=m%2BvhAV3c9uk%3D&tabid=82

Nor Akmar, A. A., Konijnendijk, C.C., Stigsdotter, U.K., & Nillson, K. (2012). Malaysian case studies on the relation between the use of green space and health promotion. *Alam Cipta*, 5, June 2012.

Nor Akmar, A. A., Konijnendijk, C. C., Sreetheran, M., & Nilsson, K. (2011). Greenspace Planning and Management in Klang Valley, Peninsular Malaysia. *Arboriculture & Urban Forestry, 37*, 99–107.

Paul, S., & Nagendra, H. (2017). Factors Influencing Perceptions and Use of Urban Nature: Surveys of Park Visitors in Delhi. *Land, 6*, 27. https://doi.org/10.3390/land6020027

Peters, K., Elands, B., & Buijs, A. (2010). Social interactions in urban parks: Stimulating social cohesion? *Urban Forestry & Urban Greening, 9*, 93–100.

Pretty, J., Barton, J., Colbeck, I., Hine, R., Mourato, S., Mackerron, G., & Wood, C. (2011). Health values from ecosystems. Chapter 32. In: National ecosystem assessment. London: Defra; p. 1153–1181.

Pretty, J., Barton, J., Bharucha, Z. P., Bragg, R., Pencheon, D., Wood, C., & Depledge, M. H. (2016). Improving health and well-being independently of GDP: Dividends of greener and prosocial economies. *International Journal of Environmental Health & Research, 26*, 11–36.

Pretty, J. (2017). Manifesto for the green mind. *Resurgence & Ecologist, 301*, 18–21.

Prezza, M., Amici, M., Roberti, T., & Tedeschi, G. (2001). Sense of community referred to the whole town: Its relations with neighboring, loneliness, life satisfaction, and area of residence. *Journal of Community Psychology, 29*, 29–52.

Project Evergreen. (2017). *Economic Benefits of Green Spaces*. http://projectevergreen.org/resources/economic-benefits-of-green-spaces/

Regeneris Consulting. (2009). The Economic Contribution of the Mersey Forest's Objective One-Funded Investments. Regeneris Consulting Ltd, Cheshire, the UK.

Richardson, E. A., Pearce, J., Mitchell, R., & Kingham, S. (2013). Role of physical activity in the relationship between urban green space and health. *Public Health, 127*, 318–324.

Roe, J. J., Ward, T. C., Aspinall, P. A., Brewer, M. J., Duff, E. I., Miller, D., Mitchell, R., & Clow, A. (2013). Green space and stress: Evidence from cortisol measures in deprived urban communities. *International Journal of Environmental Research & Public Health, 10*, 4086–4103.

Saraev, V. (2012). Economic benefits of greenspace: a critical assessment of evidence of net economic benefits Forestry Commission Research Report. Forestry Commission, Edinburgh.

Shaharuddin, M.I., Latiff, A. & Muda, Y. (2012). Forestry in the colonial Selangor and its implications to biodiversity conservation in peninsular Malaysia. In Prosiding ekspedisi saintifik kepelbagaian biology- Hutan Simpan Kota Damansara, Selangor (pp 27–41). Selangor State Forestry Department, Malaysia.

Silva, R. A., Rogers, K., & Buckley, T. J. (2018). Advancing environmental epidemiology to assess the beneficial influence of the natural environment on human health and well-being. *Environmental Science & Technology, 52*(17), 9545–9555.

Subramanian, D., & Jana, A. (2018). Assessing urban recreational open spaces for the elderly: A case of three Indian cities. *Urban Forestry & Urban Greening, 35*, 115–128.

Sugiyama, T., Francis, J., Middleton, N. J., Owen, N., & Giles-Corti, B. (2010). Associations between recreational walking and attractiveness, size, and proximity of neighborhood open spaces. *American Journal of Public Health, 100*, 1752–1757.

Sulaiman, N., Mohamad, N., Tahir, A.B.M. & Abdullah, M. (2012). Butterfly fauna (Lepidoptera:Rhopalocera) of Kota Damansara Forest Reserve, Selangor. In Prosiding ekspedisi saintifik kepelbagaian biology- Hutan Simpan Kota Damansara, Selangor (pp101–109). Selangor State Forestry Department, Malaysia.

Thompson Coon, J., Boddy, K., Stein, K., Whear, R., Barton, J., & Depledge, M. H. (2011). Does participating in physical activity in outdoor natural environments have a greater effect on physical and mental wellbeing than physical activity indoors? A systematic review. *Environmental Science & Technology, 45*, 1761–1772.

Van Hecke, L., Ghekiere, A., Veitch, J., Van Dyck, D., Van Cauwenberg, J., Clarys, P., & Deforche, B. (2018). Public open space characteristics influencing adolescents' use and physical activity: A systematic literature review of qualitative and quantitative studies. *Health & Place, 51*, 158–173.

Vaz, J., & Lim T. W. (2012). Realising our shared vision for the Kota Damansara Forest Reserve, Selangor. In Prosiding ekspedisi saintifik kepelbagaian biology- Hutan Simpan Kota Damansara, Selangor (pp 176–183). Selangor State Forestry Department, Malaysia.

Villanueva, K., Badland, H., Hooper, P., Koohsari, M. J., Mavoa, S., Davern, M., Roberts, R., Goldfeld, S., & Giles-Corti, B. (2015). Developing indicators of public open space to promote health and wellbeing in communities. *Applied Geography, 57*, 112–119.

Ward-Thompson, C., Roe, J., Aspinall, P., Mitchell, R., Clow, A., & Miller, D. (2012). More green space is linked to less stress in deprived communities: Evidence from salivary cortisol patterns. *Landscape & Urban Planning, 105*, 221–229.

White, M. P., Elliott, L. R., Taylor, T., Wheeler, B. W., Spencer, A., Bone, A., Depledge, M. H., & Fleming, L. E. (2016). Recreational physical activity in natural environments and implications for health: A population based cross-sectional study in England. *Preventive Medicine, 91*, 383–388.

White, M. P., Pahl, S., Ashbullby, K., Herbert, S., & Depledge, M. H. (2013). Feelings of restoration from recent nature visits. *Journal of Environmental Psychology, 35*, 40–51.

Wood, L., Hooper, P., Foster, S., & Bull, F. (2017). Public green spaces and positive mental health – investigating the relationship between access, quantity and types of parks and mental wellbeing. *Health & Place, 48*, 63–71.

World Data Atlas-Malaysia. (2021). Malaysia urban population as a share of total population. https://knoema.com/atlas/Malaysia/Urban-population#:~:text=In%202020%2C%20urban%20population%20for,defined%20by%20national%20statistical%20offices.

Yung, E. H. K., Conejos, S., & Chan, E. H. W. (2016). Social needs of the elderly and active aging in public open spaces in urban renewal. *Cities*, 52, 114–122.

Zhang, W., Yang, J., Ma, L., & Huang, C. (2015). Factors affecting the use of urban green spaces for physical activities: Views of young urban residents in Beijing. *Urban Forestry & Urban Greening, 14*, 851–857.

Zhu, Bo-Wei., Zhang, Jia-Rui., Tzeng, Gwo-Hshiung, Huang, Shan-Lin., & Xiong, L. (2017). Public open space development for elderly people by using the DANP-V model to establish continuous improvement strategies towards a sustainable and healthy aging society. *Sustainability*, 9, 420. https://doi.org/10.3390/su9030420.

Zupancic, T., Kingsley, M., Jason, T., & Macfarlane, R. (2015). Green City: Why Nature Matters to Health-An Evidence Review September 2015. *Toronto Public Health, 1*, 1–22.

Tapan Kumar Nath is an Interdisciplinary Researcher with experience in applying socio-ecological and sustainable livelihood frameworks for collaborative natural resources management and livelihood analysis of forest-dependent communities. His current position at the University of Nottingham (UNM) is an Associate Professor in the School of Environmental and Geographical Sciences. Prior to UNM, he worked at the University of Chittagong, the University of Tokyo and the Asian Institute of Technology at various levels. Dr. Nath has graduated from the University of Chittagong with a BSc and an MSc in Forestry and earned his Ph.D. in Forest Science from the University of Tokyo. He had a 2-year postdoctoral fellowship at the University of Tokyo. His current research focuses on the interaction between nature and society, particularly in forest-dependent communities. His research interests include Stakeholders' participation in forest management, livelihood analysis, sustainable intensification of agro-ecological practices and urban green space and human well-being.

Chapter 21
Urban Forest Ecosystem and Its Services to Human Wellbeing in Klang Valley, Malaysia

Chee Hung Foo

Abstract The population in urban areas has increased substantially in the last few decades and is expected to reach 60% of the world population by 2050. Urban forests provide a full spectrum of ecosystem services, serving as a nature-based solution to many environmental problems in cities. Identifying the ecosystem services and understanding the preferences of local communities could inform better decision-making in terms of land use planning and zoning, especially in highly urbanised areas. This chapter explores local residents' perfection and preferences for ecosystem services provided by four remnant urban forests located in the Klang Valley region, Malaysia, which exhibit different natural attributes and management approaches. This case study illustrates the potential roles of environmental issues, as well as residents' knowledge and motive to visit the urban forests in shaping their perception of ecosystem services provided by the urban forests. The findings also reveal that a lack of direct engagement with the forests and ecosystem services is unlikely to be regarded as a contributor to residents' quality of life.

Keywords Ecological services · Quality of life · Local perception · Physical activities · Naturalness

Forest Ecosystem and Its Services in the Urban Context

Urban population has increased considerably over the last century. As compared to only 30% of the world's population in 1950, it has accounted for 50% of today's population, and is expected to reach as high as 65% in 2050 (United Nations, 2014). Such a drastic rise in urban population is one of the important environmental issues across the globe (Chou et al., 2016). This is because urbanisation usually involves a high degree of land-use conversion, whereby the natural ecosystem is disturbed and replaced by the human-built environment. The most typical example is the loss of green spaces in a growing and expanding urbanised region (Fig. 21.1). Due to the

C. H. Foo (✉)
Property Business Development, MKH Berhad, Kuala Lumpur, Malaysia
e-mail: fighting_fist@yahoo.com

S. Maruthaveeran et al. (eds.), *Urban Forestry and Arboriculture in Malaysia*,
https://doi.org/10.1007/978-981-19-5418-4_21

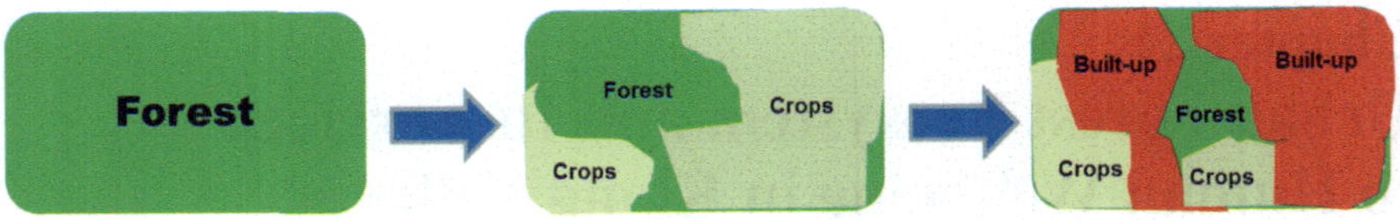

Fig. 21.1 The formation of remnant forests in urban areas—a conceptual illustration

expansion of the urban areas into the surrounding rural landscapes, natural forests which were once on a large scale, located in remote regions, possessed pristine biodiversity and functioned as a primary forest ecosystem have been significantly fragmented, degraded into remnants of secondary forest ecosystem, and become part of the urban infrastructure (Foo, 2016). These changes not just degrade most of the ecological functions, but also diminish species evenness among plant communities (Foo & Kidokoro, 2011a). Most importantly, it erodes the biodiversity of nature and might lead to a decline in the health and well-being of the urban residents.

Many studies have pointed out the negative environmental impacts brought by urbanisation. With the decrease and degradation of natural ecosystems as well as their services, coupled with the high population density and relentless development activities in urban areas, various types of "urban diseases" like traffic congestion, housing shortage, lack of amenity, environmental pollution etc. become increasingly significant in many cities (Wei et al., 2015). Empirical evidence has shown that living in cities tends to suffer directed attention fatigue (Kaplan, 1995), and is more likely to be linked to a greater risk of mental illness, since there are huge amounts of information generated by the urban environments that capture urban dwellers' attention and require further directed attention (Hartig et al., 2003; Lederbogen et al., 2011; Peen et al., 2010; Ulrich, 1984).

Moreover, greater levels of urbanisation are also associated with increased obesity, poorer mental health, and cancer mortality, owing to the increased exposure to pollution, as well as changes in behaviours and lifestyles that directly influence disease risk like smoking, sedentary time, social behaviour etc. (Ewing et al., 2003; McLafferty & Wang, 2009; Waldron, 1983); not to mention other physiological stress, negative emotions, and cardiovascular disease mortality as a result of the lack of access to green spaces also causes (Chang & Chen, 2005; Mitchell & Popham, 2008). In view of this, the existence of natural ecosystems in highly urbanised areas, such as secondary remnant forests, is indeed playing a key role to ensure the betterment of daily urban living (Chiesura, 2004; Gómez-Baggethun & Barton, 2013; Pinho et al., 2016). Even though their ecological value has often been ignored or overlooked due to their size and degree of artificiality (Foo & Kidokoro, 2011b), these remnant forests can still provide various nature-based solutions to many environmental problems in the cities (Fig. 21.2).

For example, the existence of a community forest park can provide a comfortable outdoor venue for its surrounding neighbourhood for public leisure, physical activities, social interaction, and spiritual inspiration. The experience with urban nature can not only improve cardiovascular and immune function, reduce physical

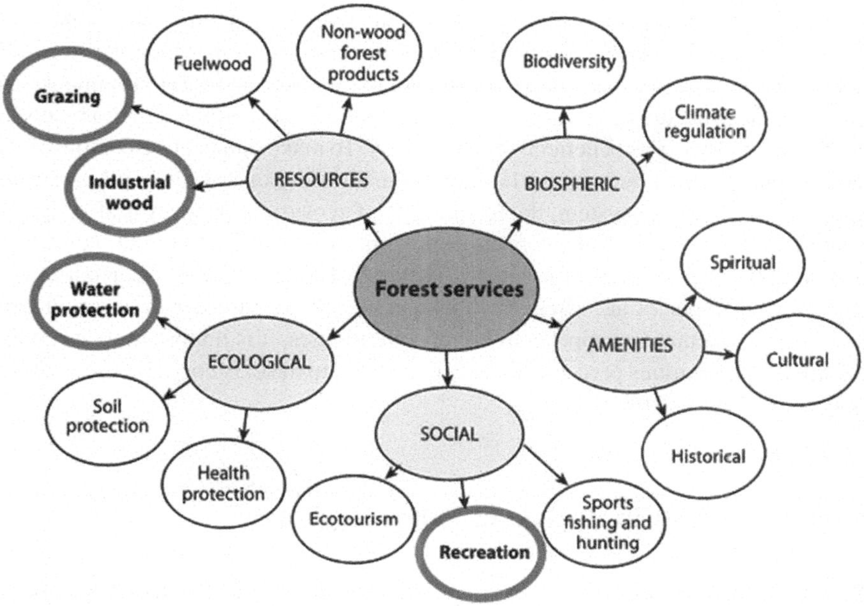

Fig. 21.2 Major classes of forest services. *Source* Millennium Ecosystem Assessment (2005)

responses to stress, promote relaxation, and reduce the negative impact of illness through conducting physical exercises (de Vries et al., 2011; Groenewegen et al., 2012; Maller et al., 2002), but also strengthen interpersonal bonds and foster other beneficial social factors (such as civic participation, trust, social relationship, etc.) through facilitating various levels of physical activity (Baur & Tynon, 2010; Hartig et al., 2011). Besides, urban nature can help reduce anxiety and evoke positive feelings through the provision of a place for relaxation and escape from the stressful rhythm of the city (Bratman et al., 2015; Chiesura, 2004; Groenewegen et al., 2012). More importantly, remnant forests in urban areas can contribute to microclimate regulation, water and air purification, wind and noise reduction, wildlife habitat, as well as enhancing environmental quality, quality of life, and sustainability of the surrounding neighbourhood (Jim & Chen, 2009; Millennium Ecosystem Assessment, 2005).

While most services are generally positive for human wellbeing, some of them may be less desirable (Delshammar et al., 2015; Döhren & Haase, 2015; Lyytimäki & Sipilä, 2009). Examples of such "less desirable" services include vegetation being a source of pests and disease vectors, or the roots of street trees damaging property (Drillet et al., 2020). Ecosystems, with varying components and environmental conditions, may provide contrasting suites of ecosystem services (Foo, 2016; Mexia et al., 2018). Moreover, people experience ecosystems in different ways. According to Russell et al. (2013), humans can interact with ecosystems through four channels: (i) knowing: thinking about an ecosystem or just the concept of an ideal ecosystem; (ii) perceiving: remote interactions with ecosystem components; (iv) interacting:

physical, active, direct multisensory interactions with ecosystem components; and (iv) living within: everyday interactions with the ecosystem in which we live. All these imply the way in which an individual interacts with the natural environment will help guide how he/she thinks about the ecosystem, thereby influencing how he/she perceives the services as beneficial or detrimental. To make things more complicated, users of ecosystem services are not homogeneous (Blazevska et al., 2012). Apart from experience with the ecosystem, the recognition of ecosystem services is also significantly relying on individual interests, beliefs, professional background, education level, access to relevant information etc. (Knez & Thorsson, 2008; Manzoor et al., 2019; Martín-López et al., 2012; Suchocka et al., 2019). In order to ensure urban nature remains intact and appeals to urban communities, it is important to ascertain how urban communities perceive the benefits related to the existence and use of these ecosystems (Foo, 2016).

Remnant Forests in the Klang Valley Region

The Klang Valley region, encompassing Kuala Lumpur and the four districts of Selangor State: Gombak, Hulu Langat, Petaling and Klang, is significantly affected by the spillover effects of the development of Kuala Lumpur (Fig. 21.3). As the highest urbanised region in Malaysia, Klang Valley has transformed rapidly in terms of demographics, economics, and physical development. The drastic change in land use, as a result of increasing urbanisation, has further seized the limited green area (Fig. 21.4). According to the Klang Valley Environmental Improvement Project, as early as 1986, almost 60% of the land in the region has undergone at least one change in land use from its original natural forest state to the commercial plantation and then followed by built-up area and mining. Today, only some small and fragmented lowland forests are found scattered within this region.

Due to its high degree of urbanity and built surface replacement that creates a significant contradiction between the built and natural ecosystem, remnant forests within the Klang Valley region are chosen as the case study to further explore how people perceive the forest ecosystem and its services in the urban context. Four remnant forests are selected as study sites: (i) Bukit Kiara Forest Park (BKPR), (ii) Bukit Gasing Forest (BGF), (iii) Kota Damansara Community Forest (KDCF), and (iv) Ayer Hitam Forest Reserve (AHFR) (Fig. 21.5). Realizing that the forest condition, forest status, as well as forest management approach will affect the way people experience nature, thereby leading to different perceptions of the local population toward the services provided by the forests; forests processing different biophysical structures are selected. For instance, BKF is initially an abandoned rubber estate, which has undergone a forest rehabilitation program to become a botanical garden. AHFR is a disturbed forest, contributing to the logging industry between 1930 and 1965, which is then gazetted as a forest reserve and leased to a university for the purposes of education, research, and extension. Both these forests represent the extreme of a man-made and natural continuum along the degree of naturalness.

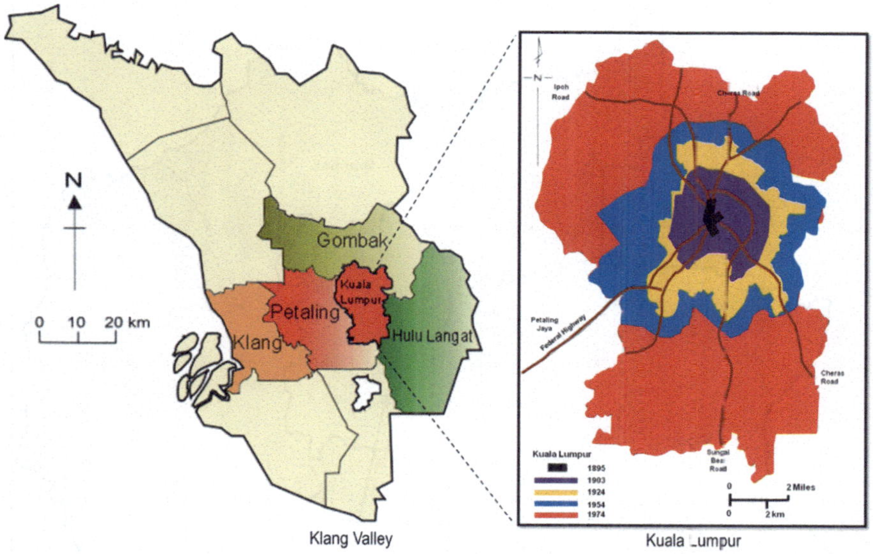

Fig. 21.3 Spillover development of Kuala Lumpur and the formation of Klang Valley conurbation

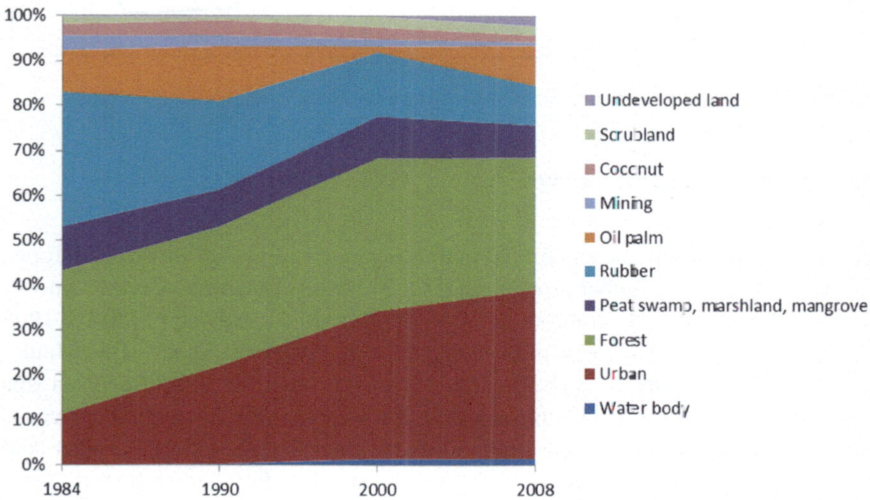

Fig. 21.4 Land use change in Klang Valley, 1984–2008

Table 21.1 summarizes the characteristics of the four selected study sites, and a more detailed description of the forests is given in the following sections.

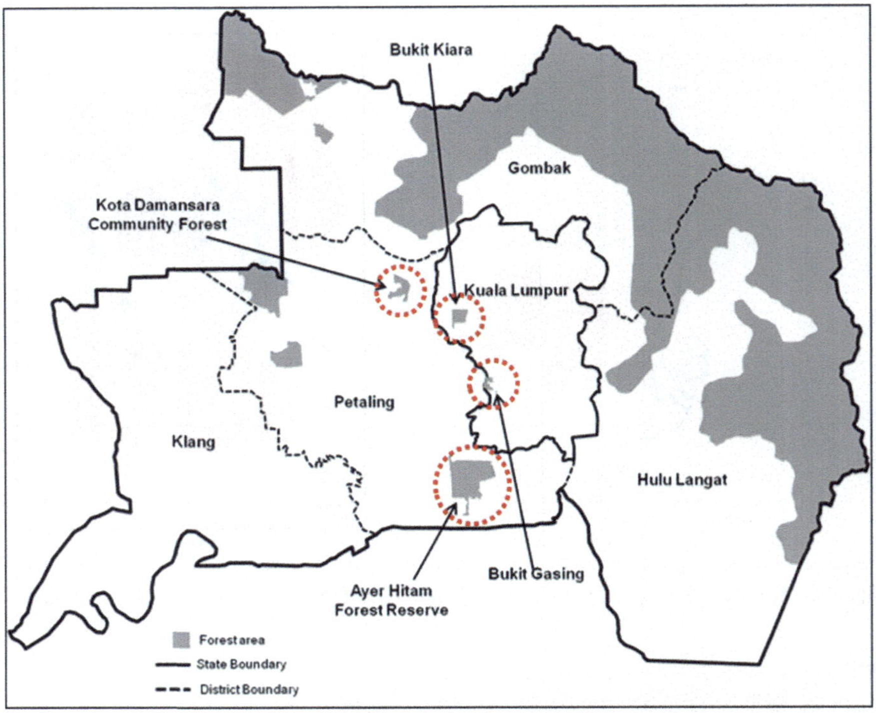

Fig. 21.5 Forested areas in Klang Valley with the location of the study sites

Ayer Hitam Forest Reserve

Ayer Hitam Forest Reserve (AHFR) is mainly covered by lowland-dipterocarp forest that has been used for timber production (Fig. 21.6). It is categorized as a secondary disturbed forest due to several rounds of logging activities—between 1930 and 1965 (Awang et al., 2000). The forest is located in Puchong, under the Subang Jaya Municipal Council's authority. It had been gazetted as a forest reserve in 1906 with an area of 4270.7 ha divided into 22 compartments. The size of AHFR was then decreased substantially at the tail end of the 1980s due to the forest degazettement for other land uses. At present, only an area of 1,176 ha remains and with 6 compartments. Since the forest is leased to Universiti Putra Malaysia (UPM) for 80 years (1996–2076) with the purpose of conducting research, education, and extension, it is not open to public access. It is specifically used for conducting academic activities, including fieldwork for courses such as dendrology, ecology, silviculture, forest mensuration, land survey, forest management, soil science and wildlife management.

Table 21.1 Summary of characteristics of the four study sites

Criteria	Forests			
	Bukit Kiara	Bukit Gasing	Kota Damansara	Ayer Hitam
Size (ha)	188.9	153.6	327	1176
Location based on local authority area	Kuala Lumpur	In between the border of Petaling Jaya and Kuala Lumpur	Petaling Jaya	Subang Jaya
Forest structure/condition	Forest rehabilitation program from an abandon rubber estate	Regenerating secondary forest, previously a rubber estate	Logged-over, regenerating secondary forest	Secondary disturbed forest
Degree of naturalness	High human input	Less human input	Less human input	Less human input
	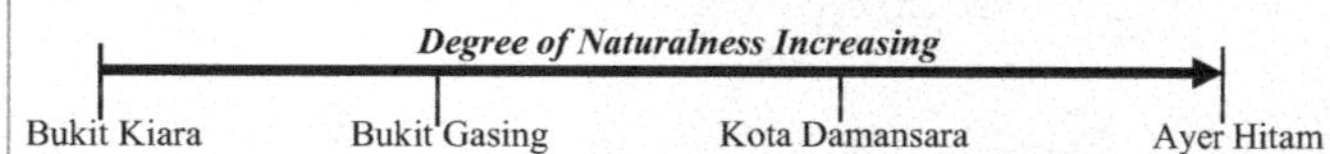			
Forest status	Federal park	Local park	State land forest	State land forest
Management system	Recreational park	Research and recreation forest park	Community forest	Research & education forest
Interaction with public	Directly contact	Directly contact	Directly contact	Restricted
Biological wealth	Plant: Mostly rubber tree Mammals: 3 species Reptiles: 2 species	Plant: 32 species Birds: about 100 species	Plant: 266 species Birds: 230 species Mammals: 10 species Reptiles & amphibians: 5 species	Plant: 430 species Timber tree: 127 species Medical plants: 98 species Moss species: 39 species Birds: 160 species Small mammals: 14 species Primates: 5 species Reptile: 10 species Amphibians: 18 species Fish: 10 species

Fig. 21.6 Photos of Ayer Hitam Forest Reserve, Puchong, Selangor Darul Ehsan

Kota Damansara Community Forest

Kota Damansara Community Forest (KDCF) is a logged-over forest under regeneration (Fig. 21.7). It is located inside a new town—Kota Damansara, under the Petaling Jaya municipality. KDCF is the remnant of the former Sungai Buloh Forest Reserve—the Peninsular Malaysia's oldest forest reserve. Being once a primary lowland-dipterocarp forest with a 5-layered canopy, it is now a secondary forest with a 1-layered canopy due to several rounds of logging activities which started in 1931 and stopped in the 1980s. In addition, the relentless development activities in recent years have reduced the forest area to just over 320 ha where many forest species have disappeared, especially the rarities and endemics. After a series of debates about whether the whole area can be converted into residential and commercial areas or not, KDCF was finally gazetted as a forest reserve in 2010. However, being designated as a forest reserve may not be able to prevent the place from being converted into other land uses, as it is still under the status of 'state land forest' (a government

Fig. 21.7 Photos of Kota Damansara Community Forest, Petaling Jaya, Selangor Darul Ehsan

land that still contains forest and can be allocated for development and other land uses such as agriculture, infrastructure and township). This land status may lead to the transformation of the forest into other land uses in the future, as it is located within a rather new township with soaring housing development in recent years.

Bukit Gasing Forest

Bukit Gasing Forest (BGF) is located between Kuala Lumpur and Petaling Jaya. Being once a rubber plantation and left idle for 50 years, the course of nature has overtaken the hillsides and produced a lush secondary forest brimming with woody herbs and large tropical trees (Fig. 21.8). Today, BGF functions as a green lung within a highly urbanised area. With a size of 153 ha, this piece of greenery is not only provides recreation services to the public but also plays an important role in soil erosion prevention. The forest is a favourite location for jogging and jungle tracking, where it has been experiencing an increase in popularity beyond the surrounding communities and even among foreign visitors. It is estimated that BGF receives almost 1,400 visitors during weekends and about 500 visitors on weekdays.

Bukit Kiara Forest Park

Bukit Kiara Forest Park (BKFP) is another typical example of a secondary forest transformed from a rubber estate. It is different from BGF in the sense that the regeneration process is dominated by human interventions. It is located within an

Fig. 21.8 Photos of Bukit Gasing Forest which is located on the border between Kuala Lumpur and Petaling Jaya, Selangor Darul Ehsan in Malaysia

area in the western part of the Kuala Lumpur city which has been well-known as an enigmatic, affluent, and exclusive area of the city. The forest has been functioning as a green lung of the city and enjoying a high number of visitors on a daily basis (Fig. 21.9). The forest that once struggled from being cleared for landscape and developed into an artificially constructed garden park, was then given to the National Landscape Department in 2010 for the management and development as a large-scale federal park. The concept of "design with nature" has been adopted by the department. The park will be equipped with all necessary amenities for the recreation purpose, while at the same time a large part of its greenery will remain. Since it is located in an affluent area, the pressure from housing developers to turn the park into a low-density, high-end property residential area has never been lessened.

Local Perception Towards Remnant Forest Ecosystem Services

To explore and describe the opinions of the local population on the benefits of remnant forests and the services provided, a semi-structured questionnaire survey was adopted. Several on-site observations were carried out prior to the questionnaire survey, in order to compile a list of services that can be provided by the forest ecosystem. The list was then refined by consultation with the local informants and was standardized; so that a comparison could be made among the four study sites (Table 21.2).

At each study site, a total of 30 respondents who live in the surrounding area of the forest, were sampled by the method next to pass, to conduct the semi-structured questionnaire survey. Apart from selecting listed ecosystem services from the list that

Fig. 21.9 Photos of Bukit Kiara Forest Park, Kuala Lumpur

Table 21.2 List of ecosystem services used in the questionnaire survey

Category of services	Ecosystem services
Social	Opportunity for recreation and leisure Opportunity for research and education Place for exercise and physical activity Place to strengthen community's relationship Importance for psychology and mental health
Ecological	Reduce the temperature of the surrounding area Air pollutant absorption Noise abatement CO_2 sequestration
Resource	Natural resources provision Space for further property development Increase property value of the surrounding area
Biospheric	Habitat for wildlife Source of biodiversity
Amenity	Scenic beauty Cultural and historical connotation

are beneficial to them, the respondents were also asked to describe: (i) how the forest services could enhance their quality of life; and (ii) their motives to visit the forest. Table 21.3 summarizes the respondents' profiles at each study site, and Fig. 21.10 shows the relative importance of forest services perceived by the respondents at each study site.

It is obvious that how people interact with the forest will influence their perception of services provided by the forest. People tend to highly rank those services to which they have direct access to and give low rank to those services which are intangible and seemingly insignificant to them. As demonstrated by AHFR, being gazetted as a

Table 21.3 Survey respondents' profile

	BKFP	BGF	KDCF	AHFR
Gender				
Male	63.4%	56.4%	61.8%	58.7%
Female	36.6%	43.6%	38.2%	41.3%
Age (year)				
<18	–	–	8.3%	–
18–25	10.7%	9.1%	27.8%	44.0%
26–35	22.1%	32.7%	30.6%	36.0%
36–50	26.0%	31.8%	21.3%	17.3%
>50	41.2%	26.4%	12.0%	2.7%
Educational background				
SPM	8.4%	10.9%	33.6%	18.9%
Diploma	13.7%	19.1%	28.2%	24.3%
STPM	5.3%	10.0%	0.9%	5.4%
Degree	49.6%	44.5%	25.5%	44.6%
Postgraduate	19.8%	11.8%	7.3%	5.4%
Others	3.1%	3.6%	4.5%	1.4%
Monthly Income (RM)				
No income	10.7%	6.4%	25.2%	5.3%
<RM2,000	9.9%	8.2%	18.9%	25.3%
RM2,000–RM3,500	10.7%	20.9%	25.5%	45.3%
RM3,501–RM5,000	18.3%	25.5%	13.2%	16.0%
RM5,001–RM10,000	30.5%	26.4%	5.7%	8.0%
>RM10,000	19.8%	12.7%	11.3%	–
Employment status				
Working	89.3%	92.7%	72.7%	94.7%
Housewife	–	0.9%	3.6%	–
Student	–	–	19.1%	5.3%
Retired	10.7%	6.4%	4.5%	–

research and educational forest, it is restricted to public accessibility. Forest services that require direct human-nature interaction such as 'place for physical activity, 'place for recreation', and 'enhance community relationship' were not chosen by the respondents. In the case of forests that people can contact directly, these services were highly ranked.

With no surprise, services including 'wildlife habitat', 'research and education, 'source of biodiversity, and 'natural resources provision' were perceived as important in AHFR and KDCF. This is because these forests are rich in flora and fauna. More importantly, the degree of human intrusion in these forests is relatively lower than in

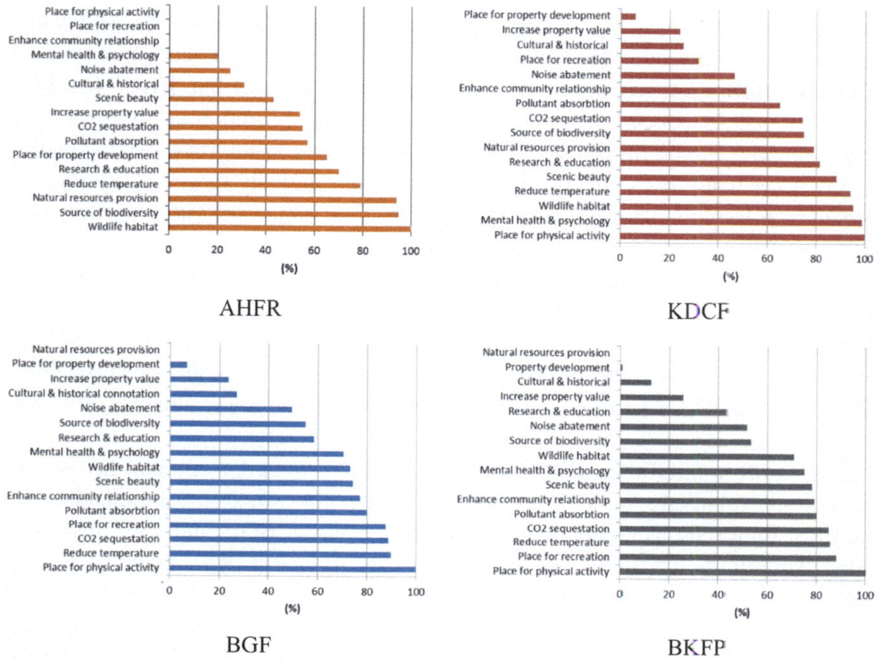

Fig. 21.10 Relative importance (%) of forest services in each study site

BGF and BKF. Therefore, both AHFR and KDCF are highly appreciated as 'natural' and 'wild'. However, it is rather surprising to find that several intangible services such as 'CO_2 sequestration', 'pollutant absorption', and 'reduce temperature' were highly rated in BGF and BKFP, as compared to AHFR and KDCR which are relatively larger in size and supposed to function better in providing these services. The possible explanation is that people's perceptions on forest services are found to be affected not only by the forest's natural attributes but also by its surrounding condition.

In fact, Klang Valley is more susceptible to air pollution as compared to other regions of the country. The worst polluted areas, including the Kuala Lumpur city, Shah Alam and Klang, lie along the Kuala Lumpur—Port Klang development axis. Being the country's capital city and the core of all the economic and investment activities, Kuala Lumpur has been experiencing a high volume of traffic, associated with people's mobility to and from the city outskirts. As reported in 2007, there are both 1.305 million and 2.125 million vehicles crossing the Middle Ring Road 1 (MRR1) and the Middle Ring Road 2 (MRR2) on a daily basis, respectively (Mohamad & Kiggundu, 2007). Such heavy inter and intra-urban movement together with the increased usage of private vehicles, as well as the rising trend of single occupancy of vehicles have inevitably led to serious congestion problems and consequently contributed to the air quality degradation within and around the city areas.

Besides, based to the Human Development Report 2007–2008, Malaysia is one of the world's fastest-growing countries in terms of carbon emissions. The country has experienced a 221% increase in CO_2 emissions since 1990. In 2006, Malaysia's carbon emissions stood at 187 million tonnes, and transportation accounted for 14% of global greenhouse gas emissions. Therefore, urban natures like BKFP and BGF located in or near Kuala Lumpur city were seen as highly important to the respondents, as these 'green lungs' were expected to function to improve the air quality. On the other hand, both the AHFR and KDCF are relatively more natural and are located in a less population density area; thus, 'CO_2 sequestration' and 'pollution absorption' were not perceived as important as other more appealing services.

In general, owing to the urban heat island effect, the forests' potential in reducing the temperature of the surrounding area was highly rated by the respondents. However, such service was deemed to be more significant in BGF and BKFP as these forests are located in a higher population density area, in which the urban heat island effect tends to be more severe as compared to the areas where AHFR and KDCF are located.

It is worth highlighting that respondent in KDCF regarded 'mental health and psychology' relatively higher than the other three study sites. It was ranked as the 2nd most important as compared to 8th in BKFP, 9th in BGF, and 13th in AHFR. This is largely related to the respondents' motives to visit KDCF. As in the case of KDCF, respondents reported that the main reason for them to visit the forest is 'to kill time', 'to relax', 'to get inspiration', and 'to change feeling', which all are important contributors to 'mental health and psychology. Meanwhile, both BGF and BKFP provide a place for the public to conduct physical activities and outdoor exercises (Fig. 21.11).

All the above ecosystem services were grouped according to the 'category of services': social, ecological, resource, biospheric and amenity. The respective mean values were calculated and compared among the four forests (Table 21.4). The rankings in BGF and BKFP were similar. This was probably due to the similar forest conditions, forest functions, as well as natural attributes of these two forests. The ranking pattern of KDCF was also quite similar to BGF and BKFP, with the only difference in ranking 'biospheric services'. It was ranked higher than 'ecological services' in KDCF. The most obvious difference was found between AHFR and the other three forests, where 'resource services' were ranked 2nd and 'social services'

Fig. 21.11 Motives to visit the forest

were ranked 5th. This was due to the different engagement that people have with the forest as discussed above. The only similarity among these four forests was that 'amenity services' such as scenic beauty and cultural & historical connotation were ranked lower. This was likely because such services were deemed ancillary, which was not as important as other services that can directly affect urban dwellers' wellbeing.

It is surprising to find that respondents, in general, ranked those intangible benefits highly, such as 'ecological services'. The possible explanation is that personal characteristics, particularly educational background, might influence people's perception of forest services. We found that respondents with higher educational backgrounds tended to recognize intangible services as a valuable benefit (Fig. 21.12). Even though establishing a statistical association between people's perceptions and their socioeconomic characteristics is beyond the scope of this study, it is undeniable that people did rely on their knowledge when perceiving forest services, particularly in assessing those intangible benefits.

Table 21.4 Ranking of the category of forest services

Category	BKFP			BGF			KDCF			AHFR		
	Mean	S.D	Rank	Mean	S.D	Rank	Mean	S.D	Rank	Mean	S.D	Rank
Social	4.65	0.61	1	4.72	0.56	1	4.38	0.88	1	4.09	1.03	5
Biospheric	4.41	0.71	3	4.50	0.74	3	4.27	1.00	2	4.20	0.94	1
Ecological	4.52	0.74	2	4.55	0.74	2	4.12	0.97	3	4.15	1.01	3
Resource	4.26	0.90	5	4.42	0.78	5	4.09	0.97	5	4.17	0.97	2
Amenity	4.33	0.72	4	4.45	0.72	4	4.10	0.86	4	4.10	0.98	4

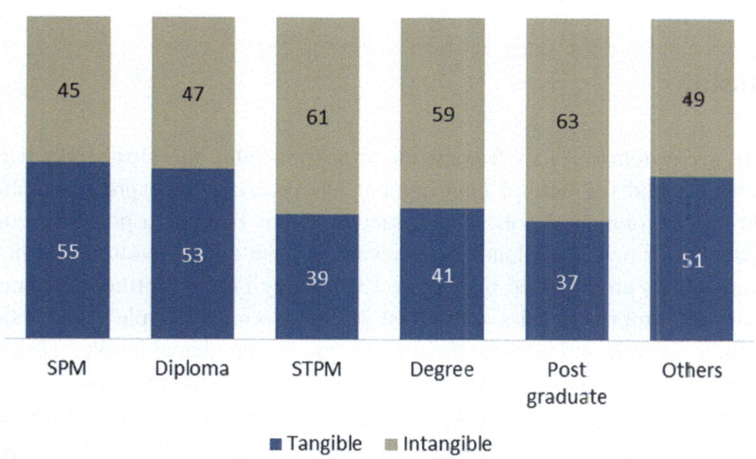

Fig. 21.12 Distribution (%) of respondents in selecting tangible/intangible services, by educational background

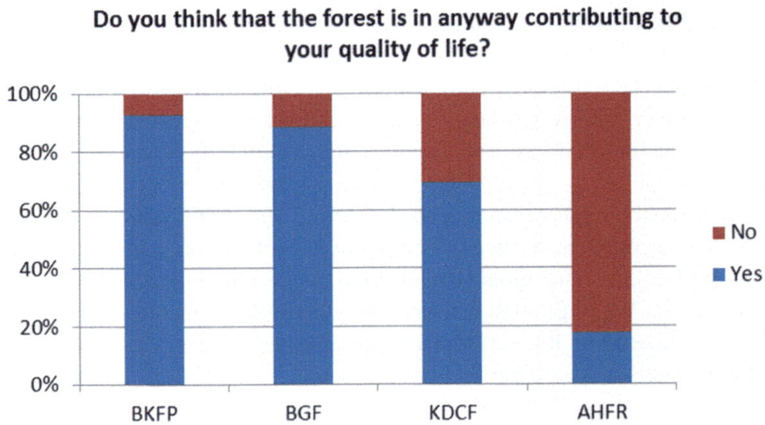

Fig. 21.13 Perception of forests' contribution to the quality of life

When examining whether and how forest services could contribute to people's quality of life, we found that people only consider the forest as a contributor to their quality of life if they have direct contacts with the forest, such as in KDCF, BGF, and BKFP (Fig. 21.13). This might be related to the degree of the forest's naturalness and human interventions (Foo, 2016). Direct engagement with nature enables residents to develop a higher attachment and sense of place to the forests, thereby becoming more aware of how the benefits can influence their everyday life (Foo, 2018). In other words, this also implies that knowing the ecosystem services is one thing but recognizing them as beneficial parts of the people's quality of life is another thing, particularly if direct human-nature interactions are lacking.

Conclusion

Using an ecosystem services framework to acknowledge the close links between human beings and the natural environment has received widespread attention in the field of environmental policy and practice. It has become a powerful conceptual approach for managing landscapes resulting from a joint evolution of humans and nature across an extended time span. Even though the scientific understanding of forestry has improved a lot in the last several decades, people's perception of forests, their services and relevant governance plays an undeniable role, particularly in conserving the declining (in size) and degrading (in biodiversity) urban natures.

The current research successfully gained insights into perceptions, preferences, and needs of the local population in the Klang Valley region toward remnant urban forests. More importantly, it has demonstrated how people's perceptions of ecosystem services are affected by the forest's natural attributes, environmental issues in the surrounding areas, as well as people's motive to visit and their educational background.

References

Awang Noor, A. G., & Mohd. Shahwahid, H. O., Rusli, M., Shukri, M., Ibrahim, F. H., & Hussin, M. Z. (2000). Economic valuation of forest goods and services of Ayer Hitam Forest Reserve, Puchong, Selangor: Implications on land use. In H. H. Mohd Nordin (Ed.), *Langat basin ecosystem health* (pp. 84–99). National University Malaysia.

Baur, J. W. R., & Tynon, J. F. (2010). Small-scale urban nature parks: Why should we care? *Leisure Science, 32*(2), 195–200.

Blazevska, A., Miceva, K., Stojanova, B., & Stojanovska, M. (2012). Perception of the local population toward urban forests in municipality of Aerodrom. *SEEFOR, 3*(2), 87–96.

Bratman, G. N., Daily, G. C., Levy, B. J., & Gross, J. J. (2015). The benefits of nature experience: Improved affect and cognition. *Landscape and Urban Planning, 138*, 41–50.

Chang, C. Y., & Chen, P. K. (2005). Human response to window views and indoor plants in the workplace. *HortScience, 40*(5), 1354–1359.

Chiesura, A. (2004). The role of urban parks for the sustainable city. *Landscape and Urban Planning, 68*(1), 129–138.

Chou, W. Y., Lee, C. H., & Chang, C. Y. (2016). Relationships between urban open spaces and humans' health benefits from an ecological perspective: A study in an urban campus. *Landscape Ecological Engineering, 12*, 255–267.

Delshammar, T., Östberg, J., & Öxell, C. (2015). Urban trees and ecosystem disservices—A pilot study using complaints records from three Swedish cities. *Arboriculture & Urban Forestry, 41*, 187–193.

de Vries, S., Classen, T., Eigenheer-Hug, S.-M., Korpela, K., Maas, J., Mitchell, R., & Schantz, P. (2011). Contributions of natural environments to physical activity. In K. Nilsson, M. Sangster, C. Gallis, T. Hartig, S. De Vries, K. Seeland, & J. Schipperijn (Eds.), *Theory and evidence base: Forests, trees and human health* (pp. 205–243). Springer.

Döhren, P., & Haase, D. (2015). Ecosystem disservices research: A review of the state of the art with a focus on cities. *Ecological Indicators, 52*, 490–497.

Drillet, Z., Fung, T. K., Leong, R. A. T., Sachidhanandam, U., Edwards, P., & Richards, D. (2020). Urban vegetation types are not perceived equally in providing ecosystem services and disservices. *Sustainability, 12*, 2076. https://doi.org/10.3390/su12052076

Ewing, R., Schmid, T., Killingsworth, R., Zlot, A., & Raudenbush, S. (2003). Relationship between urban sprawl and physical activity, obesity, and morbidity. *American Journal of Health Promotion, 18*, 47–57.

Foo, C. H. (2016). Linking forest naturalness and human wellbeing—A study on public's experiential connection to remnant forests within a highly urbanized region in Malaysia. *Urban Forestry & Urban Greening, 16*, 13–24.

Foo, C. H. (2018). Review, rethink, and reconsider human-nature interaction in the urban context. In H. F. Santos Viana & F. A. Garcia-Morote (Eds.), *New perspectives in forest science* (pp. 151–166). IntechOpen.

Foo, C. H., & Kidokoro, T. (2011a). Human-nature interaction—Understanding the role of forest's naturalness in influencing people's responses. *Journal of Habitat Engineering, 3*(2), 219–230.

Foo, C. H., & Kidokoro, T. (2011b). Issues on lowland forest conservation in a highly urbanized region in Malaysia. *Journal of Habitat Engineering, 3*(1), 131–140.

Gómez-Baggethun, E., & Barton, D. N. (2013). Classifying and valuing ecosystem services for urban planning. *Ecological Economics, 86*, 235–245.

Groenewegen, P. P., van den Berg, A. E., Mass, J., Verheji, R. A., & de Vries, S. (2012). Is a green residential environment better for health? If so, why? *Annals of the American Association of Geographers, 102*(5), 996–1003.
Hartig, T., Evans, G. W., Jamner, L. D., Davis, D. S., & Ga¨rling, T. (2003). Tracking restoration in natural and urban field settings. *Journal of Environmental Psychology, 23*(2), 109–123.
Hartig, T., Van den Berg, A. E., Hagerhall, C. M., Tomalak, M., Bauer, N., Hansmann, R., Ojala, A., Syngollitou, E., Carrus, G., van Herzele, A., Bell, S., Podesta, M. T. C., & Waaseth, C. (2011). Health benefits of nature experiences: psychological, social and cultural processes. In K. Nilsson, M. Sangster, C. Gallis, T. Hartig, S. De Vries, K. Seeland, & J. Schipperijn (Eds.), *Forests, trees and human health* (pp. 127–167). Springer.
Jim, C. Y., & Chen, W. Y. (2009). Ecosystem services and valuation of urban forests in China. *Cities, 26*(4), 187–194.
Kaplan, S. (1995). The restorative benefits of nature: Toward an integrative framework. *Journal of Environmental Psychology, 15*(3), 169–182.
Knez, I., & Thorsson, S. (2008). Thermal, emotional and perceptual evaluations of a park: Cross-Cultural and environmental attitude comparisons. *Building and Environment, 43*, 1483–1490.
Lederbogen, F., Kirsch, P., Haddad, L., Streit, F., Tost, H., Schuch, P., Wu¨st, S., Pruessner, J. C., Rietschel, M., Deuschle, M., & Meyer-Lindenberg, A. (2011). City living and urban upbringing affect neural social stress processing in humans. *Nature, 474*(7352), 498–550.
Lyytimäki, J., & Sipilä, M. (2009). Hopping on one leg—The challenge of ecosystem disservices for urban green management. *Urban Forestry & Urban Greening, 8*, 309–315.
Maller, C., Townsend, M., Brown, P., & Leger, L. S. (2002). *Healthy parks healthy people: The health benefits of contact with nature in a park context*. Parks Victoria.
Manzoor, S. A., Malik, A., Zubair, M., Griffiths, G., & Lukac, M. (2019). Linking social perception and provision of ecosystem services in a sprawling urban landscape: A case study of Multan Pakistan. *Sustainability, 11*, 654.
Martín-López, B., Iniesta-Arandia, I., García-Llorente, M., Palomo, I., Casado-Arzuaga, I., & Amo, D. G. D. (2012). Uncovering ecosystem service bundles through social preferences. *PLoS ONE, 7*, e38970.
McLafferty, S., & Wang, F. (2009). Rural reversal? Rural–urban disparities in late-stage cancer risk in Illinois. *Cancer, 115*, 2755–2764.
Mexia, T., Vieira, J., Príncipe, A., Anjos, A., Silva, P., Lopes, N., Freitas, C., Santos-Reis, M., Correia, O., & Branquinho, C. (2018). Ecosystem services: Urban parks under a magnifying glass. *Environmental Research, 160*, 469–478.
Millennium Ecosystem Assessment. (2005). *Ecosystems and human well-being—Synthesis*. Inland Press.
Mitchell, R., & Popham, F. (2008). Effect of exposure to natural environment on health inequalities: An observational population study. *The Lancet, 372*(9650), 1655–1660.
Mohamad, J., & Kiggundu, A. T. (2007). The rise of the private car in Kuala Lumpur, Malaysia—Assessing the policy options. *IATSS Research, 31*(1), 69–77.
Peen, J., Schoevers, R. A., Beekman, A. T., & Dekker, J. (2010). The current status of urban-rural differences in psychiatric disorders. *Acta Psychiatrica Scandinavica, 121*(2), 84–93.
Pinho, P., Correia, O., Lecoq, M., Munzi, S., Vasconcelos, S., Gonçalves, P., Rebelo, R., Antunes, C., Silva, P., Freitas, C., Lopes, N., Santos-Reis, M., & Branquinho, C. (2016). Evaluating green infrastructure in urban environments using a multi-taxa and functional diversity approach. *Environmental Research, 147*, 601–610.
Russell, R., Guerry, A. D., Balvanera, P., Gould, R. K., Basurto, X., Chan, K. M. A., Klain, S., Levine, J., & Tam, J. (2013). Humans and nature: How knowning and experiencing nature affect well-being. *Annual Review of Environment and Resources, 38*, 473–502.
Suchocka, M., Jankowski, P., & Błaszczyk, M. (2019). Perception of urban trees by Polish tree professionals vs. nonprofessionals. *Sustainability, 11*, 211.
Ulrich, R. S. (1984). View through a window may influence recovery from surgery. *Science, 224*, 420–421.

United Nations. (2014). *World Urbanization Prospects: The 2014 Revision, Highlights (ST/ESA/SER.A/352)*. https://doi.org/10.4054/DemRes.2005.12.9.
Waldron, H. A. (1983). A brief history of scrotal cancer. *British Journal of Industrial Medicine, 40*, 390–401.
Wei, Y., Huang, C., Lam, P. T. I., Sha, Y., & Feng, Y. (2015). Using urban-carrying capacity as a benchmark for sustainable urban development: An empirical study of Beijing. *Sustainability, 7*, 3244–3268.

Chee Hung Foo is a Professional Technologist registered under the Malaysian Board of Technologists. He is also a member of the Institution of Engineers Malaysia (IEM), a committee member of REHDA Selangor Planning Policies and Standards Sub-committee, a research committee of REHDA Institute (RI), a GreenRE manager, a Certified Sustainable INFRASTAR facilitator, and a Certified MyCREST Qualified Professional and Assessor. He is now the principal researcher of MKH Berhad, a well-known property development company based in Klang Valley, Malaysia.

Index

S. Maruthaveeran et al. (eds.), *Urban Forestry and Arboriculture in Malaysia*, https://doi.org/10.1007/978-981-19-5418-4

The manufacturer's authorised representative in the EU is Springer Nature Customer Service Centre GmbH, Europaplatz 3, 69115 Heidelberg, Germany. If you have any concerns regarding our products, please contact ProductSafety@springernature.com

Printed and bound by CPI Group (UK) Ltd, Croydon, CR0 4YY
01/10/2026
02226185-0007